Physiology

SEVENTH EDITION

Physiology

SEVENTH EDITION

Linda S. Costanzo, Ph.D.
Professor of Physiology and Biophysics
School of Medicine
Virginia Commonwealth University
Richmond, Virginia

Philadelphia • Baltimore • New York • London
Buenos Aires • Hong Kong • Sydney • Tokyo

Acquisitions Editor: Crystal Taylor
Development Editor: Andrea Vosburgh
Editorial Coordinator: Tim Rinehart
Marketing Manager: Michael McMahon
Production Project Manager: Marian Bellus
Design Coordinator: Holly McLaughlin
Manufacturing Coordinator: Margie Orzech
Prepress Vendor: SPi Global

Seventh edition

9 8 7 6 5 4 3 2

Printed in China

Library of Congress Cataloging-in-Publication Data
Names: Costanzo, Linda S., 1947- author.
Title: Physiology / Linda S. Costanzo.
Other titles: Board review series.
Description: Seventh edition. | Philadelphia : Wolters Kluwer, [2019] | Series: BRS | Includes index.
Identifiers: LCCN 2017054852 | ISBN 9781496367617
Subjects: | MESH: Physiological Phenomena | Examination Questions
Classification: LCC QP40 | NLM QT 18.2 | DDC 612.0076—dc23 LC record available at https://lccn.loc.gov/2017054852

LWW.com

CCS0119

For Richard
And
for Dan, Rebecca, and Sheila
And
for Elise and Max

Preface

The subject matter of physiology is the foundation of the practice of medicine, and a firm grasp of its principles is essential for the physician. This book is intended to aid the student preparing for the United States Medical Licensing Examination (USMLE) Step 1. It is a concise review of key physiologic principles and is intended to help the student recall material taught during the first and second years of medical school. It is not intended to substitute for comprehensive textbooks or for course syllabi, although the student may find it a useful adjunct to physiology and pathophysiology courses.

The material is organized by organ system into seven chapters. The first chapter reviews general principles of cellular physiology. The remaining six chapters review the major organ systems—neurophysiology, cardiovascular, respiratory, renal and acid–base, gastrointestinal, and endocrine physiology.

Difficult concepts are explained stepwise, concisely, and clearly, with appropriate illustrative examples and sample problems. Numerous clinical correlations are included so that the student can understand physiology in relation to medicine. An integrative approach is used, when possible, to demonstrate how the organ systems work together to maintain homeostasis. More than 130 full-color illustrations and flow diagrams and more than 50 tables help the student visualize the material quickly and aid in long-term retention. Appendices contain "Key Physiology Topics for USMLE Step 1," "Key Physiology Equations for USMLE Step 1," and "Normal Blood Values."

Questions reflecting the content and format of USMLE Step 1 are included at the end of each chapter and in a Comprehensive Examination at the end of the book. These questions, many with clinical relevance, require problem-solving skills rather than straight recall. Clear, concise explanations accompany the questions and guide the student through the correct steps of reasoning. The questions can be used as a pretest to identify areas of weakness or as a posttest to determine mastery. Special attention should be given to the Comprehensive Examination, because its questions integrate several areas of physiology and related concepts of pathophysiology and pharmacology.

New to this edition:

- Addition of new full-color figures
- Updated organization and text
- Expanded coverage of neurophysiology, and respiratory, renal, gastrointestinal, and endocrine physiology
- Addition of new multi-step questions

Best of luck in your preparation for USMLE Step 1!

Linda S. Costanzo, Ph.D.

Acknowledgments

It has been a pleasure to be a part of the Board Review Series and to work with the staff at Wolters Kluwer. Crystal Taylor and Andrea Vosburgh provided expert editorial assistance.

My sincere thanks to students in the School of Medicine at Virginia Commonwealth University/Medical College of Virginia, who have provided so many helpful suggestions for *BRS Physiology*. Thanks also to the many students from other medical schools who have taken the time to write to me about their experiences with this book.

Linda S. Costanzo, Ph.D.

Contents

Chapter 1 Cell Physiology

I. CELL MEMBRANES

- are composed primarily of phospholipids and proteins.

A. Lipid bilayer

1. **Phospholipids** have a **glycerol backbone,** which is the hydrophilic (water soluble) head, and two **fatty acid tails,** which are hydrophobic (water insoluble). The hydrophobic tails face each other and form a bilayer.
2. **Lipid-soluble substances** (e.g., O_2, CO_2, steroid hormones) cross cell membranes because they can dissolve in the hydrophobic lipid bilayer.
3. **Water-soluble substances** (e.g., Na^+, Cl^-, glucose, H_2O) cannot dissolve in the lipid of the membrane, but may cross through water-filled channels, or pores, or may be transported by carriers.

B. Proteins

1. **Integral proteins**
   - are anchored to, and imbedded in, the cell membrane through **hydrophobic** interactions.
   - may span the cell membrane.
   - include ion channels, transport proteins, receptors, and guanosine 5′-triphosphate (GTP)–binding proteins (G proteins).
2. **Peripheral proteins**
   - are *not* imbedded in the cell membrane.
   - are *not* covalently bound to membrane components.
   - are loosely attached to the cell membrane by **electrostatic** interactions.

C. Intercellular connections

1. **Tight junctions (zonula occludens)**
   - are the attachments between cells (often epithelial cells).
   - may be an intercellular pathway for solutes, depending on the size, charge, and characteristics of the tight junction.
   - may be **"tight"** (impermeable), as in the renal distal tubule, or **"leaky"** (permeable), as in the renal proximal tubule and gallbladder.
2. **Gap junctions**
   - are the attachments between cells that permit intercellular communication.
   - for example, permit current flow and electrical **coupling between myocardial cells.**

II. TRANSPORT ACROSS CELL MEMBRANES (TABLE 1.1)

A. Simple diffusion

1. Characteristics of simple diffusion

- is the only form of transport that is **not carrier mediated.**
- occurs **down an electrochemical gradient** ("downhill").
- does not require metabolic energy and therefore is passive.

2. Diffusion can be measured using the following equation:

$$J = -PA(C_1 - C_2)$$

where:

J = flux (flow) (mmol/sec)
P = permeability (cm/sec)
A = area (cm^2)
C_1 = $concentration_1$ (mmol/L)
C_2 = $concentration_2$ (mmol/L)

3. Sample calculation for diffusion

- The urea concentration of blood is 10 mg/100 mL. The urea concentration of proximal tubular fluid is 20 mg/100 mL. If the permeability to urea is 1×10^{-5} cm/sec and the surface area is 100 cm^2, what are the magnitude and direction of the urea flux?

$$\begin{aligned}\text{Flux} &= \left(\frac{1\times10^{-5}\ \text{cm}}{\text{sec}}\right)\left(100\ \text{cm}^2\right)\left(\frac{20\ \text{mg}}{100\ \text{mL}} - \frac{10\ \text{mg}}{100\ \text{mL}}\right)\\ &= \left(\frac{1\times10^{-5}\ \text{cm}}{\text{sec}}\right)\left(100\ \text{cm}^2\right)\left(\frac{10\ \text{mg}}{100\ \text{mL}}\right)\\ &= \left(\frac{1\times10^{-5}\ \text{cm}}{\text{sec}}\right)\left(100\ \text{cm}^2\right)\left(\frac{0.1\ \text{mg}}{\text{cm}^3}\right)\\ &= 1\times10^{-4}\ \text{mg/sec from lumen to blood (high to low concentration)}\end{aligned}$$

Note: The minus sign preceding the diffusion equation indicates that the direction of flux, or flow, is from high to low concentration. It can be ignored if the higher concentration is called C_1 and the lower concentration is called C_2.
Also note: 1 mL = 1 cm^3.

table 1.1 Characteristics of Different Types of Transport

Type	Electrochemical Gradient	Carrier-Mediated	Metabolic Energy	Na^+ Gradient	Inhibition of Na^+–K^+ Pump
Simple diffusion	Downhill	No	No	No	—
Facilitated diffusion	Downhill	Yes	No	No	—
Primary active transport	Uphill	Yes	Yes	—	Inhibits (if Na^+–K^+ pump)
Cotransport	Uphill*	Yes	Indirect	Yes, same direction	Inhibits (by abolishing Na^+ gradient)
Countertransport	Uphill*	Yes	Indirect	Yes, opposite direction	Inhibits (by abolishing Na^+ gradient)

*One or more solutes are transported uphill; Na^+ is transported downhill.

4. Permeability

- is the P in the equation for diffusion.
- describes the ease with which a solute diffuses through a membrane.
- depends on the characteristics of the solute and the membrane.

a. Factors that increase permeability:

- ↑ **Oil/water partition coefficient** of the solute increases solubility in the lipid of the membrane.
- ↓ **Radius (size) of the solute** increases the diffusion coefficient and speed of diffusion.
- ↓ **Membrane thickness** decreases the diffusion distance.

b. Small hydrophobic solutes (e.g., O_2, CO_2) have the highest permeabilities in lipid membranes.

c. Hydrophilic solutes (e.g., Na^+, K^+) must cross cell membranes through water-filled channels, or pores, or via transporters. If the solute is an ion (is charged), then its flux will depend on both the concentration difference and the potential difference across the membrane.

B. Carrier-mediated transport

- includes facilitated diffusion and primary and secondary active transport.
- The **characteristics** of carrier-mediated transport are

1. Stereospecificity. For example, D-glucose (the natural isomer) is transported by facilitated diffusion, but the L-isomer is not. Simple diffusion, in contrast, would not distinguish between the two isomers because it does not involve a carrier.

2. Saturation. The transport rate increases as the concentration of the solute increases, until the carriers are saturated. The **transport maximum (T_m)** is analogous to the maximum velocity **(V_{max})** in enzyme kinetics.

3. Competition. Structurally related solutes compete for transport sites on carrier molecules. For example, galactose is a competitive inhibitor of glucose transport in the small intestine.

C. Facilitated diffusion

1. Characteristics of facilitated diffusion

- occurs **down an electrochemical gradient** ("downhill"), similar to simple diffusion.
- does not require metabolic energy and therefore is **passive.**
- is more **rapid** than simple diffusion.
- is **carrier mediated** and therefore exhibits stereospecificity, saturation, and competition.

2. Example of facilitated diffusion

- Glucose transport in muscle and adipose cells is "downhill," is carrier mediated, and is inhibited by sugars such as galactose; therefore, it is categorized as facilitated diffusion. In **diabetes mellitus,** glucose uptake by muscle and adipose cells is impaired because the carriers for facilitated diffusion of glucose require **insulin.**

D. Primary active transport

1. Characteristics of primary active transport

- occurs **against an electrochemical gradient** ("uphill").
- requires **direct input of metabolic energy** in the form of adenosine triphosphate **(ATP)** and therefore is **active.**
- is **carrier mediated** and therefore exhibits stereospecificity, saturation, and competition.

2. Examples of primary active transport

a. Na^+, K^+-ATPase (or Na^+–K^+ pump) in cell membranes transports Na^+ from intracellular to extracellular fluid and K^+ from extracellular to intracellular fluid; it maintains low intracellular [Na^+] and high intracellular [K^+].

- Both **Na^+ and K^+ are transported against their electrochemical gradients.**
- Energy is provided from the terminal phosphate bond of ATP.

- The **usual stoichiometry is 3 Na^+/2 K^+.**
- Specific inhibitors of Na^+, K^+-ATPase are the cardiac glycoside drugs ouabain and **digitalis.**

b. Ca^{2+}-ATPase (or Ca^{2+} pump) in the sarcoplasmic reticulum (SR) or cell membranes transports Ca^{2+} against an electrochemical gradient.

- Sarcoplasmic and endoplasmic reticulum Ca^{2+}-ATPase is called **SERCA.**

c. H^+, K^+-ATPase (or proton pump) in gastric parietal cells and renal α-intercalated cells transports H^+ into the lumen (of the stomach or renal tubule) against its electrochemical gradient.

- It is inhibited by proton pump inhibitors, such as **omeprazole.**

E. Secondary active transport

1. Characteristics of secondary active transport

a. The transport of two or more solutes is **coupled.**

b. One of the solutes (usually Na^+) is transported "downhill" and provides energy for the "uphill" transport of the other solute(s).

c. Metabolic energy is not provided directly but indirectly from the **Na^+ gradient** that is maintained across cell membranes. Thus, inhibition of Na^+, K^+-ATPase will decrease transport of Na^+ out of the cell, decrease the transmembrane Na^+ gradient, and eventually inhibit secondary active transport.

d. If the solutes move in the same direction across the cell membrane, it is called **cotransport** or **symport.**

- Examples are **Na^+-glucose cotransport** in the small intestine and renal early proximal tubule and **Na^+–K^+–2Cl^-** cotransport in the renal thick ascending limb.

e. If the solutes move in opposite directions across the cell membranes, it is called **countertransport, exchange,** or **antiport.**

- Examples are **Na^+-Ca^{2+} exchange** and **Na^+–H^+ exchange.**

2. Example of Na^+–glucose cotransport (Figure 1.1)

a. The carrier for Na^+–glucose cotransport is located in the luminal membrane of intestinal mucosal and renal proximal tubule cells.

b. Glucose is transported "uphill"; Na^+ is transported "downhill."

c. Energy is derived from the "downhill" movement of Na^+. The inwardly directed Na^+ gradient is maintained by the Na^+–K^+ pump on the basolateral (blood side) membrane. Poisoning the Na^+–K^+ pump decreases the transmembrane Na^+ gradient and consequently inhibits Na^+–glucose cotransport.

3. Example of Na^+–Ca^{2+} countertransport or exchange (Figure 1.2)

a. Many cell membranes contain a Na^+–Ca^{2+} exchanger that transports Ca^{2+} "uphill" from low intracellular [Ca^{2+}] to high extracellular [Ca^{2+}]. Ca^{2+} and Na^+ move in opposite directions across the cell membrane.

b. The energy is derived from the "downhill" movement of Na^+. As with cotransport, the inwardly directed Na^+ gradient is maintained by the Na^+–K^+ pump. Poisoning the Na^+–K^+ pump therefore inhibits Na^+–Ca^{2+} exchange.

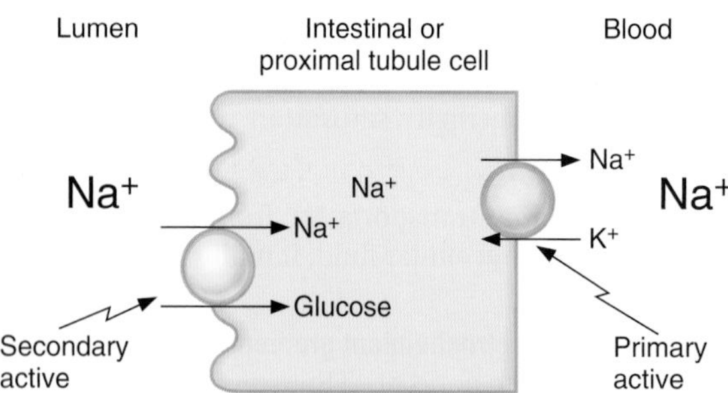

FIGURE 1.1. Na^+–glucose cotransport (symport) in intestinal or proximal tubule epithelial cell.

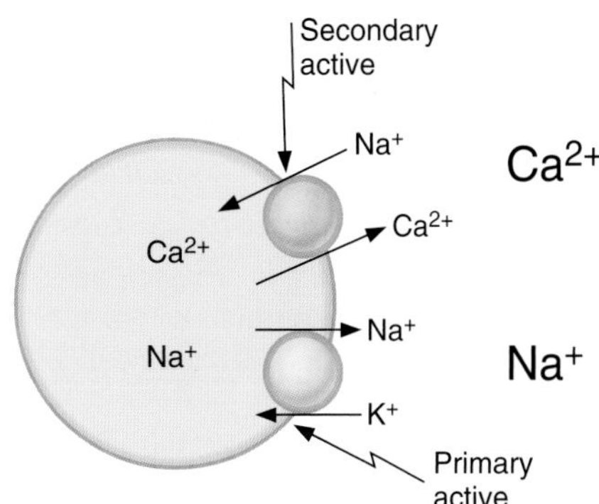

FIGURE 1.2. Na^+–Ca^{2+} countertransport (antiport).

III. OSMOSIS

A. Osmolarity

- is the concentration of osmotically active particles in a solution.
- is a colligative property that can be measured by freezing point depression.
- can be calculated using the following **equation:**

$$\textbf{Osmolarity} = \mathbf{g} \times \mathbf{C}$$

where:

Osmolarity = concentration of particles (Osm/L)
g = number of particles in solution (Osm/mol) [e.g., $g_{NaCl} = 2$; $g_{glucose} = 1$]
C = concentration (mol/L)

- Two solutions that have the same calculated osmolarity are **isosmotic.** If two solutions have different calculated osmolarities, the solution with the higher osmolarity is **hyperosmotic** and the solution with the lower osmolarity is **hyposmotic.**
- **Sample calculation:** What is the osmolarity of a 1 M NaCl solution?

$$\begin{aligned} \text{Osmolarity} &= g \times C \\ &= 2\ \text{Osm/mol} \times 1\ \text{M} \\ &= 2\ \text{Osm/L} \end{aligned}$$

B. Osmosis and osmotic pressure

- **Osmosis** is the **flow of water** across a semipermeable membrane from a solution with low solute concentration to a solution with high solute concentration.

1. Example of osmosis (Figure 1.3)

a. Solutions 1 and 2 are separated by a semipermeable membrane. Solution 1 contains a solute that is too large to cross the membrane. Solution 2 is pure water. The presence of the solute in solution 1 produces an **osmotic pressure.**

b. The osmotic pressure difference across the membrane causes water to flow from solution 2 (which has no solute and the lower osmotic pressure) to solution 1 (which has the solute and the higher osmotic pressure).

c. With time, the volume of solution 1 increases and the volume of solution 2 decreases.

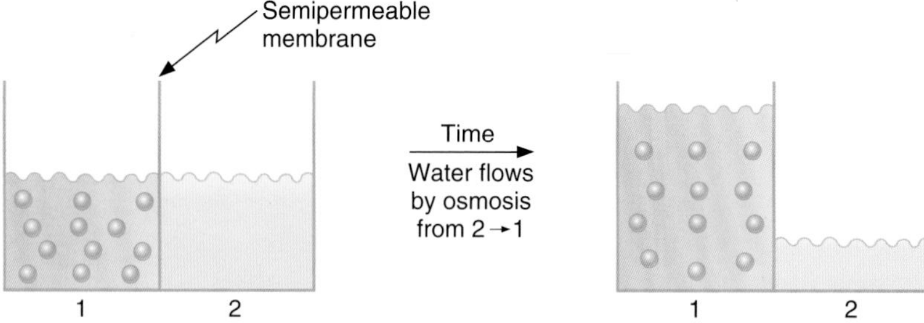

FIGURE 1.3. Osmosis of H_2O across a semipermeable membrane.

2. Calculating osmotic pressure (van't Hoff's law)

a. The **osmotic pressure** of solution 1 (see Figure 1.3) can be calculated by van't Hoff's law, which states that osmotic pressure depends on the concentration of osmotically active particles. The concentration of particles is converted to pressure according to the following **equation:**

$$\pi = g \times C \times RT$$

where:

π = osmotic pressure (mm Hg or atm)
g = number of particles in solution (osm/mol)
C = concentration (mol/L)
R = gas constant (0.082 L—atm/mol—K)
T = absolute temperature (K)

b. The osmotic pressure increases when the solute concentration increases. A solution of 1 M $CaCl_2$ has a higher osmotic pressure than does a solution of 1 M KCl because, for a given volume, the number of osmotically active particles is higher.

c. The higher the osmotic pressure of a solution, the greater the water flow into it.

d. Two solutions having the same effective osmotic pressure are **isotonic** because no water flows across a semipermeable membrane separating them. If two solutions separated by a semipermeable membrane have different effective osmotic pressures, the solution with the higher effective osmotic pressure is **hypertonic** and the solution with the lower effective osmotic pressure is **hypotonic.** Water flows from the hypotonic to the hypertonic solution.

e. Colloid osmotic pressure, or **oncotic pressure,** is the osmotic pressure created by proteins (e.g., plasma proteins).

3. Reflection coefficient (σ)

- is a number between zero and one that describes the ease with which a solute permeates a membrane.

a. If the reflection coefficient is one, the solute is impermeable. Therefore, it is retained in the original solution, it creates an osmotic pressure, and it causes water flow. **Serum albumin** (a large solute) has a reflection coefficient of nearly one.

b. If the reflection coefficient is zero, the solute is completely permeable. Therefore, it will not exert any osmotic effect, and it will not cause water flow. **Urea** (a small solute) usually has a reflection coefficient of close to zero and it is, therefore, an **ineffective osmole.**

4. Calculating effective osmotic pressure

- Effective osmotic pressure is the osmotic pressure (calculated by van't Hoff's law) multiplied by the reflection coefficient.
- If the reflection coefficient is one, the solute will exert maximal effective osmotic pressure. If the reflection coefficient is zero, the solute will exert no osmotic pressure.

IV. DIFFUSION POTENTIAL, RESTING MEMBRANE POTENTIAL, AND ACTION POTENTIAL

A. Ion channels

- are **integral proteins** that span the membrane and, when open, permit the passage of certain ions.

1. **Ion channels are selective;** they permit the passage of some ions, but not others. Selectivity is based on the size of the channel and the distribution of charges that line it.
   - For example, a small channel lined with negatively charged groups will be selective for small cations and exclude large solutes and anions. Conversely, a small channel lined with positively charged groups will be selective for small anions and exclude large solutes and cations.
2. **Ion channels may be open or closed.** When the channel is open, the ion(s) for which it is selective can flow through. When the channel is closed, ions cannot flow through.
3. **The conductance of a channel** depends on the probability that the channel is open. The higher the probability that a channel is open, the higher the conductance, or **permeability.** Opening and closing of channels are controlled by **gates.**
   a. **Voltage-gated channels** are opened or closed by changes in membrane potential.
      - The **activation gate of the Na^+ channel** in nerve is opened by depolarization.
      - The **inactivation gate of the Na^+ channel** in nerve is closed by depolarization.
      - When both the activation and inactivation gates on Na^+ channels are open, the channels are open and permeable to Na^+ (e.g., during the upstroke of the nerve action potential).
      - If either the activation or inactivation gate on the Na^+ channel is closed, the channel is closed and impermeable to Na^+. For example, at the resting potential, the activation gates are closed and thus the Na^+ channels are closed.

   b. **Ligand-gated channels** are opened or closed by hormones, second messengers, or neurotransmitters.
      - For example, the **nicotinic receptor** for acetylcholine (ACh) at the motor end plate is an ion channel that opens when ACh binds to it. When open, it is permeable to Na^+ and K^+, causing the motor end plate to depolarize.

B. Diffusion and equilibrium potentials

- A **diffusion potential** is the potential difference generated across a membrane because of a concentration difference of an ion.
- A diffusion potential can be generated only if the membrane is permeable to the ion.
- The **size of the diffusion potential** depends on the size of the concentration gradient.
- The **sign of the diffusion potential** depends on whether the diffusing ion is positively or negatively charged.
- Diffusion potentials are created by the diffusion of **very few ions** and, therefore, do not result in changes in concentration of the diffusing ions.
- The **equilibrium potential** is the potential difference that would exactly balance (oppose) the tendency for diffusion down a concentration difference. At **electrochemical equilibrium,** the chemical and electrical driving forces that act on an ion are equal and opposite, and no further net diffusion of the ion occurs.

1. **Example of a Na^+ diffusion potential** (Figure 1.4)
   a. Two solutions of NaCl are separated by a membrane that is permeable to Na^+ but not to Cl^-. The NaCl concentration of solution 1 is higher than that of solution 2.
   b. Because the membrane is permeable to Na^+, Na^+ will diffuse from solution 1 to solution 2 down its concentration gradient. Cl^- is impermeable and therefore will not accompany Na^+.
   c. As a result, a **diffusion potential** will develop at the membrane and solution 1 will become negative with respect to solution 2.

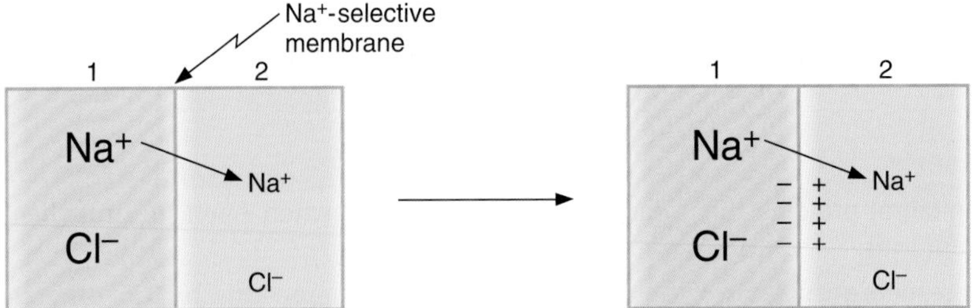

FIGURE 1.4. Generation of an Na^+ diffusion potential across a Na^+-selective membrane.

d. Eventually, the potential difference will become large enough to oppose further net diffusion of Na^+. The potential difference that exactly counterbalances the diffusion of Na^+ down its concentration gradient is the **Na^+ equilibrium potential.** At electrochemical equilibrium, the chemical and electrical driving forces on Na^+ are equal and opposite, and there is no net diffusion of Na^+.

2. Example of a Cl^- diffusion potential (Figure 1.5)

a. Two solutions identical to those shown in Figure 1.4 are now separated by a membrane that is permeable to Cl^- rather than to Na^+.

b. Cl^- will diffuse from solution 1 to solution 2 down its concentration gradient. Na^+ is impermeable and therefore will not accompany Cl^-.

c. A **diffusion potential** will be established at the membrane such that solution 1 will become positive with respect to solution 2. The potential difference that exactly counterbalances the diffusion of Cl^- down its concentration gradient is the **Cl^- equilibrium potential.** At electrochemical equilibrium, the chemical and electrical driving forces on Cl^- are equal and opposite, and there is no net diffusion of Cl^-.

3. Using the Nernst equation to calculate equilibrium potentials

a. The **Nernst equation** is used to calculate the equilibrium potential at a given concentration difference of a permeable ion across a cell membrane. It tells us what potential would exactly balance the tendency for diffusion down the concentration gradient; in other words, **at what potential would the ion be at electrochemical equilibrium?**

$$\mathbf{E = -2.3\frac{RT}{zF}\log_{10}\frac{[C_i]}{[C_e]}}$$

where:

E = equilibrium potential (mV)

$$2.3\frac{RT}{zF} = \frac{60\text{ mV}}{z}\text{ at }37°\text{C}$$

z = charge on the ion (+1 for Na^+, +2 for Ca^{2+}, −1 for Cl^-)

C_i = intracellular concentration (mM)

C_e = extracellular concentration (mM)

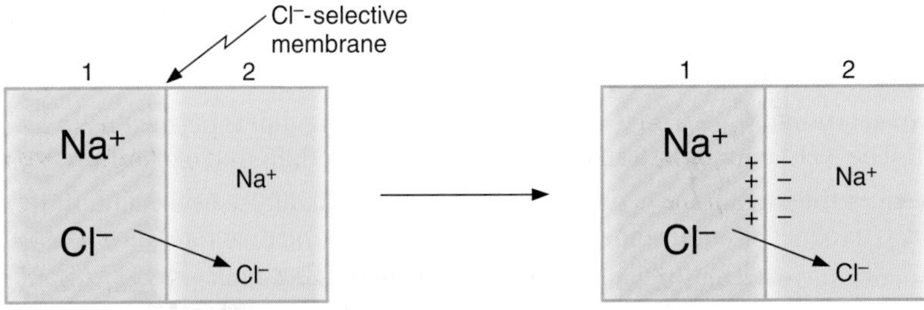

FIGURE 1.5. Generation of a Cl^- diffusion potential across a Cl^--selective membrane.

b. Sample calculation with the Nernst equation

- If the intracellular [Na^+] is 15 mM and the extracellular [Na^+] is 150 mM, what is the equilibrium potential for Na^+?

$$\begin{aligned} E_{Na^+} &= \frac{-60\text{ mV}}{z}\log_{10}\frac{[C_i]}{[C_e]} \\ &= \frac{-60\text{ mV}}{+1}\log_{10}\frac{15\text{ mM}}{150\text{ mM}} \\ &= -60\text{ mV}\log_{10}0.1 \\ &= +60\text{ mV} \end{aligned}$$

Note: You need not remember which concentration goes in the numerator. Because it is a log function, perform the calculation either way to get the absolute value of 60 mV. Then use an "intuitive approach" to determine the correct sign. (Intuitive approach: The [Na^+] is higher in extracellular fluid than in intracellular fluid, so Na^+ ions will diffuse from extracellular to intracellular, making the inside of the cell positive [i.e., +60 mV at equilibrium].)

c. Approximate values for equilibrium potentials in nerve and skeletal muscle

E_{Na^+}	+65 mV
$E_{Ca^{2+}}$	+120 mV
E_{K^+}	–85 mV
E_{Cl^-}	–85 mV

C. Driving force and current flow

- The **driving force** on an ion is the difference between the actual membrane potential (E_m) and the ion's equilibrium potential (calculated with the Nernst equation). In other words, the driving force is the difference between the actual membrane potential and what the ion would "like" the membrane potential to be; the ion would "like" the membrane potential to be its equilibrium potential, as calculated by the Nernst equation.
- **Current flow** occurs if there is a driving force on the ion and the membrane is permeable to the ion. The *direction* of current flow is in the same direction as the driving force. The *magnitude* of current flow is determined by the size of the driving force and the permeability (or conductance) of the ion. If there is no driving force on the ion, no current flow can occur. If the membrane is impermeable to the ion, no current flow can occur.

D. Resting membrane potential

- is expressed as the measured potential difference across the cell membrane in millivolts (mV).
- is, by convention, expressed as the intracellular potential relative to the extracellular potential. Thus, a resting membrane potential of –70 mV means **70 mV, cell negative.**

1. **The resting membrane potential is established by diffusion potentials** that result from concentration differences of permeant ions.
2. **Each permeable ion attempts to drive the membrane potential toward its equilibrium potential.** Ions with the highest permeabilities, or conductances, will make the greatest contributions to the resting membrane potential, and those with the lowest permeabilities will make little or no contribution.
3. **For example,** the resting membrane potential of nerve is –70 mV, which is close to the calculated K^+ equilibrium potential of –85 mV, but far from the calculated Na^+ equilibrium potential of +65 mV. **At rest, the nerve membrane is far more permeable to K^+ than to Na^+.**
4. **The Na^+–K^+ pump contributes only indirectly** to the resting membrane potential by maintaining, across the cell membrane, the Na^+ and K^+ concentration gradients that then produce diffusion potentials. The direct **electrogenic** contribution of the pump (3 Na^+ pumped out of the cell for every 2 K^+ pumped into the cell) is small.

E. Action potentials

1. Definitions

a. **Depolarization** makes the membrane potential **less negative** (the cell interior becomes less negative).

b. **Hyperpolarization** makes the membrane potential **more negative** (the cell interior becomes more negative).

c. **Inward current** is the flow of positive charge into the cell. Inward current **depolarizes** the membrane potential.

d. **Outward current** is the flow of positive charge out of the cell. Outward current **hyperpolarizes** the membrane potential.

e. **Action potential** is a property of excitablecells (i.e., nerve, muscle) that consists of a rapid depolarization, or upstroke, followed by repolarization of the membrane potential. Action potentials have **stereotypical size and shape,** are **propagating,** and are **all-or-none.**

f. **Threshold** is the membrane potential at which the action potential is inevitable. At threshold potential, net inward current becomes larger than net outward current. The resulting depolarization becomes self-sustaining and gives rise to the upstroke of the action potential. If net inward current is less than net outward current, no action potential will occur (i.e., all-or-none response).

2. Ionic basis of the nerve action potential (Figure 1.6)

a. **Resting membrane potential**

- is approximately –70 mV, cell negative.
- is the result of the **high resting conductance to K^+**, which drives the membrane potential toward the K^+ equilibrium potential.
- At rest, although the inactivation gates on Na^+ channels are open (having been opened by repolarization from the preceding action potential), the activation gates on Na^+ channels are closed and thus the Na^+ channels are closed and Na^+ conductance is low.

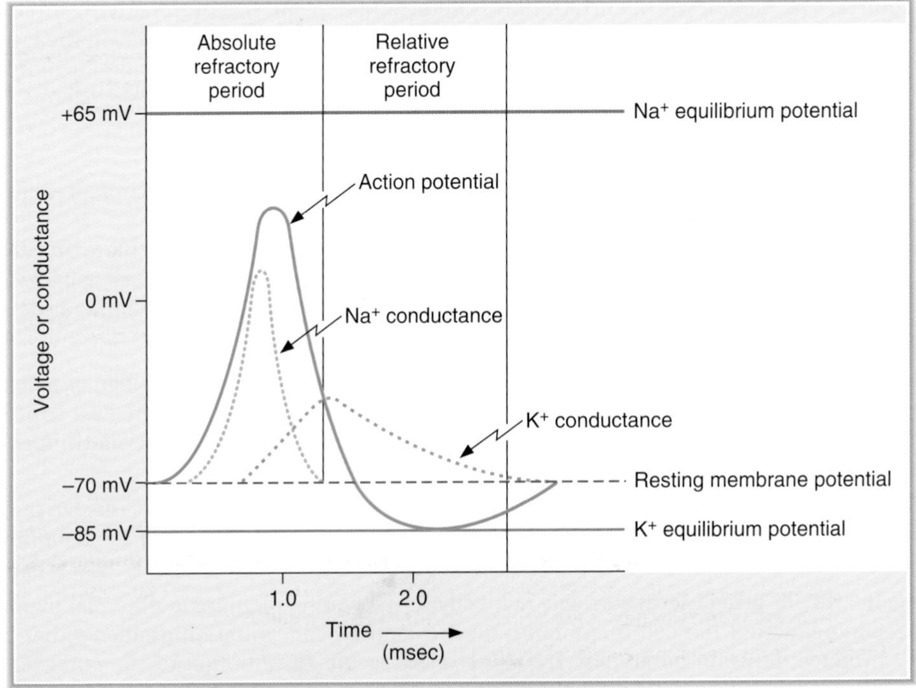

FIGURE 1.6. Nerve action potential and associated changes in Na^+ and K^+ conductance.

b. Upstroke of the action potential

(1) Inward current depolarizes the membrane potential to threshold.

(2) **Depolarization causes rapid opening of the activation gates of the Na^+ channels.** Now, both activation and inactivation gates are open and the Na^+ conductance of the membrane promptly increases.

(3) The Na^+ conductance becomes higher than the K^+ conductance, and the membrane potential is driven toward (but does not quite reach) the Na^+ equilibrium potential of +65 mV. Thus, the rapid depolarization during the upstroke is caused by an **inward Na^+ current.**

(4) The **overshoot** is the brief portion at the peak of the action potential when the membrane potential is positive.

(5) **Tetrodotoxin (TTX)** and **lidocaine** block these voltage-sensitive Na^+ channels and abolish action potentials.

c. Repolarization of the action potential

(1) **Depolarization also closes the inactivation gates of the Na^+ channels** (but more slowly than it opens the activation gates). Closure of the inactivation gates results in closure of the Na^+ channels, and the Na^+ conductance returns toward zero.

(2) **Depolarization slowly opens K^+ channels and increases K^+ conductance** to even higher levels than at rest. **Tetraethylammonium (TEA)** blocks these voltage-gated K^+ channels.

(3) The combined effect of closing the Na^+ channels and greater opening of the K^+ channels makes the K^+ conductance higher than the Na^+ conductance, and the membrane potential is repolarized. Thus, repolarization is caused by an **outward K^+ current.**

d. Undershoot (hyperpolarizing afterpotential)

- The K^+ conductance remains higher than at rest for some time after closure of the Na^+ channels. During this period, the membrane potential is driven very close to the K^+ equilibrium potential.

3. Refractory periods (see Figure 1.6)

a. Absolute refractory period

- is the period during which another action potential cannot be elicited, no matter how large the stimulus.
- coincides with almost the entire duration of the action potential.
- **Explanation:** Recall that the inactivation gates of the Na^+ channels are closed when the membrane potential is depolarized. They remain closed until repolarization occurs. No action potential can occur until the inactivation gates open.

b. Relative refractory period

- begins at the end of the absolute refractory period and continues until the membrane potential returns to the resting level.
- An action potential can be elicited during this period only if a larger than usual inward current is provided.
- **Explanation:** The K^+ conductance is higher than at rest, and the membrane potential is closer to the K^+ equilibrium potential and, therefore, farther from threshold; more inward current is required to bring the membrane to threshold.

c. Accommodation

- occurs when the cell membrane is held at a depolarized level such that the threshold potential is passed without firing an action potential.
- occurs because depolarization closes inactivation gates on the Na^+ channels.
- is demonstrated in **hyperkalemia,** in which skeletal muscle membranes are depolarized by the high serum K^+ concentration. Although the membrane potential is closer to threshold, action potentials do not occur because inactivation gates on Na^+ channels are closed by depolarization, causing **muscle weakness.**

4. Propagation of action potentials (Figure 1.7)

- occurs by the spread of **local currents** to adjacent areas of membrane, which are then depolarized to threshold and generate action potentials.

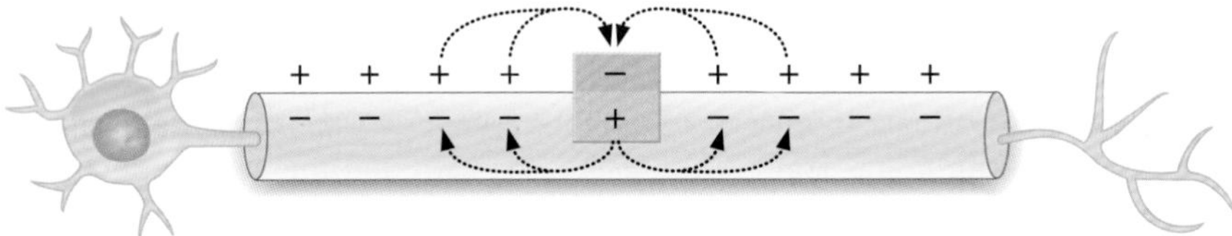

FIGURE 1.7. Unmyelinated axon showing spread of depolarization by local current flow. *Box* shows active zone where action potential had reversed the polarity.

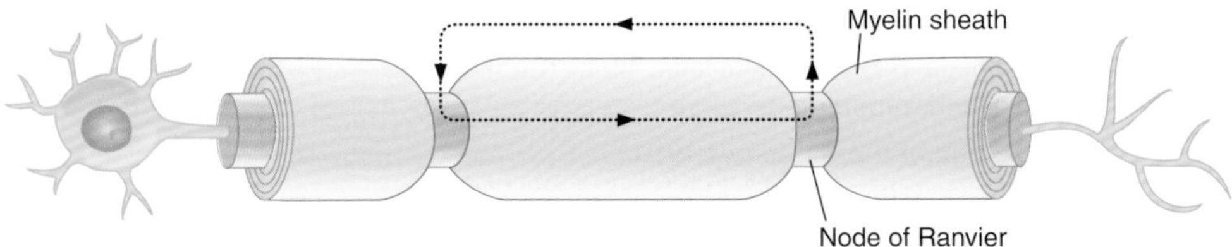

FIGURE 1.8. Myelinated axon. Action potentials can occur at nodes of Ranvier.

- **Conduction velocity is increased by:**

a. ↑ fiber size. Increasing the diameter of a nerve fiber results in decreased internal resistance; thus, conduction velocity down the nerve is faster.

b. Myelination. Myelin acts as an insulator around nerve axons and increases conduction velocity. Myelinated nerves exhibit **saltatory conduction** because action potentials can be generated only at the **nodes of Ranvier,** where there are gaps in the myelin sheath (Figure 1.8).

V. NEUROMUSCULAR AND SYNAPTIC TRANSMISSION

A. General characteristics of chemical synapses

1. **An action potential in the presynaptic cell** causes depolarization of the presynaptic terminal.
2. As a result of the depolarization, **Ca^{2+} enters the presynaptic terminal,** causing **release of neurotransmitter** into the synaptic cleft.
3. Neurotransmitter diffuses across the synaptic cleft and combines with **receptors on the postsynaptic cell membrane,** causing a change in its permeability to ions and, consequently, a change in its membrane potential.
4. **Inhibitory neurotransmitters** hyperpolarize the postsynaptic membrane: **excitatory neurotransmitters** depolarize the postsynaptic membrane.

B. Neuromuscular junction (Figure 1.9 and Table 1.2)

- is the synapse between axons of motoneurons and skeletal muscle.
- The neurotransmitter released from the presynaptic terminal is **ACh,** and the postsynaptic membrane contains a **nicotinic receptor.**

1. Synthesis and storage of ACh in the presynaptic terminal

- **Choline acetyltransferase** catalyzes the formation of ACh from acetyl coenzyme A (CoA) and choline in the presynaptic terminal.
- ACh is stored in **synaptic vesicles** with ATP and proteoglycan for later release.

2. Depolarization of the presynaptic terminal and Ca^{2+} uptake

- Action potentials are conducted down the motoneuron. Depolarization of the presynaptic terminal **opens Ca^{2+} channels.**

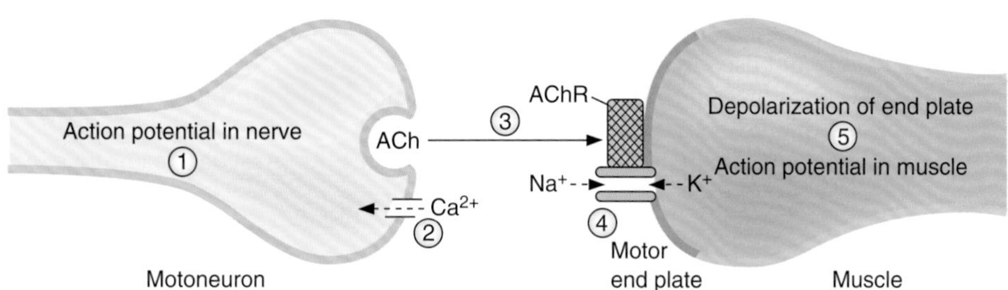

FIGURE 1.9. Neuromuscular junction. ACh = acetylcholine; AChR = acetylcholine receptor.

- When Ca^{2+} permeability increases, Ca^{2+} rushes into the presynaptic terminal down its electrochemical gradient.

3. **Ca^{2+} uptake causes release of ACh into the synaptic cleft**
   - The synaptic vesicles fuse with the plasma membrane and empty their contents into the cleft by **exocytosis.**
4. **Diffusion of ACh to the postsynaptic membrane (muscle end plate) and binding of ACh to nicotinic receptors**
   - The nicotinic ACh receptor is also a **Na^+ and K^+ ion channel.**
   - Binding of ACh to α subunits of the receptor causes a conformational change that opens the central core of the channel and increases its conductance to Na^+ and K^+. These are examples of **ligand-gated channels.**
5. **End plate potential (EPP) in the postsynaptic membrane**
   - Because the channels opened by ACh conduct both Na^+ and K^+ ions, the postsynaptic membrane potential is depolarized to a value halfway between the Na^+ and K^+ equilibrium potentials (approximately 0 mV).
   - The contents of one synaptic vesicle (one quantum) produce a **miniature end plate potential** (MEPP), the smallest possible EPP.
   - MEPPs summate to produce a full-fledged EPP. **The EPP is not an action potential,** but simply a depolarization of the specialized muscle end plate.
6. **Depolarization of adjacent muscle membrane to threshold**
   - Once the end plate region is depolarized, local currents cause depolarization and action potentials in the adjacent muscle tissue. Action potentials in the muscle are followed by contraction.
7. **Degradation of ACh**
   - The EPP is transient because ACh is degraded to acetyl CoA and choline by **acetylcholinesterase** (AChE) on the muscle end plate.

table 1.2 Agents Affecting Neuromuscular Transmission

Example	Action	Effect on Neuromuscular Transmission
Botulinum toxin	Blocks release of ACh from presynaptic terminals	Total blockade
Curare	Competes with ACh for receptors on motor end plate	Decreases size of EPP; maximal doses produce paralysis of respiratory muscles and death
Neostigmine	Inhibits acetylcholinesterase	Prolongs and enhances action of ACh at muscle end plate
Hemicholinium	Blocks reuptake of choline into presynaptic terminal	Depletes ACh stores from presynaptic terminal

ACh = acetylcholine; EPP = end plate potential.

- One-half of the choline is taken back into the presynaptic ending by Na^+-choline cotransport and used to synthesize new ACh.
- **AChE inhibitors (neostigmine)** block the degradation of ACh, prolong its action at the muscle end plate, and increase the size of the EPP.
- **Hemicholinium** blocks choline reuptake and depletes the presynaptic endings of ACh stores.

8. Disease—myasthenia gravis

- is caused by the presence of antibodies to the ACh receptor.
- is characterized by skeletal muscle weakness and fatigability resulting from a **reduced number of ACh receptors** on the muscle end plate.
- The size of the EPP is reduced; therefore, it is more difficult to depolarize the muscle membrane to threshold and to produce action potentials.
- **Treatment with AChE inhibitors (e.g., neostigmine)** prevents the degradation of ACh and prolongs the action of ACh at the muscle end plate, partially compensating for the reduced number of receptors.

C. Synaptic transmission

1. Types of arrangements

a. One-to-one synapses (such as those found at the neuromuscular junction)

- An action potential in the presynaptic element (the motor nerve) produces an action potential in the postsynaptic element (the muscle).

b. Many-to-one synapses (such as those found on spinal motoneurons)

- An action potential in a single presynaptic cell is insufficient to produce an action potential in the postsynaptic cell. Instead, many cells synapse on the postsynaptic cell to depolarize it to threshold. The presynaptic input may be excitatory or inhibitory.

2. Input to synapses

- The postsynaptic cell integrates excitatory and inhibitory inputs.
- When the sum of the input brings the membrane potential of the postsynaptic cell to threshold, it fires an action potential.

a. Excitatory postsynaptic potentials (EPSPs)

- are inputs that **depolarize** the postsynaptic cell, bringing it closer to threshold and closer to firing an action potential.
- are caused by **opening of channels that are permeable to Na^+ and K^+**, similar to the ACh channels. The membrane potential depolarizes to a value halfway between the equilibrium potentials for Na^+ and K^+ (approximately 0 mV).
- **Excitatory neurotransmitters** include ACh, norepinephrine, epinephrine, dopamine, glutamate, and serotonin.

b. Inhibitory postsynaptic potentials (IPSPs)

- are inputs that **hyperpolarize** the postsynaptic cell, moving it away from threshold and farther from firing an action potential.
- are caused by **opening Cl^- channels.** The membrane potential is hyperpolarized toward the Cl^- equilibrium potential (–90 mV).
- **Inhibitory neurotransmitters** are γ-aminobutyric acid **(GABA)** and **glycine.**

3. Summation at synapses

a. Spatial summation occurs when two excitatory inputs arrive at a postsynaptic neuron simultaneously. Together, they produce greater depolarization.

b. Temporal summation occurs when two excitatory inputs arrive at a postsynaptic neuron in rapid succession. Because the resulting postsynaptic depolarizations overlap in time, they add in stepwise fashion.

c. Facilitation, augmentation, and posttetanic potentiation occur after tetanic stimulation of the presynaptic neuron. In each of these, depolarization of the postsynaptic neuron is greater than expected because greater than normal amounts of neurotransmitter are released, possibly because of the accumulation of Ca^{2+} in the presynaptic terminal.

- **Long-term potentiation** (memory) involves new protein synthesis.

4. Neurotransmitters

a. ACh (see V B)

b. Norepinephrine, epinephrine, and dopamine (Figure 1.10)

(1) Norepinephrine

- is the primary transmitter released from **postganglionic sympathetic neurons.**
- is synthesized in the nerve terminal and released into the synapse to bind with **α or β receptors** on the postsynaptic membrane.
- is removed from the synapse by **reuptake** or is metabolized in the presynaptic terminal by monoamine oxidase **(MAO)** and catechol-*O*-methyltransferase **(COMT).** The **metabolites** are:

(a) 3,4-Dihydroxymandelic acid (DOMA)
(b) Normetanephrine (NMN)
(c) 3-Methoxy-4-hydroxyphenylglycol (MOPEG)
(d) 3-Methoxy-4-hydroxymandelic acid or vanillylmandelic acid (VMA)

- In **pheochromocytoma,** a tumor of the adrenal medulla that secretes catecholamines, urinary excretion of **VMA** is increased.

(2) Epinephrine

- is synthesized from norepinephrine by the action of phenylethanolamine-*N*-methyltransferase in the **adrenal medulla**
- a methyl group is transferred to norepinephrine from *S*-adenosylmethionine

(3) Dopamine

- is prominent in **midbrain** neurons.
- is released from the hypothalamus and **inhibits prolactin secretion;** in this context, it is called prolactin-inhibiting factor (PIF).
- is metabolized by MAO and COMT.

(a) D_1 **receptors** activate adenylate cyclase via a G_s protein.
(b) D_2 **receptors** inhibit adenylate cyclase via a G_i protein.
(c) **Parkinson disease** involves degeneration of dopaminergic neurons that use the D_2 receptors.
(d) **Schizophrenia** involves increased levels of D_2 receptors.

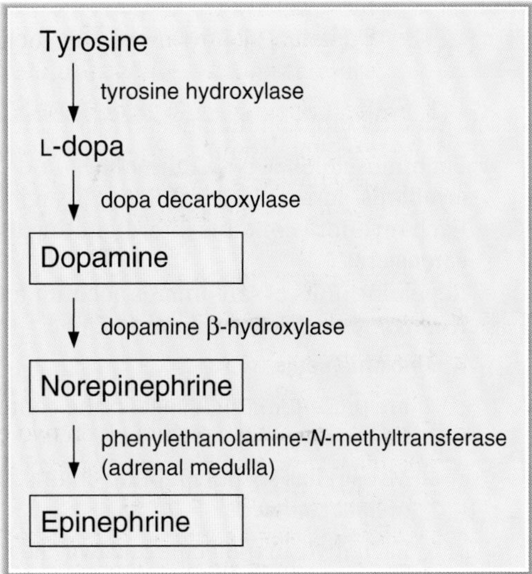

FIGURE 1.10. Synthetic pathway for dopamine, norepinephrine, and epinephrine.

c. Serotonin

- is present in high concentrations in the **brain stem.**
- is formed from tryptophan.
- is converted to melatonin in the pineal gland.

d. Histamine

- is formed from histidine.
- is present in the neurons of the **hypothalamus.**

e. Glutamate

- is the **most prevalent excitatory neurotransmitter** in the brain.
- There are four subtypes of glutamate receptors.
- Three subtypes are **ionotropic receptors** (ligand-gated ion channels) including the **NMDA** (*N*-methyl-D-aspartate) receptor.
- One subtype is a **metabotropic receptor,** which is coupled to ion channels via a heterotrimeric G protein.

f. GABA

- is an **inhibitory neurotransmitter.**
- is synthesized from glutamate by glutamate decarboxylase.
- has two types of receptors:

(1) The **$GABA_A$ receptor** increases Cl^- conductance and is the site of action of **benzodiazepines** and **barbiturates.**

(2) The **$GABA_B$ receptor** increases K^+ conductance.

g. Glycine

- is an **inhibitory neurotransmitter** found primarily in the spinal cord and brain stem.
- increases Cl^- conductance.

h. Nitric oxide (NO)

- is a short-acting **inhibitory neurotransmitter** in the gastrointestinal tract, blood vessels, and the central nervous system.
- is synthesized in presynaptic nerve terminals, where **NO synthase** converts arginine to citrulline and NO.
- is a permeant gas that diffuses from the presynaptic terminal to its target cell.
- also functions in signal transduction of guanylyl cyclase in a variety of tissues, including vascular smooth muscle.

VI. SKELETAL MUSCLE

A. Muscle structure and filaments (Figure 1.11)

- Each muscle fiber is multinucleate and behaves as a single unit. It contains bundles of **myofibrils,** surrounded by **SR** and invaginated by **transverse tubules (T tubules).**
- Each myofibril contains interdigitating **thick and thin filaments** arranged longitudinally in **sarcomeres.**
- Repeating units of sarcomeres account for the unique banding pattern in striated muscle. A sarcomere runs from **Z line to Z line.**

1. Thick filaments

- are present in the **A band** in the center of the sarcomere.
- contain **myosin.**

a. Myosin has six polypeptide chains, including one pair of **heavy chains** and two pairs of **light chains.**

b. Each myosin molecule has **two "heads"** attached to a single "tail." The myosin heads bind ATP and actin and are involved in cross-bridge formation.

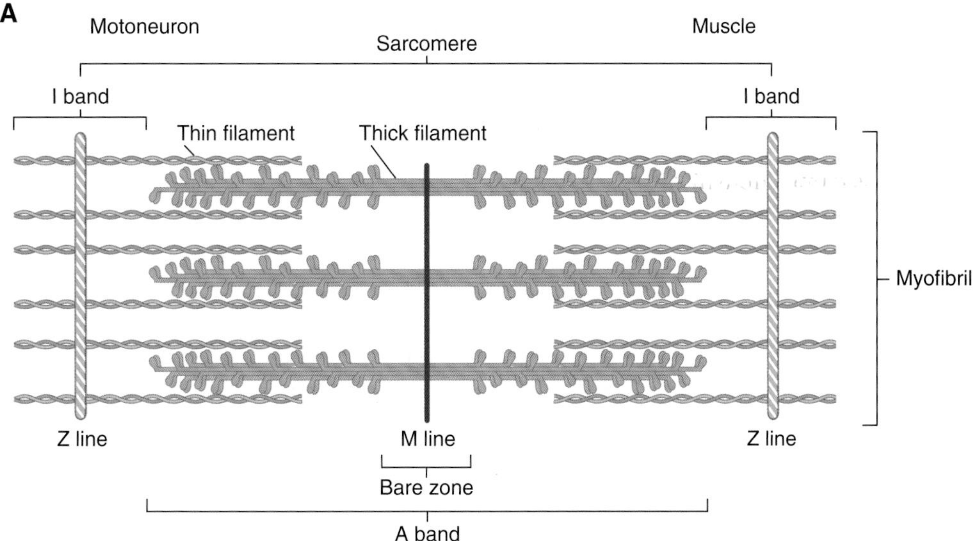

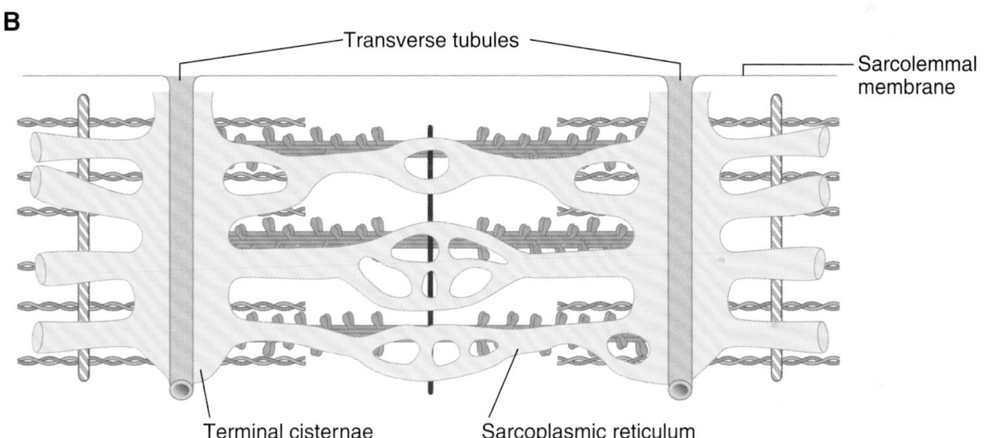

FIGURE 1.11. Structure of the sarcomere in skeletal muscle. **A:** Arrangement of thick and thin filaments. **B:** Transverse tubules and sarcoplasmic reticulum.

2. Thin filaments

- are anchored at the Z lines.
- are present in the **I bands.**
- interdigitate with the thick filaments in a portion of the A band.
- contain **actin, tropomyosin,** and **troponin.**

a. Troponin is the regulatory protein that permits cross-bridge formation when it binds Ca^{2+}.

b. **Troponin** is a complex of three globular proteins:

- **Troponin T** ("T" for tropomyosin) attaches the troponin complex to tropomyosin.
- **Troponin I** ("I" for inhibition) inhibits the interaction of actin and myosin.
- **Troponin C** ("C" for Ca^{2+}) is the Ca^{2+}-binding protein that, when bound to Ca^{2+}, permits the interaction of actin and myosin.

3. **T tubules**
   - are an extensive tubular network, open to the extracellular space, that carry the depolarization from the sarcolemmal membrane to the cell interior.
   - are located at the junctions of A bands and I bands.
   - contain a voltage-sensitive protein called the **dihydropyridine receptor;** depolarization causes a conformational change in the dihydropyridine receptor.
4. **SR**
   - is the internal tubular structure that is the **site of Ca^{2+} storage and release** for excitation–contraction coupling.
   - has **terminal cisternae** that make intimate contact with the T tubules in a triad arrangement.
   - membrane contains **Ca^{2+}-ATPase (Ca^{2+} pump),** which transports Ca^{2+} from intracellular fluid into the SR interior, keeping intracellular [Ca^{2+}] low.
   - contains Ca^{2+} bound loosely to **calsequestrin.**
   - contains a Ca^{2+} release channel called the **ryanodine receptor.**

B. Steps in excitation–contraction coupling in skeletal muscle (Figures 1.12 to 1.14)

1. **Action potentials** in the muscle cell membrane initiate depolarization of the T tubules.
2. **Depolarization of the T tubules** causes a conformational change in its dihydropyridine receptor, which opens **Ca^{2+} release channels** (ryanodine receptors) in the nearby **SR,** causing release of Ca^{2+} from the SR into the intracellular fluid.
3. **Intracellular [Ca^{2+}] increases.**

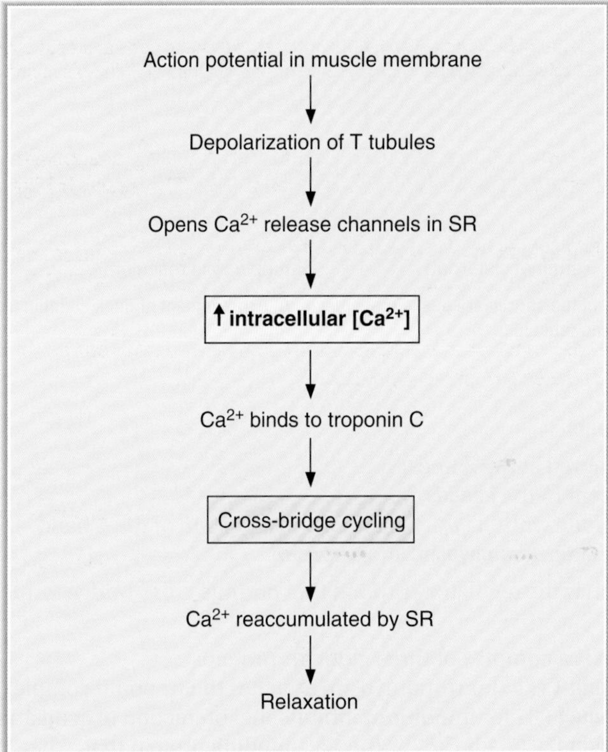

FIGURE 1.12. Steps in excitation–contraction coupling in skeletal muscle. SR = sarcoplasmic reticulum.

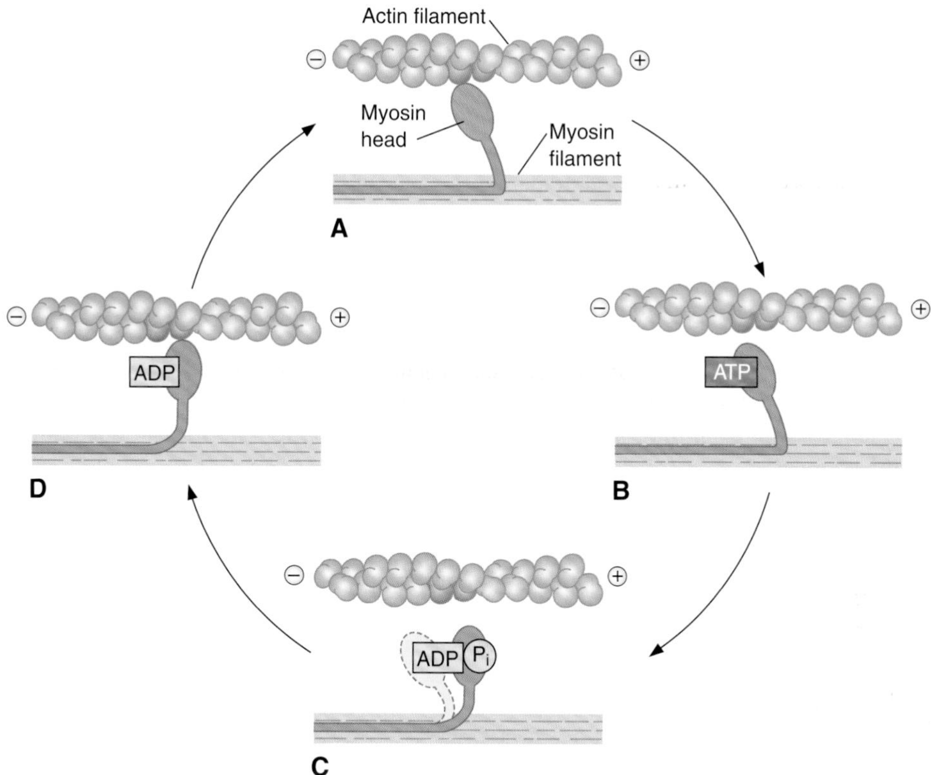

FIGURE 1.13. Cross-bridge cycle. Myosin "walks" toward the plus end of actin to produce shortening and force generation. ADP = adenosine diphosphate; ATP = adenosine triphosphate; P_i = inorganic phosphate.

4. **Ca^{2+} binds to troponin C** on the thin filaments, causing a conformational change in troponin that moves tropomyosin out of the way. The **cross-bridge cycle** begins (see Figure 1.13):
   - a. At first, **no ATP is bound** to myosin **(A)** and myosin is tightly attached to actin. In rapidly contracting muscle, this stage is brief. In the absence of ATP, this state is permanent (i.e., **rigor**).
   - b. **ATP then binds to myosin (B)** producing a conformational change in myosin that causes myosin to be released from actin.

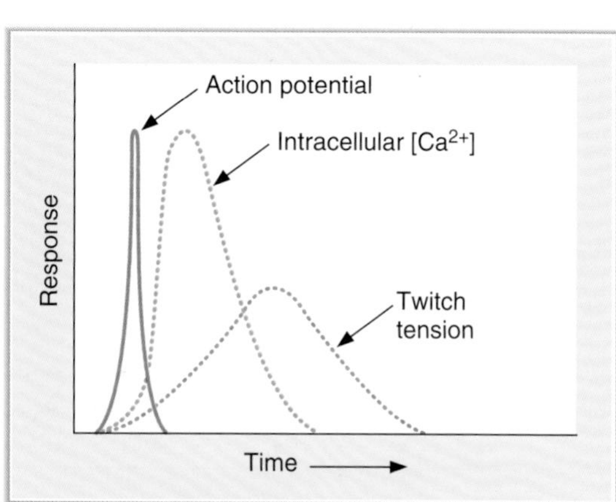

FIGURE 1.14. Relationship of the action potential, the increase in intracellular [Ca^{2+}], and muscle contraction in skeletal muscle.

c. **Myosin is displaced toward the plus end of actin.** There is hydrolysis of ATP to ADP and inorganic phosphate (P_i). ADP remains attached to myosin **(C)**

d. Myosin attaches to a new site on actin, which constitutes the **power (force-generating) stroke (D)** ADP is then released, returning myosin to its rigor state.

e. The cycle repeats as long as Ca^{2+} is bound to troponin C. Each cross-bridge cycle "walks" myosin further along the actin filament.

5. **Relaxation** occurs when Ca^{2+} is reaccumulated by the **SR Ca^{2+}-ATPase** (SERCA). Intracellular Ca^{2+} concentration decreases, Ca^{2+} is released from troponin C, and tropomyosin again blocks the myosin-binding site on actin. As long as intracellular Ca^{2+} concentration is low, cross-bridge cycling cannot occur.

6. **Mechanism of tetanus.** A single action potential causes the release of a standard amount of Ca^{2+} from the SR and produces a single twitch. However, if the muscle is stimulated repeatedly, more Ca^{2+} is released from the SR and there is a cumulative increase in intracellular [Ca^{2+}], extending the time for cross-bridge cycling. The muscle does not relax (tetanus).

C. Length–tension and force–velocity relationships in muscle

- **Isometric contractions** are measured when **length is held constant.** Muscle length **(preload)** is fixed, the muscle is stimulated to contract, and the developed tension is measured. There is ***no* shortening.**
- **Isotonic contractions** are measured when **load is held constant.** The load against which the muscle contracts **(afterload)** is fixed, the muscle is stimulated to contract, and **shortening** is measured.

1. **Length–tension relationship** (Figure 1.15)

   - measures tension developed during **isometric contractions** when the muscle is set to fixed lengths (preload).

   a. **Passive tension** is the tension developed by stretching the muscle to different lengths.

   b. **Total tension** is the tension developed when the muscle is stimulated to contract at different lengths.

   c. **Active tension** is the difference between total tension and passive tension.

   - Active tension represents the active force developed from contraction of the muscle. It can be explained by the cross-bridge cycle model.
   - **Active tension is proportional to the number of cross-bridges formed.** Tension will be maximum when there is maximum overlap of thick and thin filaments. When the muscle is stretched to greater lengths, the number of cross-bridges is reduced because there is less overlap. When muscle length is decreased, the thin filaments collide and tension is reduced.

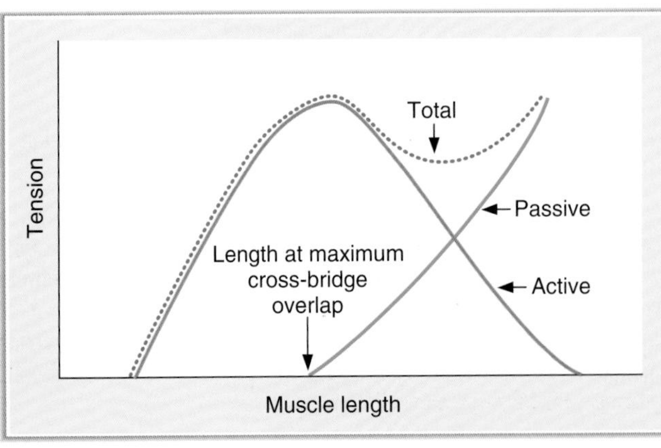

FIGURE 1.15. Length–tension relationship in skeletal muscle.

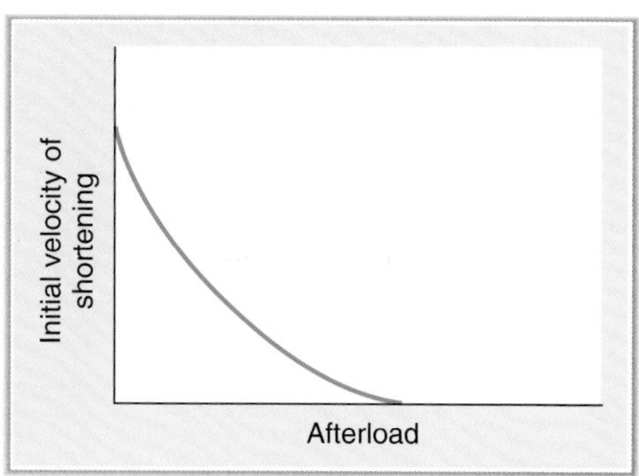

FIGURE 1.16. Force–velocity relationship in skeletal muscle.

2. **Force–velocity relationship** (Figure 1.16)
   - measures the velocity of shortening of **isotonic contractions** when the muscle is challenged with different afterloads (the load against which the muscle must contract).
   - The **velocity of shortening decreases as the afterload increases.**

VII. SMOOTH MUSCLE

- has thick and thin filaments that are not arranged in sarcomeres; therefore, they appear homogeneous rather than striated.

A. Types of smooth muscle

1. **Multiunit smooth muscle**
   - is present in the **iris, ciliary muscle of the lens,** and **vas deferens.**
   - behaves as separate motor units.
   - has little or no electrical coupling between cells.
   - is **densely innervated;** contraction is controlled by neural innervation (e.g., autonomic nervous system).
2. **Unitary (single-unit) smooth muscle**
   - is the most common type and is present in the **uterus, gastrointestinal tract, ureter,** and **bladder.**
   - is spontaneously active (exhibits **slow waves**) and exhibits "pacemaker" activity (see Chapter 6 III A), which is modulated by hormones and neurotransmitters.
   - has a high degree of electrical coupling between cells and, therefore, permits coordinated contraction of the organ (e.g., bladder).
3. **Vascular smooth muscle**
   - has properties of both multiunit and single-unit smooth muscle.

B. Steps in excitation–contraction coupling in smooth muscle (Figure 1.17)

- The mechanism of excitation–contraction coupling is different from that in skeletal muscle.
- There is ***no*** **troponin;** instead, Ca^{2+} regulates myosin on the thick filaments.

1. **Depolarization of the cell membrane opens voltage-gated Ca^{2+} channels** and Ca^{2+} flows into the cell down its electrochemical gradient, increasing the intracellular [Ca^{2+}]. **Hormones and neurotransmitters** may open ligand-gated Ca^{2+} channels in the cell membrane. Ca^{2+} entering the cell causes release of more Ca^{2+} from the SR in a process called **Ca^{2+}-induced Ca^{2+} release**. Hormones and neurotransmitters also directly release Ca^{2+} from the SR through inositol 1,4,5-trisphosphate **(IP_3)–gated Ca^{2+} channels.**

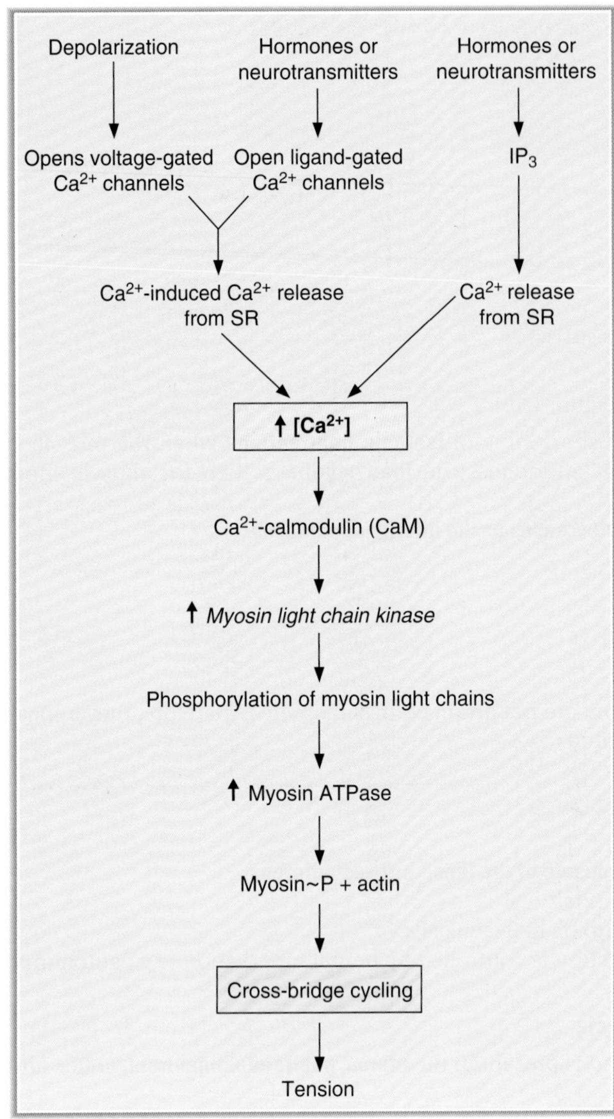

FIGURE 1.17. Sequence of events in contraction of smooth muscle.

2. **Intracellular [Ca^{2+}] increases.**
3. Ca^{2+} binds to **calmodulin.** The Ca^{2+}–calmodulin complex binds to and activates **myosin light chain kinase.** When activated, myosin light chain kinase **phosphorylates myosin** and allows it to bind to actin, thus initiating cross-bridge cycling. The amount of tension produced is proportional to the intracellular Ca^{2+} concentration.
4. A decrease in intracellular [Ca^{2+}] produces relaxation.

VIII. COMPARISON OF SKELETAL MUSCLE, SMOOTH MUSCLE, AND CARDIAC MUSCLE

- Table 1.3 compares the ionic basis for the action potential and mechanism of contraction in skeletal muscle, smooth muscle, and cardiac muscle.
- Cardiac muscle is discussed in Chapter 3.

table 1.3 Comparison of Skeletal, Smooth, and Cardiac Muscles

Feature	Skeletal Muscle	Smooth Muscle	Cardiac Muscle
Appearance	Striated	No striations	Striated
Upstroke of action potential	Inward Na^+ current	Inward Ca^{2+} current	Inward Ca^{2+} current (SA node) Inward Na^+ current (atria, ventricles, Purkinje fibers)
Plateau	No	No	No (SA node) Yes (atria, ventricles, Purkinje fibers; due to inward Ca^{2+} current)
Duration of action potential	~1 msec	~10 msec	150 msec (SA node, atria) 250–300 msec (ventricles and Purkinje fibers)
Excitation–contraction coupling	Action potential → T tubules	Action potential opens voltage-gated Ca^{2+} channels in cell membrane	Inward Ca^{2+} current during plateau of action potential
	Ca^{2+} released from nearby SR		Ca^{2+}-induced Ca^{2+} release from SR
	↑ $[Ca^{2+}]_i$	Hormones and transmitters open IP_3-gated Ca^{2+} channels in SR	↑ $[Ca^{2+}]_i$
Molecular basis for contraction	Ca^{2+}–troponin C	Ca^{2+}–calmodulin ↑ myosin-light-chain kinase	Ca^{2+}–troponin C

IP_3 = inositol 1,4,5-triphosphate; SA = sinoatrial; SR = sarcoplasmic reticulum.

Review Test

1. Which of the following characteristics is shared by simple and facilitated diffusion of glucose?

(A) Occurs down an electrochemical gradient
(B) Is saturable
(C) Requires metabolic energy
(D) Is inhibited by the presence of galactose
(E) Requires a Na^+ gradient

2. During the upstroke of the nerve action potential

(A) there is net outward current and the cell interior becomes more negative
(B) there is net outward current and the cell interior becomes less negative
(C) there is net inward current and the cell interior becomes more negative
(D) there is net inward current and the cell interior becomes less negative

3. Solutions A and B are separated by a semipermeable membrane that is permeable to K^+ but not to Cl^-. Solution A is 100 mM KCl, and solution B is 1 mM KCl. Which of the following statements about solution A and solution B is true?

(A) K^+ ions will diffuse from solution A to solution B until the $[K^+]$ of both solutions is 50.5 mM
(B) K^+ ions will diffuse from solution B to solution A until the $[K^+]$ of both solutions is 50.5 mM
(C) KCl will diffuse from solution A to solution B until the [KCl] of both solutions is 50.5 mM
(D) K^+ will diffuse from solution A to solution B until a membrane potential develops with solution A negative with respect to solution B
(E) K^+ will diffuse from solution A to solution B until a membrane potential develops with solution A positive with respect to solution B

4. The correct temporal sequence for events at the neuromuscular junction is

(A) action potential in the motor nerve; depolarization of the muscle end plate; uptake of Ca^{2+} into the presynaptic nerve terminal
(B) uptake of Ca^{2+} into the presynaptic terminal; release of acetylcholine (ACh); depolarization of the muscle end plate
(C) release of ACh; action potential in the motor nerve; action potential in the muscle
(D) uptake of Ca^{2+} into the motor end plate; action potential in the motor end plate; action potential in the muscle
(E) release of ACh; action potential in the muscle end plate; action potential in the muscle

5. Which characteristic or component is shared by skeletal muscle and smooth muscle?

(A) Thick and thin filaments arranged in sarcomeres
(B) Troponin
(C) Elevation of intracellular $[Ca^{2+}]$ for excitation–contraction coupling
(D) Spontaneous depolarization of the membrane potential
(E) High degree of electrical coupling between cells

6. Repeated stimulation of a skeletal muscle fiber causes a sustained contraction (tetanus). Accumulation of which solute in intracellular fluid is responsible for the tetanus?

(A) Na^+
(B) K^+
(C) Cl^-
(D) Mg^{2+}
(E) Ca^{2+}
(F) Troponin
(G) Calmodulin
(H) Adenosine triphosphate (ATP)

7. Solutions A and B are separated by a membrane that is permeable to Ca^{2+} and impermeable to Cl^-. Solution A contains 10 mM $CaCl_2$, and solution B contains 1 mM $CaCl_2$. Assuming that $2.3\ RT/F = 60$ mV, Ca^{2+} will be at electrochemical equilibrium when

(A) solution A is +60 mV
(B) solution A is +30 mV

(C) solution A is –60 mV
(D) solution A is –30 mV
(E) solution A is +120 mV
(F) solution A is –120 mV
(G) the Ca^{2+} concentrations of the two solutions are equal
(H) the concentrations of the two solutions are equal

8. A 42-year-old man with myasthenia gravis notes increased muscle strength when he is treated with an acetylcholinesterase (AChE) inhibitor. The basis for his improvement is increased

(A) amount of acetylcholine (ACh) released from motor nerves
(B) levels of ACh at the muscle end plates
(C) number of ACh receptors on the muscle end plates
(D) amount of norepinephrine released from motor nerves
(E) synthesis of norepinephrine in motor nerves

9. In a hospital error, a 60-year-old woman is infused with large volumes of a solution that causes lysis of her red blood cells (RBCs). The solution was most likely

(A) 150 mM NaCl
(B) 300 mM mannitol
(C) 350 mM mannitol
(D) 300 mM urea
(E) 150 mM $CaCl_2$

10. During a nerve action potential, a stimulus is delivered as indicated by the arrow shown in the following figure. In response to the stimulus, a second action potential

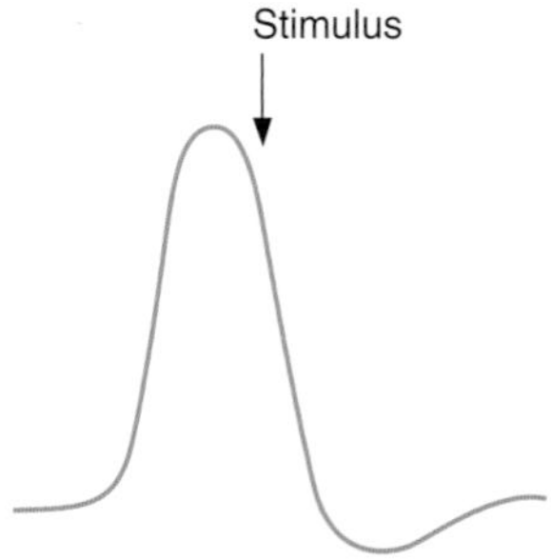

(A) of smaller magnitude will occur
(B) of normal magnitude will occur
(C) of normal magnitude will occur but will be delayed
(D) will occur but will not have an overshoot
(E) will not occur

11. Solutions A and B are separated by a membrane that is permeable to urea. Solution A is 10 mM urea, and solution B is 5 mM urea. If the concentration of urea in solution A is doubled, the flux of urea across the membrane will

(A) double
(B) triple
(C) be unchanged
(D) decrease to one-half
(E) decrease to one-third

12. A muscle cell has an intracellular [Na^+] of 14 mM and an extracellular [Na^+] of 140 mM. Assuming that 2.3 RT/F = 60 mV, what would the membrane potential be if the muscle cell membrane were permeable only to Na^+?

(A) –80 mV
(B) –60 mV
(C) 0 mV
(D) +60 mV
(E) +80 mV

QUESTIONS 13–15

The following diagram of a nerve action potential applies to Questions 13–15.

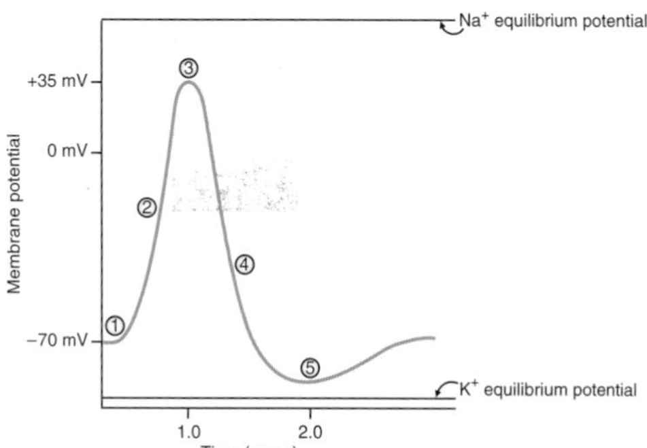

13. At which labeled point on the action potential is K^+ closest to electrochemical equilibrium?

(A) 1
(B) 2
(C) 3
(D) 4
(E) 5

14. What process is responsible for the change in membrane potential that occurs between point 1 and point 3?

(A) Movement of Na^+ into the cell
(B) Movement of Na^+ out of the cell
(C) Movement of K^+ into the cell

(D) Movement of K^+ out of the cell
(E) Activation of the Na^+–K^+ pump
(F) Inhibition of the Na^+–K^+ pump

15. What process is responsible for the change in membrane potential that occurs between point 3 and point 4?

(A) Movement of Na^+ into the cell
(B) Movement of Na^+ out of the cell
(C) Movement of K^+ into the cell
(D) Movement of K^+ out of the cell
(E) Activation of the Na^+–K^+ pump
(F) Inhibition of the Na^+–K^+ pump

16. The velocity of conduction of action potentials along a nerve will be increased by

(A) stimulating the Na^+–K^+ pump
(B) inhibiting the Na^+–K^+ pump
(C) decreasing the diameter of the nerve
(D) myelinating the nerve
(E) lengthening the nerve fiber

17. Solutions A and B are separated by a semipermeable membrane. Solution A contains 1 mM sucrose and 1 mM urea. Solution B contains 1 mM sucrose. The reflection coefficient for sucrose is one, and the reflection coefficient for urea is zero. Which of the following statements about these solutions is correct?

(A) Solution A has a higher effective osmotic pressure than does solution B
(B) Solution A has a lower effective osmotic pressure than does solution B
(C) Solutions A and B are isosmotic
(D) Solution A is hyperosmotic with respect to solution B, and the solutions are isotonic
(E) Solution A is hyposmotic with respect to solution B, and the solutions are isotonic

18. Transport of D- and L-glucose proceeds at the same rate down an electrochemical gradient by which of the following processes?

(A) Simple diffusion
(B) Facilitated diffusion
(C) Primary active transport
(D) Cotransport
(E) Countertransport

19. Which of the following will double the permeability of a solute in a lipid bilayer?

(A) Doubling the molecular radius of the solute
(B) Doubling the oil/water partition coefficient of the solute
(C) Doubling the thickness of the bilayer
(D) Doubling the concentration difference of the solute across the bilayer

20. A newly developed local anesthetic blocks Na^+ channels in nerves. Which of the following effects on the action potential would it be expected to produce?

(A) Decrease the rate of rise of the upstroke of the action potential
(B) Shorten the absolute refractory period
(C) Abolish the hyperpolarizing afterpotential
(D) Increase the Na^+ equilibrium potential
(E) Decrease the Na^+ equilibrium potential

21. At the muscle end plate, acetylcholine (ACh) causes the opening of

(A) Na^+ channels and depolarization toward the Na^+ equilibrium potential
(B) K^+ channels and depolarization toward the K^+ equilibrium potential
(C) Ca^{2+} channels and depolarization toward the Ca^{2+} equilibrium potential
(D) Na^+ and K^+ channels and depolarization to a value halfway between the Na^+ and K^+ equilibrium potentials
(E) Na^+ and K^+ channels and hyperpolarization to a value halfway between the Na^+ and K^+ equilibrium potentials

22. An inhibitory postsynaptic potential

(A) depolarizes the postsynaptic membrane by opening Na^+ channels
(B) depolarizes the postsynaptic membrane by opening K^+ channels
(C) hyperpolarizes the postsynaptic membrane by opening Ca^{2+} channels
(D) hyperpolarizes the postsynaptic membrane by opening Cl^- channels

23. Which of the following would occur as a result of the inhibition of Na^+, K^+-ATPase?

(A) Decreased intracellular Na^+ concentration
(B) Increased intracellular K^+ concentration
(C) Increased intracellular Ca^{2+} concentration
(D) Increased Na^+–glucose cotransport
(E) Increased Na^+–Ca^{2+} exchange

24. Which of the following temporal sequences is correct for excitation–contraction coupling in skeletal muscle?

(A) Increased intracellular [Ca^{2+}]; action potential in the muscle membrane; cross-bridge formation
(B) Action potential in the muscle membrane; depolarization of the T tubules; release of Ca^{2+} from the sarcoplasmic reticulum (SR)

(C) Action potential in the muscle membrane; splitting of adenosine triphosphate (ATP); binding of Ca^{2+} to troponin C
(D) Release of Ca^{2+} from the SR; depolarization of the T tubules; binding of Ca^{2+} to troponin C

25. Which of the following transport processes is involved if transport of glucose from the intestinal lumen into a small intestinal cell is inhibited by abolishing the usual Na^+ gradient across the cell membrane?

(A) Simple diffusion
(B) Facilitated diffusion
(C) Primary active transport
(D) Cotransport
(E) Countertransport

26. In skeletal muscle, which of the following events occurs before depolarization of the T tubules in the mechanism of excitation–contraction coupling?

(A) Depolarization of the sarcolemmal membrane
(B) Opening of Ca^{2+} release channels on the sarcoplasmic reticulum (SR)
(C) Uptake of Ca^{2+} into the SR by Ca^{2+}-adenosine triphosphatase (ATPase)
(D) Binding of Ca^{2+} to troponin C
(E) Binding of actin and myosin

27. Which of the following is an inhibitory neurotransmitter in the central nervous system (CNS)?

(A) Norepinephrine
(B) Glutamate
(C) γ-Aminobutyric acid (GABA)
(D) Serotonin
(E) Histamine

28. Adenosine triphosphate (ATP) is used indirectly for which of the following processes?

(A) Accumulation of Ca^{2+} by the sarcoplasmic reticulum (SR)
(B) Transport of Na^+ from intracellular to extracellular fluid
(C) Transport of K^+ from extracellular to intracellular fluid
(D) Transport of H^+ from parietal cells into the lumen of the stomach
(E) Absorption of glucose by intestinal epithelial cells

29. Which of the following causes rigor in skeletal muscle?

(A) Lack of action potentials in motoneurons
(B) An increase in intracellular Ca^{2+} level
(C) A decrease in intracellular Ca^{2+} level
(D) An increase in adenosine triphosphate (ATP) level
(E) A decrease in ATP level

30. Degeneration of dopaminergic neurons has been implicated in

(A) schizophrenia
(B) Parkinson disease
(C) myasthenia gravis
(D) curare poisoning

31. Assuming complete dissociation of all solutes, which of the following solutions would be hyperosmotic to 1 mM NaCl?

(A) 1 mM glucose
(B) 1.5 mM glucose
(C) 1 mM $CaCl_2$
(D) 1 mM sucrose
(E) 1 mM KCl

32. A new drug is developed that blocks the transporter for H^+ secretion in gastric parietal cells. Which of the following transport processes is being inhibited?

(A) Simple diffusion
(B) Facilitated diffusion
(C) Primary active transport
(D) Cotransport
(E) Countertransport

33. A 56-year-old woman with severe muscle weakness is hospitalized. The only abnormality in her laboratory values is an elevated serum K^+ concentration. The elevated serum K^+ causes muscle weakness because

(A) the resting membrane potential is hyperpolarized
(B) the K^+ equilibrium potential is hyperpolarized
(C) the Na^+ equilibrium potential is hyperpolarized
(D) K^+ channels are closed by depolarization
(E) K^+ channels are opened by depolarization
(F) Na^+ channels are closed by depolarization
(G) Na^+ channels are opened by depolarization

34. In contraction of gastrointestinal smooth muscle, which of the following events occurs after binding of Ca^{2+} to calmodulin?

(A) Depolarization of the sarcolemmal membrane
(B) Ca^{2+}-induced Ca^{2+} release
(C) Increased myosin light chain kinase
(D) Increased intracellular Ca^{2+} concentration
(E) Opening of ligand-gated Ca^{2+} channels

35. In an experimental preparation of a nerve axon, membrane potential (E_m), K^+ equilibrium potential, and K^+ conductance can be measured. Which combination of values will create the largest outward current flow?

	E_m (mV)	E_K (mV)	K conductance (relative units)
(A)	−90	−90	1
(B)	−100	−90	1
(C)	−50	−90	1
(D)	0	−90	1
(E)	+20	−90	1
(F)	−90	−90	2

36. A 68-year-old man with oat cell carcinoma of the lung has a grand mal seizure at home. In the emergency room, based on measurement of the man's plasma osmolarity, the physician diagnosed him with syndrome of inappropriate antidiuretic hormone (SIADH) and treated him immediately with hypertonic saline to prevent another seizure. Which of the following is the most likely value of the man's plasma osmolarity before treatment?

(A) 235 mOSm/L
(B) 290 mOsm/L
(C) 300 mOsm/L
(D) 320 mOSm/L
(E) 330 mOSm/L

Answers and Explanations

1. **The answer is A** [II A 1, C]. Both types of transport occur down an electrochemical gradient ("downhill") and do not require metabolic energy. Saturability and inhibition by other sugars are characteristic only of carrier-mediated glucose transport; thus, facilitated diffusion is saturable and inhibited by galactose, whereas simple diffusion is not.
2. **The answer is D** [IV E 1 a, b, 2 b]. During the upstroke of the action potential, the cell depolarizes or becomes less negative. The depolarization is caused by inward current, which is, by definition, the movement of positive charge into the cell. In nerve and in most types of muscle, this inward current is carried by Na^+.
3. **The answer is D** [IV B]. Because the membrane is permeable only to K^+ ions, K^+ will diffuse down its concentration gradient from solution A to solution B, leaving some Cl^- ions behind in solution A. A diffusion potential will be created, with solution A negative with respect to solution B. Generation of a diffusion potential involves movement of only a few ions and, therefore, does not cause a change in the concentration of the bulk solutions.
4. **The answer is B** [V B 1–6]. Acetylcholine (ACh) is stored in vesicles and is released when an action potential in the motor nerve opens Ca^{2+} channels in the presynaptic terminal. ACh diffuses across the synaptic cleft and opens Na^+ and K^+ channels in the muscle end plate, depolarizing it (but not producing an action potential). Depolarization of the muscle end plate causes local currents in adjacent muscle membrane, depolarizing the membrane to threshold and producing action potentials.
5. **The answer is C** [VI A, B 1–4; VII B 1–4]. An elevation of intracellular [Ca^{2+}] is common to the mechanism of excitation–contraction coupling in skeletal and smooth muscle. In skeletal muscle, Ca^{2+} binds to troponin C, initiating the cross-bridge cycle. In smooth muscle, Ca^{2+} binds to calmodulin. The Ca^{2+}–calmodulin complex activates myosin light chain kinase, which phosphorylates myosin so that shortening can occur. The striated appearance of the sarcomeres and the presence of troponin are characteristic of skeletal, not smooth, muscle. Spontaneous depolarizations and gap junctions are characteristics of unitary smooth muscle but not skeletal muscle.
6. **The answer is E** [VI B 6]. During repeated stimulation of a muscle fiber, Ca^{2+} is released from the sarcoplasmic reticulum (SR) more quickly than it can be reaccumulated; therefore, the intracellular [Ca^{2+}] does not return to resting levels as it would after a single twitch. The increased [Ca^{2+}] allows more cross-bridges to form and, therefore, produces increased tension (tetanus). Intracellular Na^+ and K^+ concentrations do not change during the action potential. Very few Na^+ or K^+ ions move into or out of the muscle cell, so bulk concentrations are unaffected. Adenosine triphosphate (ATP) levels would, if anything, decrease during tetanus.
7. **The answer is D** [IV B]. The membrane is permeable to Ca^{2+} but impermeable to Cl^-. Although there is a concentration gradient across the membrane for both ions, only Ca^{2+} can diffuse down this gradient. Ca^{2+} will diffuse from solution A to solution B, leaving negative charge behind in solution A. The magnitude of this voltage can be calculated for electrochemical equilibrium with the Nernst equation as follows: $E_{Ca^{2+}} = 2.3\,RT / zF \log C_A / C_B = 60\text{ mV} / +2 \log 10\text{ mM} / 1\text{ mM} = 30\text{ mV} \log 10 = 30\text{ mV}$. The sign is determined with an intuitive approach—Ca^{2+} diffuses from solution A to solution B, so solution A develops a negative voltage (−30 mV). Net diffusion of Ca^{2+} will cease when this voltage is achieved, that is, when the chemical driving force is exactly balanced by the electrical driving force (not when the Ca^{2+} concentrations of the solutions become equal).

8. The answer is B [V B 8]. Myasthenia gravis is characterized by a decreased density of acetylcholine (ACh) receptors at the muscle end plate. An acetylcholinesterase (AChE) inhibitor blocks degradation of ACh in the neuromuscular junction, so levels at the muscle end plate remain high, partially compensating for the deficiency of receptors.

9. The answer is D [III B 2 d]. Lysis of the patient's red blood cells (RBCs) was caused by entry of water and swelling of the cells to the point of rupture. Water would flow into the RBCs if the extracellular fluid became hypotonic (had a lower osmotic pressure) relative to the intracellular fluid. By definition, isotonic solutions do not cause water to flow into or out of cells because the osmotic pressure is the same on both sides of the cell membrane. Hypertonic solutions would cause shrinkage of the RBCs. 150 mM NaCl and 300 mM mannitol are isotonic. 350 mM mannitol and 150 mM $CaCl_2$ are hypertonic. Because the reflection coefficient of urea is <1.0, 300 mM urea is hypotonic.

10. The answer is E [IV E 3 a]. Because the stimulus was delivered during the absolute refractory period, no action potential occurs. The inactivation gates of the Na^+ channel were closed by depolarization and remain closed until the membrane is repolarized. As long as the inactivation gates are closed, the Na^+ channels cannot be opened to allow for another action potential.

11. The answer is B [II A]. Flux is proportional to the concentration difference across the membrane, $J = -PA\ (C_A - C_B)$. Originally, $C_A - C_B = 10\text{ mM} - 5\text{ mM} = 5\text{ mM}$. When the urea concentration was doubled in solution A, the concentration difference became 20 mM – 5 mM = 15 mM or three times the original difference. Therefore, the flux would also triple. Note that the negative sign preceding the equation is ignored if the lower concentration is subtracted from the higher concentration.

12. The answer is D [IV B 3 a, b]. The Nernst equation is used to calculate the equilibrium potential for a single ion. In applying the Nernst equation, we assume that the membrane is freely permeable to that ion alone. $E_{Na^+} = 2.3\ RT/zF \log C_e/C_i = 60\text{ mV} \log 140/14 = 60\text{ mV} \log 10 = 60\text{ mV}$. Notice that the signs were ignored and that the higher concentration was simply placed in the numerator to simplify the log calculation. To determine whether E_{Na^+} is +60 mV or –60 mV, use the intuitive approach—Na^+ will diffuse from extracellular to intracellular fluid down its concentration gradient, making the cell interior positive.

13. The answer is E [IV E 2 d]. The hyperpolarizing afterpotential represents the period during which K^+ permeability is highest, and the membrane potential is closest to the K^+ equilibrium potential. At that point, K^+ is closest to electrochemical equilibrium. The force driving K^+ movement out of the cell down its chemical gradient is balanced by the force driving K^+ into the cell down its electrical gradient.

14. The answer is A [IV E 2 b (1)–(3)]. The upstroke of the nerve action potential is caused by opening of the Na^+ channels (once the membrane is depolarized to threshold). When the Na^+ channels open, Na^+ moves into the cell down its electrochemical gradient, driving the membrane potential toward the Na^+ equilibrium potential.

15. The answer is D [IV E 2 c]. The process responsible for repolarization is the opening of K^+ channels. The K^+ permeability becomes very high and drives the membrane potential toward the K^+ equilibrium potential by flow of K^+ out of the cell.

16. The answer is D [IV E 4 b]. Myelin insulates the nerve, thereby increasing conduction velocity; action potentials can be generated only at the nodes of Ranvier, where there are breaks in the insulation. Activity of the Na^+–K^+ pump does not directly affect the formation or conduction of action potentials. Decreasing nerve diameter would increase internal resistance and, therefore, slow the conduction velocity.

17. The answer is D [III A, B 4]. Solution A contains both sucrose and urea at concentrations of 1 mM, whereas solution B contains only sucrose at a concentration of 1 mM. The calculated osmolarity of solution A is 2 mOsm/L, and the calculated osmolarity of solution B is 1 mOsm/L. Therefore, solution A, which has a higher osmolarity, is hyperosmotic with respect to solution B. Actually, solutions A and B have the same effective osmotic pressure (i.e., they are isotonic) because the only "effective" solute is sucrose, which has the same concentration in both solutions. Urea is not an effective solute because its reflection coefficient is zero.

18. **The answer is A** [II A 1, C 1]. Only two types of transport occur "downhill"—simple and facilitated diffusion. If there is no stereospecificity for the D- or L-isomer, one can conclude that the transport is not carrier mediated and, therefore, must be simple diffusion.

19. **The answer is B** [II A 4 a–c]. Increasing the oil/water partition coefficient increases solubility in a lipid bilayer and therefore increases permeability. Increasing molecular radius and increased membrane thickness decrease permeability. The concentration difference of the solute has no effect on permeability.

20. **The answer is A** [IV E 1–3]. Blockade of the Na^+ channels would prevent action potentials. The upstroke of the action potential depends on the entry of Na^+ into the cell through these channels and therefore would also be reduced or abolished. The absolute refractory period would be lengthened because it is based on the availability of the Na^+ channels. The hyperpolarizing afterpotential is related to increased K^+ permeability. The Na^+ equilibrium potential is calculated from the Nernst equation and is the theoretical potential at electrochemical equilibrium (and does not depend on whether the Na^+ channels are open or closed).

21. **The answer is D** [V B 5]. Binding of acetylcholine (ACh) to receptors in the muscle end plate opens channels that allow passage of both Na^+ and K^+ ions. Na^+ ions will flow into the cell down its electrochemical gradient, and K^+ ions will flow out of the cell down its electrochemical gradient. The resulting membrane potential will be depolarized to a value that is approximately halfway between their respective equilibrium potentials.

22. **The answer is D** [V C 2 b]. An inhibitory postsynaptic potential hyperpolarizes the postsynaptic membrane, taking it farther from threshold. Opening Cl^- channels would hyperpolarize the postsynaptic membrane by driving the membrane potential toward the Cl^- equilibrium potential (about −90 mV). Opening Ca^{2+} channels would depolarize the postsynaptic membrane by driving it toward the Ca^{2+} equilibrium potential.

23. **The answer is C** [II D 2 a]. Inhibition of Na^+, K^+-adenosine triphosphatase (ATPase) leads to an increase in intracellular Na^+ concentration. Increased intracellular Na^+ concentration decreases the Na^+ gradient across the cell membrane, thereby inhibiting Na^+–Ca^{2+} exchange and causing an increase in intracellular Ca^{2+} concentration. Increased intracellular Na^+ concentration also inhibits Na^+–glucose cotransport.

24. **The answer is B** [VI B 1–4]. The correct sequence is action potential in the muscle membrane; depolarization of the T tubules; release of Ca^{2+} from the sarcoplasmic reticulum (SR); binding of Ca^{2+} to troponin C; cross-bridge formation; and splitting of adenosine triphosphate (ATP).

25. **The answer is D** [II D 2 a, E 1]. In the "usual" Na^+ gradient, the $[Na^+]$ is higher in extracellular than in intracellular fluid (maintained by the Na^+–K^+ pump). Two forms of transport are energized by this Na^+ gradient—cotransport and countertransport. Because glucose is moving in the same direction as Na^+, one can conclude that it is cotransport.

26. **The answer is A** [VI A 3]. In the mechanism of excitation–contraction coupling, excitation always precedes contraction. Excitation refers to the electrical activation of the muscle cell, which begins with an action potential (depolarization) in the sarcolemmal membrane that spreads to the T tubules. Depolarization of the T tubules then leads to the release of Ca^{2+} from the nearby sarcoplasmic reticulum (SR), followed by an increase in intracellular Ca^{2+} concentration, binding of Ca^{2+} to troponin C, and then contraction.

27. **The answer is C** [V C 2 a, b]. γ-Aminobutyric acid (GABA) is an inhibitory neurotransmitter. Norepinephrine, glutamate, serotonin, and histamine are excitatory neurotransmitters.

28. **The answer is E** [II D 2]. All of the processes listed are examples of primary active transport (and therefore use adenosine triphosphate [ATP] directly), except for absorption of glucose by intestinal epithelial cells, which occurs by secondary active transport

(i.e., cotransport). Secondary active transport uses the Na^+ gradient as an energy source and, therefore, uses ATP indirectly (to maintain the Na^+ gradient).

29. The answer is E [VI B]. Rigor is a state of permanent contraction that occurs in skeletal muscle when adenosine triphosphate (ATP) levels are depleted. With no ATP bound, myosin remains attached to actin and the cross-bridge cycle cannot continue. If there were no action potentials in motoneurons, the muscle fibers they innervate would not contract at all, since action potentials are required for release of Ca^{2+} from the sarcoplasmic reticulum (SR). When intracellular Ca^{2+} concentration increases, Ca^{2+} binds troponin C, permitting the cross-bridge cycle to occur. Decreases in intracellular Ca^{2+} concentration cause relaxation.

30. The answer is B [V C 4 b (3)]. Dopaminergic neurons and D_2 receptors are deficient in people with Parkinson disease. Schizophrenia involves increased levels of D_2 receptors. Myasthenia gravis and curare poisoning involve the neuromuscular junction, which uses acetylcholine (ACh) as a neurotransmitter.

31. The answer is C [III A]. Osmolarity is the concentration of particles (osmolarity = g × C). When two solutions are compared, that with the higher osmolarity is hyperosmotic. The 1-mM $CaCl_2$ solution (osmolarity = 3 mOsm/L) is hyperosmotic to 1 mM NaCl (osmolarity = 2 mOsm/L). The 1-mM glucose, 1.5-mM glucose, and 1-mM sucrose solutions are hyposmotic to 1 mM NaCl, whereas 1 mM KCl is isosmotic.

32. The answer is C [II D c]. H^+ secretion by gastric parietal cells occurs by H^+–K^+ adenosine triphosphatase (ATPase), a primary active transporter.

33. The answer is F [IV E 2]. Elevated serum K^+ concentration causes depolarization of the K^+ equilibrium potential and therefore depolarization of the resting membrane potential in skeletal muscle. Sustained depolarization closes the inactivation gates on Na^+ channels and prevents the occurrence of action potentials in the muscle.

34. The answer is C [VII B]. The steps that produce contraction in smooth muscle occur in the following order: various mechanisms that raise intracellular Ca^{2+} concentration, including depolarization of the sarcolemmal membrane, which opens voltage-gated Ca^{2+} channels, and opening of ligand-gated Ca^{2+} channels; Ca^{2+}-induced Ca^{2+} released from SR; increased intracellular Ca^{2+} concentration; binding of Ca^{2+} to calmodulin; increased myosin-light-chain kinase; phosphorylation of myosin; binding of myosin to actin; and cross-bridge cycling, which produces contraction.

35. The answer is E [IV C]. Data sets A and F have no difference between membrane potential (E_m) and E_K and thus have no driving force or current flow; although data set F has the higher K^+ conductance, this is irrelevant since the driving force is zero. Data sets C, D, and E all will have outward K^+ current, since E_m is less negative than E_K; of these, data set E will have the largest outward K^+ current because it has the highest driving force. Data set B will have inward K^+ current since E_m is more negative than E_K.

36. The answer is A [III B]. The man has syndrome of inappropriate antidiuretic hormone (SIADH) and suffers a grand mal seizure, suggesting that inappropriately high ADH levels led to excessive water reabsorption in the collecting ducts of the kidney. The excess water retention diluted his extracellular osmolarity and caused decreased extracellular osmotic pressure. The decreased extracellular osmotic pressure caused osmotic water flow into all cells, including brain cells, increasing intracellular volume and causing cell swelling. Since the brain is encased in a fixed structure, the skull, swelling of brain cells led to the seizure. The other piece of evidence suggesting that the problem is decreased extracellular osmolarity (and osmotic pressure) is that the treatment used to prevent another seizure is hypertonic saline; the hypertonic saline treatment would increase extracellular fluid osmolarity, thus reducing or eliminating the osmotic driving force that caused water flow into cells. Of the answers, the only value for pretreatment plasma osmolarity that is hyposmotic and hypotonic is 235 mOsm/L. The other values are either essentially isosmotic (290 and 300 mOsm/L) or hyperosmotic (320 and 330 mOsm/L).

Chapter 2

Neurophysiology

I. AUTONOMIC NERVOUS SYSTEM (ANS)

- is a set of pathways to and from the central nervous system (CNS) that innervates and regulates **smooth muscle**, **cardiac muscle**, and **glands**.
- is distinct from the somatic nervous system, which innervates skeletal muscle.
- has three divisions: **sympathetic**, **parasympathetic**, and **enteric** (the enteric division is discussed in Chapter 6).

A. Organization of the ANS (Table 2.1 and Figure 2.1)

1. **Synapses between neurons are made in the autonomic ganglia.**
   a. **Parasympathetic ganglia** are located in or near the effector organs.
   b. **Sympathetic ganglia** are located in the paravertebral chain.
2. **Preganglionic neurons** have their cell bodies in the CNS and synapse in autonomic ganglia.
   - Preganglionic neurons of the **sympathetic nervous system** originate in spinal cord segments T1–L3 or the **thoracolumbar** region.
   - Preganglionic neurons of the **parasympathetic nervous system** originate in the nuclei of cranial nerves and in spinal cord segments S2–S4 or the **craniosacral** region.
3. **Postganglionic neurons** of both divisions have their cell bodies in the autonomic ganglia and synapse on effector organs (e.g., heart, blood vessels, sweat glands).
4. **Adrenal medulla** is a specialized ganglion of the sympathetic nervous system.
   - Preganglionic fibers synapse directly on **chromaffin cells** in the adrenal medulla.
   - The chromaffin cells secrete **epinephrine** (80%) and norepinephrine (20%) into the circulation (see Figure 2.1).
   - **Pheochromocytoma** is a tumor of the adrenal medulla that secretes excessive amounts of catecholamines and is associated with increased excretion of 3-methoxy-4-hydroxymandelic acid **(VMA)**.

B. Neurotransmitters of the ANS

- **Adrenergic neurons** release **norepinephrine** as the neurotransmitter.
- **Cholinergic neurons**, whether in the sympathetic or parasympathetic nervous system, release **acetylcholine (ACh)** as the neurotransmitter.
- **Nonadrenergic, noncholinergic neurons** include *some* postganglionic parasympathetic neurons of the gastrointestinal (GI) tract, which release substance P, vasoactive intestinal peptide (VIP), or nitric oxide (NO).

table 2.1 Organization of the Autonomic Nervous System

Characteristic	Sympathetic	Parasympathetic	Somatic*
Origin of preganglionic nerve	Nuclei of spinal cord segments T1–T12; L1–L3 (thoracolumbar)	Nuclei of cranial nerves III, VII, IX, and X; spinal cord segments S2–S4 (craniosacral)	
Length of preganglionic nerve axon	Short	Long	
Neurotransmitter in ganglion	ACh	ACh	
Receptor type in ganglion	Nicotinic	Nicotinic	
Length of postganglionic nerve axon	Long	Short	
Effector organs	Smooth and cardiac muscle; glands	Smooth and cardiac muscle; glands	Skeletal muscle
Neurotransmitter in effector organs	Norepinephrine (except sweat glands, which use ACh)	ACh	ACh (synapse is neuromuscular junction)
Receptor types in effector organs	α_1, α_2, β_1, and β_2	Muscarinic	Nicotinic

*Somatic nervous system has been included for comparison.
ACh = acetylcholine.

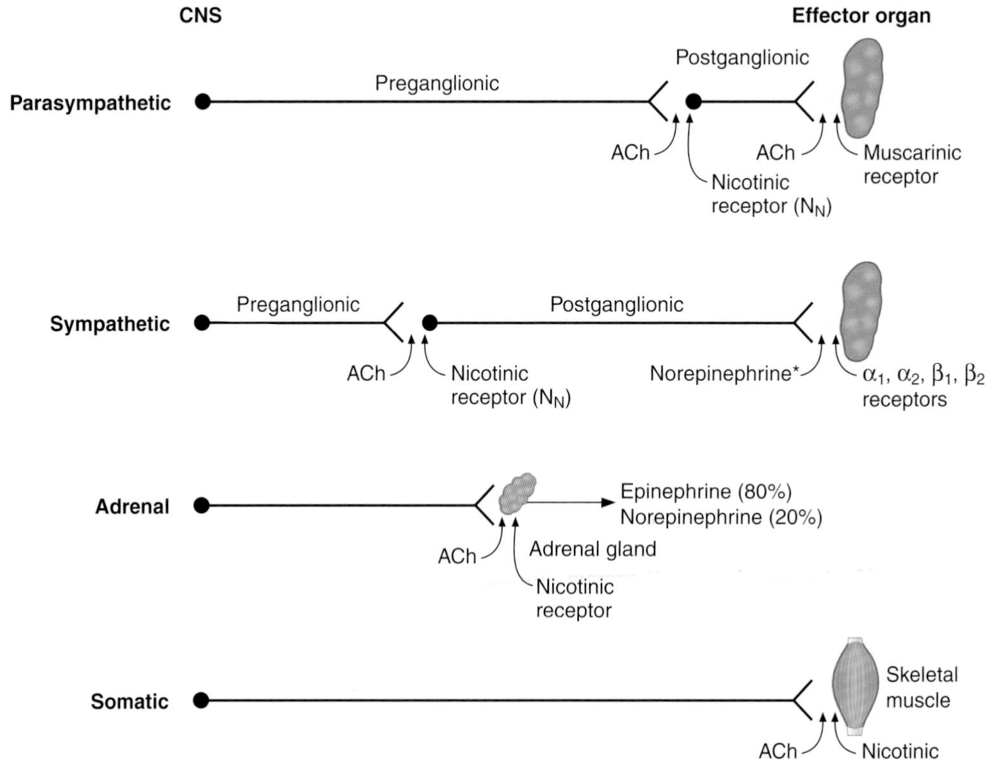

FIGURE 2.1. Organization of the autonomic nervous system. ACh = acetylcholine; CNS = central nervous system.

table 2.2 Signaling Pathways and Mechanisms for Autonomic Receptors

Receptor	Location	G Protein	Mechanism
Adrenergic			
α_1	Smooth muscle	G_q	↑ IP_3/Ca^{2+}
α_2	Gastrointestinal tract	G_i	↓ cAMP
β_1	Heart	G_s	↑ cAMP
β_2	Smooth muscle	G_s	↑ cAMP
Cholinergic			
N_M (N_1)	Skeletal muscle	—	Opening Na^+/K^+ channels
N_N (N_2)	Autonomic ganglia	—	Opening Na^+/K^+ channels
M_1	CNS	G_q	↑ IP_3/Ca^{2+}
M_2	Heart	G_i	↓ cAMP
M_3	Glands, smooth muscle	G_q	↑ IP_3/Ca^{2+}

IP_3 = inositol 1,4,5-triphosphate; cAMP = cyclic adenosine monophosphate.

C. Receptor types in the ANS (Table 2.2)

1. Adrenergic receptors (adrenoreceptors)

a. α_1 Receptors

- are located on vascular smooth muscle of the skin and splanchnic regions, the GI and bladder sphincters, and the radial muscle of the iris.
- produce **excitation** (e.g., contraction or constriction).
- are equally sensitive to norepinephrine and epinephrine. However, only norepinephrine released from adrenergic neurons is present in high enough concentrations to activate α_1 receptors.
- **Mechanism of action: G_q protein**, stimulation of phospholipase C and increase in inositol 1,4,5-triphosphate **(IP_3)** and intracellular [Ca^{2+}].

b. α_2 Receptors

- are located on sympathetic postganglionic nerve terminals (autoreceptors), platelets, fat cells, and the walls of the GI tract (heteroreceptors).
- often produce inhibition (e.g., relaxation or dilation).
- **Mechanism of action: G_i protein**, inhibition of adenylate cyclase and **decrease in cyclic adenosine monophosphate (cAMP).**

c. β_1 Receptors

- are located in the sinoatrial (SA) node, atrioventricular (AV) node, and ventricular muscle of the **heart**.
- produce **excitation** (e.g., increased heart rate, increased conduction velocity, increased contractility).
- are sensitive to both norepinephrine and epinephrine, and are more sensitive than the α_1 receptors.
- **Mechanism of action: G_s protein**, stimulation of adenylate cyclase and **increase in cAMP**.

d. β_2 Receptors

- are located on vascular smooth muscle of skeletal muscle, on bronchial smooth muscle, and in the walls of the GI tract and bladder.
- produce **relaxation** (e.g., dilation of vascular smooth muscle, dilation of bronchioles, relaxation of the bladder wall).
- are more sensitive to epinephrine than to norepinephrine.
- are more sensitive to epinephrine than the α_1 receptors.
- **Mechanism of action: G_s protein**, stimulation of adenylate cyclase and **increase in cAMP**.

2. Cholinergic receptors (cholinoreceptors)

a. Nicotinic receptors

- are located in the **autonomic ganglia** (N_N) of the sympathetic and parasympathetic nervous systems, at the **neuromuscular junction** (N_M), and in the **adrenal medulla** (N_N). The receptors at these locations are similar, but not identical.
- are **activated by ACh or nicotine.**
- produce **excitation.**
- are blocked by **ganglionic blockers (e.g., hexamethonium)** in the autonomic ganglia, but not at the neuromuscular junction.
- **Mechanism of action:** ACh binds to α subunits of the nicotinic ACh receptor. The nicotinic ACh receptors are also ion channels for Na^+ and K^+.

b. Muscarinic receptors

- are located in the **heart** (M_2), **smooth muscle** (M_3), and **glands** (M_3).
- are **inhibitory in the heart** (e.g., decreased heart rate, decreased conduction velocity in AV node).
- are **excitatory in smooth muscle** and **glands** (e.g., increased GI motility, increased secretion).
- are **activated by ACh** and **muscarine.**
- are blocked by **atropine.**
- **Mechanism of action:**

(1) *heart SA node:* **G_i protein**, inhibition of adenylate cyclase, which leads to opening of K^+ channels, slowing of the rate of spontaneous phase 4 depolarization, and decreased heart rate.

(2) *smooth muscle and glands:* **G_q protein**, stimulation of phospholipase C, and increase in IP_3 and intracellular $[Ca^{2+}]$.

3. Drugs that act on the ANS (Table 2.3)

table 2.3 Prototypes of Drugs that Affect Autonomic Activity

Type of Receptor	Agonist	Antagonist
Adrenergic		
α_1	Norepinephrine Phenylephrine	Phenoxybenzamine Phentolamine Prazosin
α_2	Clonidine	Yohimbine
β_1	Norepinephrine Isoproterenol Dobutamine	Propranolol Metoprolol
β_2	Isoproterenol Albuterol	Propranolol Butoxamine
Cholinergic		
Nicotinic	ACh Nicotine Carbachol	Curare (neuromuscular junction N_1 receptors) Hexamethonium (ganglionic N_2 receptors)
Muscarinic	ACh Muscarine Carbachol	Atropine

ACh = acetylcholine.

table 2.4 Effect of the Autonomic Nervous System on Organ Systems

Organ	Sympathetic Action	Sympathetic Receptor	Parasympathetic Action	Parasympathetic Receptor
Heart	↑ heart rate	β_1	↓ heart rate	M_2
	↑ contractility	β_1	↓ contractility (atria)	M_2
	↑ AV node conduction	β_1	↓ AV node conduction	M_2
Vascular smooth muscle	Constricts blood vessels in skin; splanchnic	α_1	—	
	Dilates blood vessels in skeletal muscle	β_2	—	
Gastrointestinal tract	↓ motility	α_2, β_2	↑ motility	M_3
	Constricts sphincters	α_1	Relaxes sphincters	M_3
Bronchioles	Dilates bronchiolar smooth muscle	β_2	Constricts bronchiolar smooth muscle	M_3
Male sex organs	Ejaculation	α	Erection	M
Bladder	Relaxes bladder wall	β_2	Contracts bladder wall	M_3
	Constricts sphincter	α_1	Relaxes sphincter	M_3
Sweat glands	↑ sweating	M (sympathetic cholinergic)	—	
Eye				
Radial muscle, iris	Dilates pupil (mydriasis)	α_1	—	
Circular sphincter muscle, iris	—		Constricts pupil (miosis)	M
Ciliary muscle	Dilates (far vision)	β	Contracts (near vision)	M
Kidney	↑ renin secretion	β_1	—	
Fat cells	↑ lipolysis	β_1	—	

AV = atrioventricular; M = muscarinic.

D. Effects of the ANS on various organ systems (Table 2.4)

E. Autonomic centers—brain stem and hypothalamus

1. Medulla

- Vasomotor center
- Respiratory center
- Swallowing, coughing, and vomiting centers

2. Pons

- Pneumotaxic center

3. Midbrain

- Micturition center

4. Hypothalamus

- Temperature regulation center
- Thirst and food intake regulatory centers

II. ORGANIZATION OF THE NERVOUS SYSTEM

A. Divisions of the nervous system

- The nervous system is composed of the CNS and the peripheral nervous system (PNS).
- The CNS includes the brain and spinal cord.
- The major divisions of the CNS are the spinal cord, brain stem (medulla, pons, and midbrain), cerebellum, diencephalon (thalamus and hypothalamus), and cerebral hemispheres (cerebral cortex, basal ganglia, hippocampus, and amygdala).
- Sensory or afferent nerves bring information into the nervous system.
- Motor or efferent nerves carry information out of the nervous system.

B. Cells of the nervous system

1. Structure of the neuron

a. Cell body surrounds the nucleus and is responsible for protein synthesis.

b. Dendrites arise from the cell body and receive information from adjacent neurons.

c. Axon projects from the axon hillock, where action potentials originate and send information to other neurons or muscle.

2. Glial cells function as support cells for neurons

a. Astrocytes supply metabolic fuels to neurons, secrete trophic factors, and synthesize neurotransmitters.

b. Oligodendrocytes synthesize myelin in the CNS (whereas Schwann cells synthesize myelin in the PNS).

c. Microglial cells proliferate following neuronal injury and serve as scavengers for cellular debris.

III. SENSORY SYSTEMS

A. Sensory receptors—general

- are specialized epithelial cells or neurons that **transduce environmental signals** into neural signals.
- The environmental signals that can be detected include **mechanical force**, **light**, **sound**, **chemicals**, and **temperature**.

1. Types of sensory transducers

a. Mechanoreceptors

- Pacinian corpuscles
- Joint receptors
- Stretch receptors in muscle
- Hair cells in auditory and vestibular systems
- Baroreceptors in carotid sinus

b. Photoreceptors

- Rods and cones of the retina

c. Chemoreceptors

- Olfactory receptors
- Taste receptors
- Osmoreceptors
- Carotid body O_2 receptors

d. Extremes of temperature and pain

- Nociceptors

table 2.5 Characteristics of Nerve Fiber Types

General Fiber Type and Example	Sensory Fiber Type and Example	Diameter	Conduction Velocity
A-alpha	**Ia**	Largest	Fastest
Large α-motoneurons	Muscle spindle afferents		
	Ib	Largest	Fastest
	Golgi tendon organs		
A-beta	**II**	Medium	Medium
Touch, pressure	Secondary afferents of muscle spindles; touch and pressure		
A-gamma	—		
γ-Motoneurons to muscle spindles (intrafusal fibers)		Medium	Medium
A-delta	**III**	Small	Medium
Touch, pressure, temperature, and pain	Touch, pressure, fast pain, and temperature		
B	—	Small	Medium
Preganglionic autonomic fibers			
C	**IV**	Smallest	Slowest
Slow pain; postganglionic autonomic fibers	Pain and temperature (unmyelinated)		

2. **Fiber types and conduction velocity** (Table 2.5)
3. **Receptive field**
   - is an area of the body that, when stimulated, changes the firing rate of a sensory neuron. If the firing rate of the sensory neuron is increased, the receptive field is **excitatory**. If the firing rate of the sensory neuron is decreased, the receptive field is **inhibitory**.
4. **Steps in sensory transduction**
   a. **Stimulus arrives at the sensory receptor**. The stimulus may be a photon of light on the retina, a molecule of NaCl on the tongue, a depression of the skin, and so forth.
   b. **Ion channels are opened in the sensory receptor**, allowing current to flow.
      - Usually, the current is inward, which produces **depolarization** of the receptor.
      - The exception is in the **photoreceptor**, where light causes *decreased* inward current and **hyperpolarization**.
   c. The change in membrane potential produced by the stimulus is the **receptor potential**, or **generator potential** (Figure 2.2).

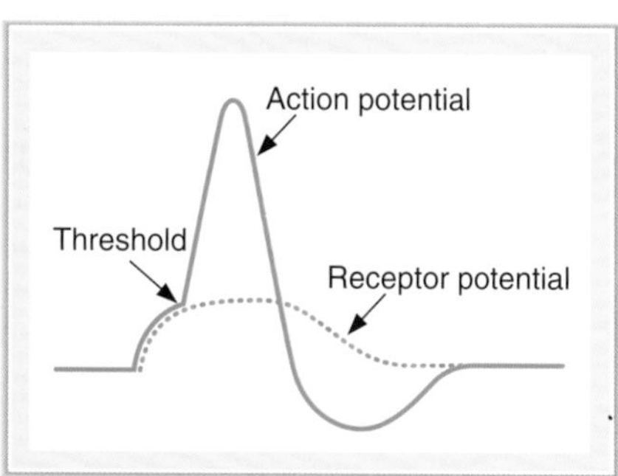

FIGURE 2.2. Receptor (generator) potential and how it may lead to an action potential.

- If the receptor potential is depolarizing, it brings the membrane potential closer to threshold. If the receptor potential is large enough, the membrane potential will exceed threshold, and an action potential will fire in the sensory neuron.
- Receptor potentials are **graded in size** depending on the size of the stimulus.

5. Adaptation of sensory receptors

a. Slowly adapting, or **tonic**, receptors (muscle spindle; pressure; slow pain)

- respond repetitively to a prolonged stimulus.
- detect a **steady stimulus.**

b. Rapidly adapting, or **phasic**, receptors (pacinian corpuscle; light touch)

- show a decline in action potential frequency with time in response to a constant stimulus.
- primarily detect **onset** and **offset** of a stimulus.

6. Sensory pathways from the sensory receptor to the cerebral cortex

a. Sensory receptors

- are activated by environmental stimuli.
- may be specialized epithelial cells (e.g., photoreceptors, taste receptors, auditory hair cells).
- may be primary afferent neurons (e.g., olfactory chemoreceptors).
- **transduce** the stimulus into **electrical energy** (i.e., receptor potential).

b. First-order neurons

- are the **primary afferent neurons** that receive the transduced signal and send the information to the CNS. Cell bodies of the primary afferent neurons are in **dorsal root** or **spinal cord ganglia.**

c. Second-order neurons

- are located in the spinal cord or brain stem.
- receive information from one or more primary afferent neurons in **relay nuclei** and transmit it to the **thalamus.**
- Axons of second-order neurons may **cross the midline** in a relay nucleus in the spinal cord before they ascend to the thalamus. Therefore, sensory **information originating on one side of the body ascends to the contralateral thalamus.**

d. Third-order neurons

- are located in the relay nuclei of the thalamus. From there, encoded sensory information ascends to the cerebral cortex.

e. Fourth-order neurons

- are located in the appropriate sensory area of the cerebral cortex. The information received results in a **conscious perception** of the stimulus.

B. Somatosensory system

- includes the sensations of touch, movement, temperature, and pain.

1. Pathways in the somatosensory system

a. Dorsal column system

- processes sensations of **fine touch, pressure, two-point discrimination, vibration**, and **proprioception.**
- consists primarily of **group II fibers.**
- **Course:** primary afferent neurons have cell bodies in the dorsal root. Their axons ascend ipsilaterally to the **nucleus gracilis** and **nucleus cuneatus** of the medulla. From the medulla, the second-order neurons cross the midline and ascend to the contralateral thalamus, where they synapse on third-order neurons. Third-order neurons ascend to the somatosensory cortex, where they synapse on fourth-order neurons.

table 2.6 Types of Mechanoreceptors

Type of Mechanoreceptor	Description	Sensation Encoded	Adaptation
Pacinian corpuscle	Onion-like structures in the subcutaneous skin (surrounding unmyelinated nerve endings)	Vibration; tapping	Rapidly adapting
Meissner corpuscle	Present in nonhairy skin	Velocity	Rapidly adapting
Ruffini corpuscle	Encapsulated	Pressure	Slowly adapting
Merkel disk	Transducer is on epithelial cells	Location	Slowly adapting

b. **Anterolateral system**

- processes sensations of **temperature**, **pain**, and **light touch**.
- consists primarily of **group III and IV fibers**, which enter the spinal cord and terminate in the dorsal horn.
- **Course:** second-order neurons cross the midline to the anterolateral quadrant of the spinal cord and ascend to the contralateral thalamus, where they synapse on third-order neurons. Third-order neurons ascend to the somatosensory cortex, where they synapse on fourth-order neurons.

2. **Mechanoreceptors for touch and pressure** (Table 2.6)

3. **Thalamus**

- Information from different parts of the body is arranged somatotopically.
- **Destruction of the thalamic nuclei** results in loss of sensation on the contralateral side of the body.

4. **Somatosensory cortex—the sensory homunculus**

- The major somatosensory areas of the cerebral cortex are **SI** and **SII**.
- SI has a somatotopic representation similar to that in the thalamus.
- This "map" of the body is called the **sensory homunculus.**
- The largest areas represent the **face**, **hands**, and **fingers**, where precise localization is most important.

5. **Pain**

- is associated with the detection and perception of noxious stimuli **(nociception)**.
- The receptors for pain are **free nerve endings** in the skin, muscle, and viscera.
- Neurotransmitters for nociceptors include **substance P**. Inhibition of the release of substance P is the basis of pain relief by **opioids**.

a. **Fibers for fast pain and slow pain**

- **Fast pain** is carried by group III fibers. It has a rapid onset and offset, and is localized.
- **Slow pain** is carried by C fibers. It is characterized as aching, burning, or throbbing that is poorly localized.

b. **Referred pain**

- Pain of visceral origin is referred to sites on the skin and follows the **dermatome rule**. These sites are innervated by nerves that arise from the same segment of the spinal cord.
- **For example**, ischemic heart pain is referred to the chest and shoulder.

C. Vision

1. **Optics**

a. **Refractive power of a lens**

- is measured in **diopters**.
- equals the reciprocal of the focal distance in meters.
- **Example:** 10 diopters = 1/10 m = 10 cm

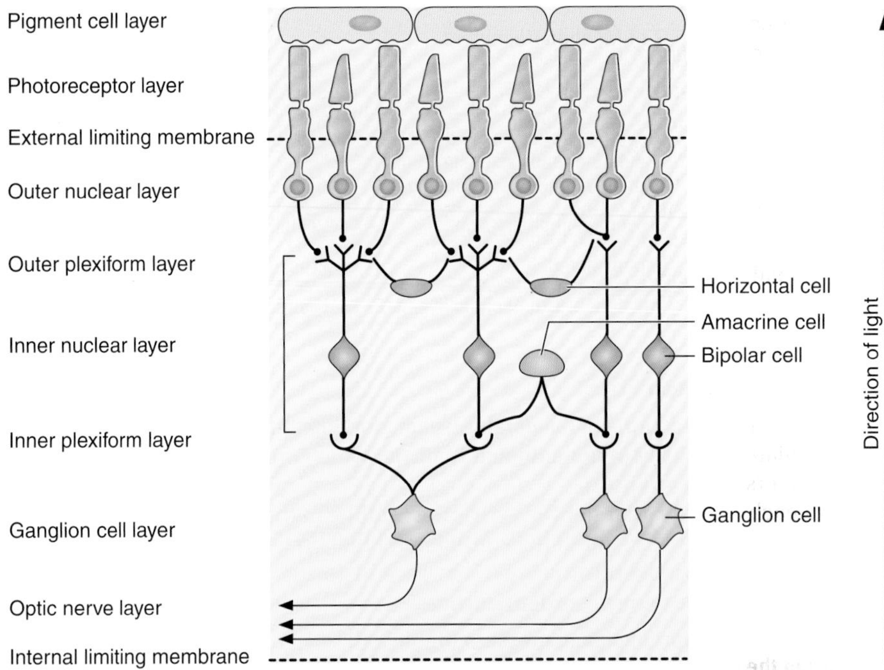

FIGURE 2.3. Cellular layers of the retina. (Reprinted with permission from Bullock J, Boyle J III, Wang MB. Physiology. 4th ed. Baltimore: Lippincott Williams & Wilkins, 2001:77.)

b. Refractive errors

(1) *Emmetropia—normal.* Light focuses on the retina.

(2) *Hyperopia—farsighted.* Light focuses behind the retina and is corrected with a **convex lens**.

(3) *Myopia—nearsighted.* Light focuses in front of the retina and is corrected with a **biconcave lens**.

(4) *Astigmatism.* Curvature of the lens is not uniform and is corrected with a **cylindric lens**.

(5) *Presbyopia* is a result of loss of the accommodation power of the lens that occurs with aging. The **near point** (closest point on which one can focus by accommodation of the lens) moves farther from the eye and is corrected with a **convex lens**.

2. Layers of the retina (Figure 2.3)

a. Pigment epithelial cells

- absorb stray light and prevent scatter of light.
- convert 11-*cis* retinal to all-*trans* retinal.

b. Receptor cells are **rods and cones** (Table 2.7).

- Rods and cones are not present on the optic disk; the result is a **blind spot**.

table 2.7 Functions of Rods and Cones

Function	Rods	Cones
Sensitivity to light	Sensitive to low-intensity light; night vision	Sensitive to high-intensity light; day vision
Acuity	Lower visual acuity Not present in fovea	Higher visual acuity Present in fovea
Dark adaptation	Rods adapt later	Cones adapt first
Color vision	No	Yes

c. **Bipolar cells.** The receptor cells (i.e., rods and cones) synapse on bipolar cells, which synapse on the ganglion cells.

(1) *Few cones synapse on a single bipolar cell,* which synapses on a single ganglion cell. This arrangement is the basis for the **high acuity** and **low sensitivity** of the cones. In the fovea, where acuity is highest, the ratio of cones to bipolar cells is 1:1.

(2) *Many rods synapse on a single bipolar cell.* As a result, there is **less acuity** in the rods than in the cones. There is also **greater sensitivity** in the rods because light striking any one of the rods will activate the bipolar cell.

d. **Horizontal and amacrine cells** form local circuits with the bipolar cells.

e. **Ganglion cells** are the output cells of the retina.

- Axons of ganglion cells form the optic nerve.

3. **Optic pathways and lesions** (Figure 2.4)

- Axons of the ganglion cells form the optic nerve and optic tract, ending in the lateral geniculate body of the thalamus.
- The fibers from each **nasal hemiretina cross** at the **optic chiasm**, whereas the fibers from each temporal hemiretina remain ipsilateral. Therefore, fibers from the **left nasal hemiretina** and fibers from the **right temporal hemiretina** form the **right optic tract** and synapse on the right lateral geniculate body.
- Fibers from the lateral geniculate body form the **geniculocalcarine tract** and pass to the **occipital lobe of the cortex**.

a. **Cutting the optic nerve** causes blindness in the ipsilateral eye.

b. **Cutting the optic chiasm** causes heteronymous bitemporal hemianopia.

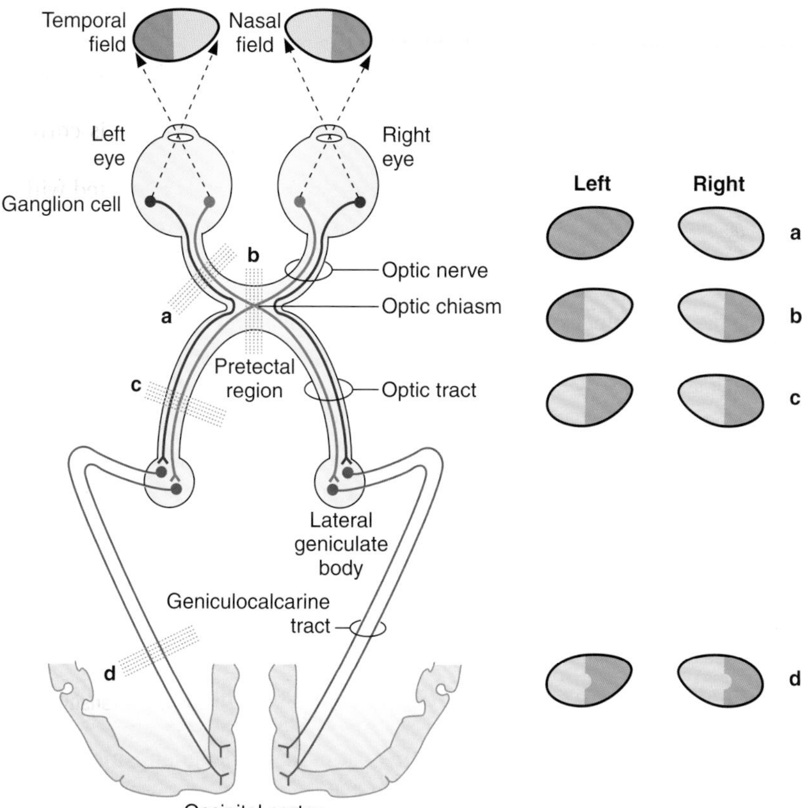

FIGURE 2.4. Effects of lesions at various levels of the optic pathway. (Modified with permission from Ganong WF. Review of Medical Physiology. 20th ed. New York: McGraw-Hill, 2001:147.)

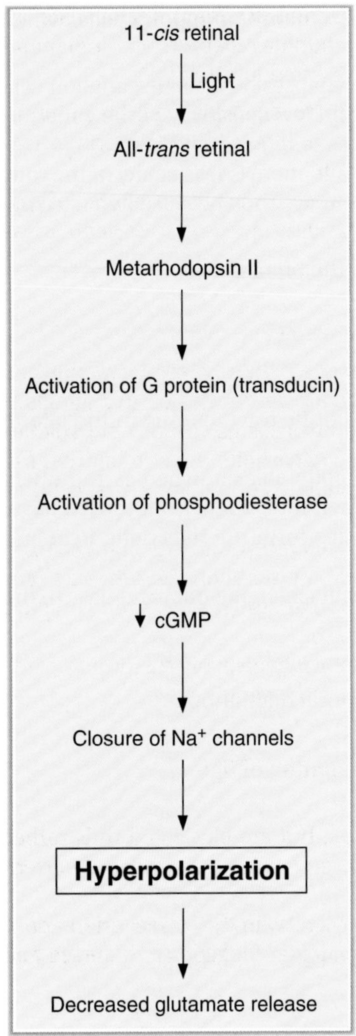

FIGURE 2.5. Steps in photoreception in rods. cGMP = cyclic guanosine monophosphate.

c. **Cutting the optic tract** causes homonymous contralateral hemianopia.

d. **Cutting the geniculocalcarine tract** causes homonymous hemianopia with **macular sparing.**

4. **Steps in photoreception in the rods** (Figure 2.5)

- The photosensitive element is **rhodopsin**, which is composed of **opsin** (a protein) belonging to the superfamily of G-protein–coupled receptors and **retinal** (an aldehyde of vitamin A).

a. **Light** on the retina converts **11-*cis*** retinal to **all-*trans*** retinal, a process called photoisomerization. A series of intermediates is then formed, one of which is **metarhodopsin II.**

- **Vitamin A** is necessary for the regeneration of 11-*cis* retinal. Deficiency of vitamin A causes **night blindness.**

b. Metarhodopsin II activates a G protein called **transducin (G_t)**, which in turn activates a **phosphodiesterase.**

c. Phosphodiesterase catalyzes the conversion of cyclic guanosine monophosphate (cGMP) to 5′-GMP, and **cGMP levels decrease.**

d. Decreased levels of cGMP cause **closure of Na^+** channels, decreased inward Na^+ current, and, as a result, **hyperpolarization** of the photoreceptor cell membrane. Increasing light intensity increases the degree of hyperpolarization.

e. When the photoreceptor is hyperpolarized, there is ***decreased* release of glutamate**, an excitatory neurotransmitter. There are two types of glutamate receptors on bipolar and horizontal cells, which determine whether the cell is excited or inhibited.

(1) Ionotropic glutamate receptors are excitatory. Therefore, decreased release of glutamate from the photoreceptors acting on ionotropic receptors causes hyperpolarization (inhibition) because there is *decreased excitation*.

(2) Metabotropic glutamate receptors are inhibitory. Therefore, decreased release of glutamate from photoreceptors acting on metabotropic receptors causes depolarization (excitation) because there is *decreased inhibition*.

5. Receptive visual fields

a. Receptive fields of the ganglion cells and lateral geniculate cells

(1) Each bipolar cell receives input from many receptor cells. In turn, each ganglion cell receives input from many bipolar cells. The receptor cells connected to a ganglion cell form the **center of its receptor field**. The receptor cells connected to ganglion cells via horizontal cells form the **surround of its receptive field**. (Remember that the response of bipolar and horizontal cells to light depends on whether that cell has ionotropic or metabotropic receptors.)

(2) On-center, off-surround is one pattern of a ganglion cell receptive field. Light striking the center of the receptive field depolarizes (excites) the ganglion cell, whereas light striking the surround of the receptive field hyperpolarizes (inhibits) the ganglion cell. **Off-center, on-surround** is another possible pattern.

(3) Lateral geniculate cells of the thalamus retain the on-center, off-surround or off-center, on-surround pattern that is transmitted from the ganglion cell.

b. Receptive fields of the visual cortex

- Neurons in the visual cortex detect shape and orientation of figures.
- Three cortical cell types are involved:

(1) *Simple cells* have center-surround, on-off patterns, but are elongated rods rather than concentric circles. They respond best to **bars of light** that have the correct **position** and **orientation**.

(2) *Complex cells* respond best to **moving bars** or **edges of light** with the correct orientation.

(3) *Hypercomplex cells* respond best to lines with particular **length** and to **curves and angles**.

D. Audition

1. Sound waves

- **Frequency** is measured in **hertz (Hz)**.
- **Intensity** is measured in **decibels (dB)**, a log scale.

$$dB = 20 \log \frac{P}{P_0}$$

where:
dB = decibel
P = sound pressure being measured
P_0 = reference pressure measured at the threshold frequency

2. Structure of the ear

a. Outer ear

- directs the sound waves into the auditory canal.

b. Middle ear

- is air filled.
- contains the **tympanic membrane** and the **auditory ossicles** (malleus, incus, and stapes). The stapes inserts into the **oval window**, a membrane between the middle ear and the inner ear.

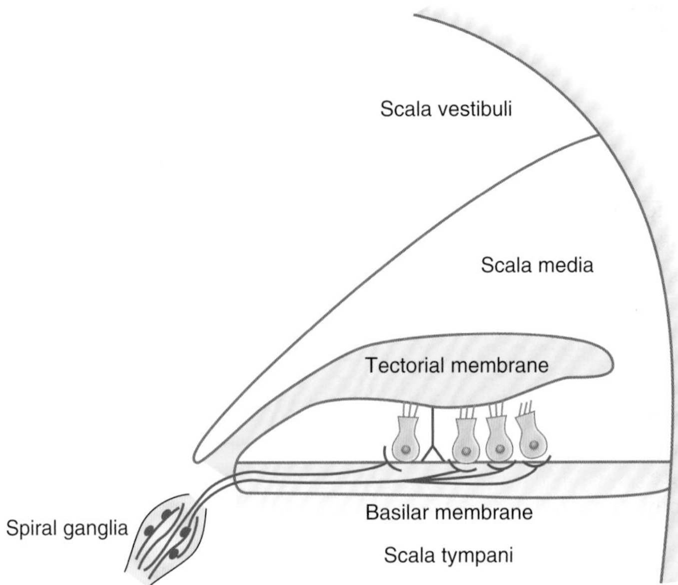

FIGURE 2.6. Organ of Corti and auditory transduction.

- Sound waves cause the tympanic membrane to vibrate. In turn, the **ossicles vibrate**, pushing the stapes into the oval window and **displacing fluid** in the **inner ear** (see II D 2 c).
- **Sound is amplified** by the lever action of the ossicles and the concentration of sound waves from the large tympanic membrane onto the smaller oval window.

c. Inner ear (Figure 2.6)

- is fluid filled.
- consists of a bony labyrinth (**semicircular canals**, **cochlea**, and **vestibule**) and a series of ducts called the membranous labyrinth. The fluid outside the ducts is **perilymph**; the fluid inside the ducts is **endolymph**.

(1) *Structure of the cochlea: three tubular canals*

(a) The scala vestibuli and scala tympani contain **perilymph**, which has a **high [Na$^+$]**.

(b) The scala media contains **endolymph**, which has a **high [K$^+$]**.

- The scala media is bordered by the **basilar membrane**, which is the site of the **organ of Corti**.

(2) *Location and structure of the organ of Corti*

- The organ of Corti is located on the basilar membrane.
- It contains the **receptor cells** (inner and outer hair cells) for auditory stimuli. **Cilia** protrude from the hair cells and are embedded in the tectorial membrane.
- **Inner hair cells** are arranged in single rows and are **few in number**.
- **Outer hair cells** are arranged in parallel rows and are **greater in number** than the inner hair cells.
- The **spiral ganglion** contains the cell bodies of the auditory nerve (cranial nerve [CN] VIII), which synapse on the hair cells.

3. Steps in auditory transduction by the organ of Corti (see Figure 2.6)

- The cell bodies of hair cells contact the **basilar membrane**. The cilia of hair cells are embedded in the **tectorial membrane**.

a. Sound waves cause **vibration** of the organ of Corti. Because the basilar membrane is more elastic than the tectorial membrane, vibration of the basilar membrane causes the hair cells to bend by a shearing force as they push against the tectorial membrane.

b. **Bending of the cilia** causes changes in **K^+ conductance** of the hair cell membrane. Bending in one direction causes depolarization; bending in the other direction causes hyperpolarization. The oscillating potential that results is the **cochlear microphonic potential**.

c. The oscillating potential of the hair cells causes intermittent firing of the cochlear nerves.

4. **How sound is encoded**

- The frequency that activates a particular hair cell depends on the location of the hair cell along the basilar membrane.

a. The **base of the basilar membrane** (near the oval and round windows) is narrow and stiff. It responds best to **high frequencies**.

b. The **apex of the basilar membrane** (near the helicotrema) is wide and compliant. It responds best to **low frequencies**.

5. **Central auditory pathways**

- Fibers ascend through the lateral lemniscus to the **inferior colliculus** to the medial geniculate nucleus of the thalamus to the **auditory cortex**.
- Fibers may be **crossed or uncrossed**. As a result, a mixture of ascending auditory fibers represents both ears at all higher levels. Therefore, lesions of the cochlea of one ear cause unilateral deafness, but more central unilateral lesions do not.
- There is **tonotopic representation** of frequencies at all levels of the central auditory pathway.
- Discrimination of complex features (e.g., recognizing a patterned sequence) is a property of the cerebral cortex.

E. Vestibular system

- detects angular and linear acceleration of the head.
- Reflex adjustments of the head, eyes, and postural muscles provide a stable visual image and steady posture.

1. **Structure of the vestibular organ**

a. It is a membranous labyrinth consisting of **three perpendicular semicircular canals**, a **utricle**, and a **saccule**. The semicircular canals detect angular acceleration or rotation. The utricle and saccule detect linear acceleration.

b. The canals are filled with **endolymph** and are bathed in perilymph.

c. The **receptors are hair cells** located at the end of each semicircular canal. Cilia on the hair cells are embedded in a gelatinous structure called the **cupula**. A single long cilium is called the **kinocilium**; smaller cilia are called **stereocilia** (Figure 2.7).

2. **Steps in vestibular transduction—angular acceleration** (see Figure 2.7)

a. During **counterclockwise (left) rotation** of the head, the horizontal semicircular canal and its attached cupula also rotate to the left. Initially, the cupula moves more quickly than the endolymph fluid. Thus, the cupula is dragged through the endolymph; as a result, the cilia on the hair cells bend.

b. If the **stereocilia are bent toward the kinocilium**, the hair cell **depolarizes** (excitation). If the **stereocilia are bent away from the kinocilium**, the hair cell **hyperpolarizes** (inhibition). Therefore, during the initial counterclockwise (left) rotation, the left horizontal canal is excited and the right horizontal canal is inhibited.

c. After several seconds, the endolymph "catches up" with the movement of the head and the cupula. The cilia return to their upright position and are no longer depolarized or hyperpolarized.

d. **When the head suddenly stops moving**, the endolymph continues to move counterclockwise (left), dragging the cilia in the opposite direction. Therefore, if the hair cell was depolarized with the initial rotation, it now will hyperpolarize. If it was hyperpolarized initially, it now will depolarize. Therefore, when the head stops moving, the left horizontal canal will be inhibited and the right horizontal canal will be excited.

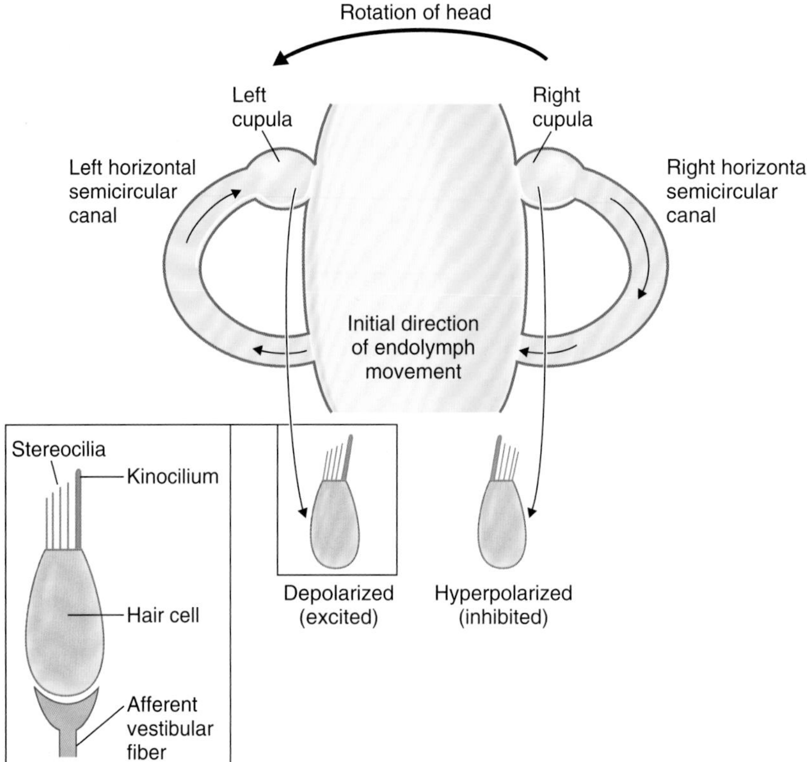

FIGURE 2.7. The semicircular canals and vestibular transduction during counterclockwise rotation.

3. Vestibular–ocular reflexes

a. Nystagmus

- An initial rotation of the head causes the eyes to move slowly in the opposite direction to maintain visual fixation. When the limit of eye movement is reached, the eyes rapidly snap back (nystagmus) and then move slowly again.
- The **direction of the nystagmus** is defined as the direction of the fast (rapid eye) movement. Therefore, the nystagmus occurs in the **same direction as the head rotation**.

b. Postrotatory nystagmus

- occurs in the **opposite direction of the head rotation**.

F. Olfaction

1. Olfactory pathway

a. Receptor cells

- are located in the olfactory epithelium.
- are **true neurons** that conduct action potentials into the CNS.
- Basal cells of the olfactory epithelium are undifferentiated stem cells that **continuously turn over** and replace the olfactory receptor cells (neurons). These are the only neurons in the adult human that replace themselves.

b. CN I (olfactory)

- carries information from the olfactory receptor cells to the olfactory bulb.
- The axons of the olfactory nerves are **unmyelinated C fibers** and are among the **smallest** and **slowest** in the nervous system.
- Olfactory epithelium is also innervated by CN V (trigeminal), which detects **noxious** or **painful stimuli**, such as ammonia.

- The olfactory nerves pass through the cribriform plate on their way to the olfactory bulb. **Fractures of the cribriform plate** sever input to the olfactory bulb and reduce **(hyposmia)** or eliminate **(anosmia)** the sense of smell. The response to ammonia, however, will be intact after fracture of the cribriform plate because this response is carried on CN V.

c. Mitral cells in the olfactory bulb

- are second-order neurons.
- Output of the mitral cells forms the olfactory tract, which projects to the **prepiriform cortex.**

2. Steps in transduction in the olfactory receptor neurons

a. Odorant molecules bind to specific **olfactory receptor proteins** located on cilia of the olfactory receptor cells.

b. When the receptors are activated, they **activate G proteins** (G_{olf}), which in turn activate adenylate cyclase.

c. There is an **increase in intracellular cAMP** that opens Na^+ channels in the olfactory receptor membrane and produces a **depolarizing receptor potential.**

d. The receptor potential depolarizes the initial segment of the axon to threshold, and **action potentials** are generated and propagated.

G. Taste

1. Taste pathways

a. Taste receptor cells line the taste buds that are located on specialized **papillae.** The receptor cells are covered with microvilli, which increase the surface area for binding taste chemicals. In contrast to olfactory receptor cells, taste receptors are **not neurons.**

b. The **anterior two-thirds of the tongue**

- has **fungiform papillae.**
- detects **salty**, **sweet**, and **umami** sensations.
- is innervated by CN VII (chorda tympani).

c. The **posterior one-third of the tongue**

- has **circumvallate** and **foliate papillae.**
- detects **sour** and **bitter** sensations.
- is innervated by **CN IX** (glossopharyngeal).
- The back of the throat and the epiglottis are innervated by **CN X.**

d. CN VII, CN IX, and CN X enter the medulla, ascend in the **solitary tract,** and terminate on second-order taste neurons in the **solitary nucleus.** They project, primarily ipsilaterally, to the ventral posteromedial nucleus of the thalamus and, finally, to the taste cortex.

2. Steps in taste transduction

- **Taste chemicals** (sour, sweet, salty, bitter, and umami) bind to taste receptors on the microvilli and produce a depolarizing receptor potential in the receptor cell.

IV. MOTOR SYSTEMS

A. Motor unit

- consists of a **single motoneuron and the muscle fibers that it innervates.** For **fine control** (e.g., muscles of the eye), a single motoneuron innervates only a few muscle fibers. For **larger movements** (e.g., postural muscles), a single motoneuron may innervate thousands of muscle fibers.
- The **motoneuron pool** is the group of motoneurons that innervates fibers within the same muscle.

- The force of muscle contraction is graded by **recruitment** of additional motor units (size principle). The **size principle** states that as additional motor units are recruited, more motoneurons are involved and more tension is generated.

1. Small motoneurons

- **innervate a few muscle fibers.**
- have the lowest thresholds and, therefore, **fire first**.
- generate the **smallest force**.

2. Large motoneurons

- **innervate many muscle fibers.**
- have the highest thresholds and, therefore, **fire last**.
- generate the **largest force**.

B. Muscle sensor

1. Types of muscle sensors (see Table 2.5)

a. Muscle spindles (groups Ia and II afferents) are arranged in parallel with extrafusal fibers. They detect both **static and dynamic changes in muscle length**.

b. Golgi tendon organs (group Ib afferents) are arranged in series with extrafusal muscle fibers. They detect **muscle tension**.

c. Pacinian corpuscles (group II afferents) are distributed throughout muscle. They detect **vibration**.

d. Free nerve endings (groups III and IV afferents) detect **noxious stimuli**.

2. Types of muscle fibers

a. Extrafusal fibers

- make up the bulk of muscle.
- are **innervated by α-motoneurons**.
- provide the **force for muscle contraction**.

b. Intrafusal fibers

- are smaller than extrafusal muscle fibers.
- are **innervated by γ-motoneurons**.
- are encapsulated in sheaths to form **muscle spindles**.
- run in parallel with extrafusal fibers, but not for the entire length of the muscle.
- are too small to generate significant force.

3. Muscle spindles

- are distributed throughout muscle.
- consist of small, encapsulated intrafusal fibers connected in parallel with large (force-generating) extrafusal fibers.
- The finer the movement required, the greater the number of muscle spindles in a muscle.

a. Types of intrafusal fibers in muscle spindles (Figure 2.8)

(1) *Nuclear bag fibers*

- detect the rate of change in muscle length (fast, **dynamic** changes).
- are innervated by group Ia afferents.
- have nuclei collected in a central "bag" region.

(2) *Nuclear chain fibers*

- detect **static** changes in muscle length.
- are innervated by group II afferents.
- are more numerous than nuclear bag fibers.
- have nuclei arranged in rows.

b. How the muscle spindle works (see Figure 2.8)

- Muscle spindle reflexes oppose (correct for) increases in muscle length (stretch).

(1) Sensory information about muscle length is received by group Ia (velocity) and group II (static) afferent fibers.

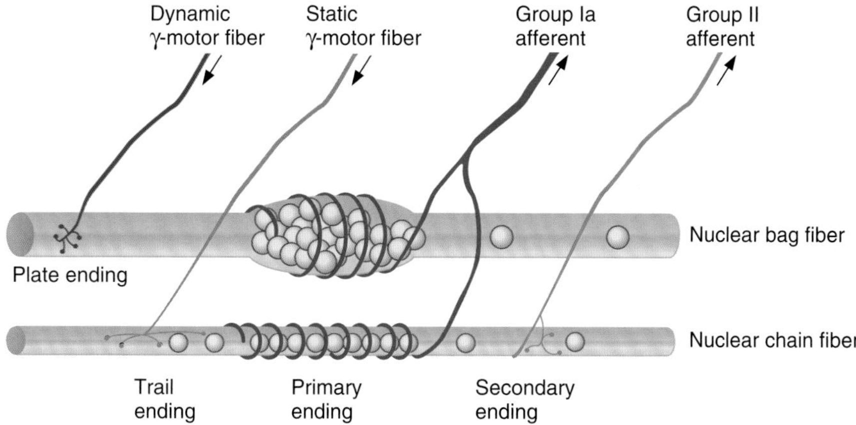

FIGURE 2.8. Organization of the muscle spindle. (Modified with permission from Matthews PBC. Muscle spindles and their motor control. Physiol Rev 1964;44:232.)

(2) When a muscle is stretched (lengthened), the muscle spindle is also stretched, stimulating group Ia and group II afferent fibers.

(3) Stimulation of group Ia afferents stimulates α-motoneurons in the spinal cord. This stimulation in turn causes contraction and shortening of the muscle. Thus, the original stretch is opposed and muscle length is maintained.

c. Function of γ-motoneurons

- innervate intrafusal muscle fibers.
- adjust the sensitivity of the muscle spindle so that it will respond appropriately during muscle contraction.
- **α-Motoneurons and γ-motoneurons are coactivated** so that muscle spindles remain sensitive to changes in muscle length during contraction.

C. Muscle reflexes (Table 2.8)

1. Stretch (myotatic) reflex—knee jerk (Figure 2.9)

- is **monosynaptic.**

a. Muscle is stretched, and the stretching stimulates **group Ia** afferent fibers.

b. Group Ia afferents synapse directly on **α-motoneurons** in the spinal cord. The pool of α-motoneurons that is activated innervates the homonymous muscle.

c. Stimulation of α-motoneurons causes **contraction in the muscle that was stretched.** As the muscle contracts, it shortens, decreasing the stretch on the muscle spindle and returning it to its original length.

d. At the same time, synergistic muscles are activated and antagonistic muscles are inhibited.

table 2.8 Summary of Muscle Reflexes

Reflex	Number of Synapses	Stimulus	Afferent Fibers	Response
Stretch reflex (knee-jerk)	Monosynaptic	Muscle is stretched.	Ia	Contraction of the muscle
Golgi tendon reflex (clasp-knife)	Disynaptic	Muscle contracts	Ib	Relaxation of the muscle
Flexor withdrawal reflex (after touching a hot stove)	Polysynaptic	Pain	II, III, and IV	Ipsilateral flexion; contralateral extension

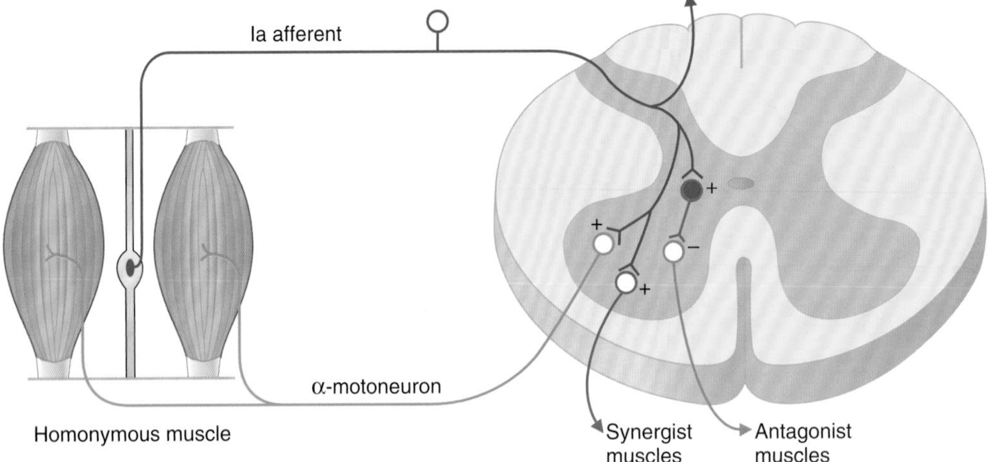

FIGURE 2.9. The stretch reflex.

e. Example of the knee-jerk reflex. Tapping on the patellar tendon causes the quadriceps to stretch. Stretch of the quadriceps stimulates group Ia afferent fibers, which activate α-motoneurons that make the quadriceps contract. Contraction of the quadriceps forces the lower leg to extend.

- **Increases in γ-motoneuron activity** increase the sensitivity of the muscle spindle and therefore exaggerate the knee-jerk reflex.

2. Golgi tendon reflex (inverse myotatic)

- is **disynaptic.**
- is the opposite, or inverse, of the stretch reflex.

a. Active muscle contraction stimulates the Golgi tendon organs and **group Ib** afferent fibers.

b. The group Ib afferents stimulate **inhibitory interneurons** in the spinal cord. These interneurons **inhibit α-motoneurons** and cause relaxation of the muscle that was originally contracted.

c. At the same time, antagonistic muscles are excited.

d. Clasp-knife reflex, an exaggerated form of the Golgi tendon reflex, can occur with **disease of the corticospinal tracts** (hypertonicity or spasticity).

- **For example**, if the arm is hypertonic, the increased sensitivity of the muscle spindles in the extensor muscles (triceps) causes resistance to flexion of the arm. Eventually, tension in the triceps increases to the point at which it activates the Golgi tendon reflex, causing the triceps to relax and the arm to flex closed like a jackknife.

3. Flexor withdrawal reflex

- is **polysynaptic.**
- results in **flexion on the ipsilateral side** and **extension on the contralateral side.** Somatosensory and pain afferent fibers elicit withdrawal of the stimulated body part from the noxious stimulus.

a. Pain (e.g., touching a hot stove) stimulates the flexor reflex afferents of **groups II, III, and IV.**

b. The afferent fibers synapse polysynaptically (via interneurons) onto motoneurons in the spinal cord.

c. On the **ipsilateral side** of the pain stimulus, flexors are stimulated (they contract) and extensors are inhibited (they relax), and the arm is jerked away from the stove. On the **contralateral side,** flexors are inhibited and extensors are stimulated **(crossed extension reflex)** to maintain balance.

d. As a result of persistent neural activity in the polysynaptic circuits, an **afterdischarge** occurs. The afterdischarge prevents the muscle from relaxing for some time.

D. Spinal organization of motor systems

1. Convergence

- occurs when a single α-motoneuron receives its input from many muscle spindle group Ia afferents in the homonymous muscle.
- produces **spatial summation** because although a single input would not bring the muscle to threshold, multiple inputs will.
- also can produce **temporal summation** when inputs arrive in rapid succession.

2. Divergence

- occurs when the muscle spindle group Ia afferent fibers project to all of the α-motoneurons that innervate the homonymous muscle.

3. Recurrent inhibition (Renshaw cells)

- Renshaw cells are inhibitory cells in the ventral horn of the spinal cord.
- They receive input from collateral axons of motoneurons and, when stimulated, negatively feedback (inhibit) on the motoneuron.

E. Brain stem control of posture

1. Motor centers and pathways

- **Pyramidal tracts** (corticospinal and corticobulbar) pass through the medullary pyramids.
- All others are **extrapyramidal tracts** and originate primarily in the following structures of the brain stem:

a. Rubrospinal tract

- originates in the red nucleus and projects to interneurons in the lateral spinal cord.
- Stimulation of the red nucleus produces **stimulation of flexors** and **inhibition of extensors**.

b. Pontine reticulospinal tract

- originates in the nuclei in the pons and projects to the ventromedial spinal cord.
- Stimulation has a general **stimulatory effect on both extensors and flexors**, with the predominant effect on extensors.

c. Medullary reticulospinal tract

- originates in the medullary reticular formation and projects to spinal cord interneurons in the intermediate gray area.
- Stimulation has a general **inhibitory effect on both extensors and flexors**, with the predominant effect on extensors.

d. Lateral vestibulospinal tract

- originates in Deiters nucleus and projects to ipsilateral motoneurons and interneurons.
- Stimulation causes a powerful **stimulation of extensors** and **inhibition of flexors**.

e. Tectospinal tract

- originates in the superior colliculus and projects to the cervical spinal cord.
- is involved in the **control of neck muscles**.

2. Effects of transections of the spinal cord

a. Paraplegia

- is the loss of voluntary movements below the level of the lesion.
- results from interruption of the descending pathways from the motor centers in the brain stem and higher centers.

b. **Loss of conscious sensation** below the level of the lesion

c. Initial loss of reflexes—spinal shock

- Immediately after transection, there is loss of the excitatory influence from α- and γ-motoneurons. **Limbs become flaccid**, and **reflexes are absent**. With time, partial recovery and return of reflexes (or even hyperreflexia) will occur.

(1) If the **lesion is at C7**, there will be loss of sympathetic tone to the heart. As a result, heart rate and arterial pressure will decrease.
(2) If the **lesion is at C3**, breathing will stop because the respiratory muscles have been disconnected from control centers in the brain stem.
(3) If the **lesion is at C1** (e.g., as a result of hanging), death occurs.

3. Effects of transections above the spinal cord

a. Lesions above the lateral vestibular nucleus

- cause **decerebrate rigidity** because of the removal of inhibition from higher centers, resulting in excitation of α- and γ-motoneurons and rigid posture.

b. Lesions above the pontine reticular formation but below the midbrain

- cause **decerebrate rigidity** because of the removal of central inhibition from the pontine reticular formation, resulting in excitation of α- and γ-motoneurons and rigid posture.

c. Lesions above the red nucleus

- result in **decorticate posturing** and intact tonic neck reflexes.

F. Cerebellum—central control of movement

1. Functions of the cerebellum

a. Vestibulocerebellum—control of balance and eye movement.

b. Pontocerebellum—planning and initiation of movement.

c. Spinocerebellum—synergy, which is control of rate, force, range, and direction of movement.

2. Layers of the cerebellar cortex

a. Granular layer

- is the innermost layer.
- contains granule cells, Golgi type II cells, and glomeruli.
- In the **glomeruli**, axons of mossy fibers form synaptic connections on dendrites of granular and Golgi type II cells.

b. Purkinje cell layer

- is the middle layer.
- contains Purkinje cells.
- **Output is always inhibitory.**

c. Molecular layer

- is the outermost layer.
- contains stellate and basket cells, dendrites of Purkinje and Golgi type II cells, and parallel fibers (axons of granule cells).
- The **parallel fibers** synapse on dendrites of Purkinje cells, basket cells, stellate cells, and Golgi type II cells.

3. Connections in the cerebellar cortex

a. Input to the cerebellar cortex

(1) *Climbing fibers*

- originate from a **single region** of the medulla (inferior olive).
- make multiple synapses onto Purkinje cells, resulting in high-frequency bursts, or **complex spikes**.
- "condition" the Purkinje cells.
- play a role in cerebellar **motor learning.**

(2) *Mossy fibers*

- originate from **many centers** in the brain stem and spinal cord.
- include vestibulocerebellar, spinocerebellar, and pontocerebellar afferents.
- make multiple synapses on Purkinje fibers via interneurons. Synapses on Purkinje cells result in **simple spikes**.

- synapse on granule cells in **glomeruli**.
- The axons of granule cells bifurcate and give rise to **parallel cells**. The parallel fibers excite multiple Purkinje cells as well as inhibitory interneurons (basket, stellate, Golgi type II).

b. Output of the cerebellar cortex

- **Purkinje cells** are the ***only output* of the cerebellar cortex**.
- Output of the Purkinje cells is **always inhibitory**; the neurotransmitter is **γ-aminobutyric acid (GABA)**.
- The output projects to deep cerebellar nuclei and to the vestibular nucleus. This inhibitory output **modulates** the output of the cerebellum and regulates rate, range, and direction of movement **(synergy)**.

c. Clinical disorders of the cerebellum—ataxia

- result in **lack of coordination**, including delay in initiation of movement, poor execution of a sequence of movements, and inability to perform rapid alternating movements **(dysdiadochokinesia)**.

(1) *Intention tremor* occurs during attempts to perform voluntary movements.
(2) *Rebound phenomenon* is the inability to stop a movement.

G. Basal ganglia—control of movement

- consists of the striatum, globus pallidus, subthalamic nuclei, and substantia nigra.
- modulates thalamic outflow to the motor cortex to **plan and execute smooth movements**.
- Many synaptic connections are inhibitory and use **GABA** as their neurotransmitter.
- The striatum communicates with the thalamus and the cerebral cortex by two opposing pathways.
- **Indirect pathway** is, overall, inhibitory.
- **Direct pathway** is, overall, excitatory.
- Connections between the striatum and the substantia nigra use **dopamine** as their neurotransmitter. Dopamine is inhibitory on the indirect pathway (**D_2** receptors) and excitatory on the direct pathway (**D_1** receptors). Thus, the action of dopamine is, overall, excitatory.
- Lesions of the basal ganglia include the following:

1. Lesions of the globus pallidus

- result in inability to maintain postural support.

2. Lesions of the subthalamic nucleus

- are caused by the release of inhibition on the contralateral side.
- result in wild, flinging movements (e.g., hemiballismus).

3. Lesions of the striatum

- are caused by the release of inhibition.
- result in quick, continuous, and uncontrollable movements.
- occur in patients with **Huntington disease**.

4. Lesions of the substantia nigra

- are caused by **destruction of dopaminergic neurons**.
- occur in patients with **Parkinson disease**.
- Since dopamine inhibits the indirect (inhibitory) pathway and excites the direct (excitatory) pathway, destruction of dopaminergic neurons is, **overall, inhibitory**.
- Symptoms include **lead-pipe rigidity**, **tremor**, and **reduced voluntary movement**.

H. Motor cortex

1. Premotor cortex and supplementary motor cortex (area 6)

- are responsible for **generating a plan for movement**, which is transferred to the primary motor cortex for execution.
- The supplementary motor cortex programs complex motor sequences and is active during **"mental rehearsal"** for a movement.

2. Primary motor cortex (area 4)

- is responsible for the **execution of movement.** Programmed patterns of motoneurons are activated in the motor cortex. Excitation of upper motoneurons in the motor cortex is transferred to the brain stem and spinal cord, where the lower motoneurons are activated and cause voluntary movement.
- is somatotopically organized **(motor homunculus).** Epileptic events in the primary motor cortex cause **jacksonian seizures,** which illustrate the somatotopic organization.

V. HIGHER FUNCTIONS OF THE CEREBRAL CORTEX

A. Electroencephalographic (EEG) findings

- **EEG waves** consist of alternating excitatory and inhibitory synaptic potentials in the pyramidal cells of the cerebral cortex.
- A **cortical evoked potential** is an EEG change. It reflects synaptic potentials evoked in large numbers of neurons.
- In awake adults with eyes open, **beta waves** predominate.
- In awake adults with eyes closed, **alpha waves** predominate.
- During sleep, **slow waves** predominate, muscles relax, and heart rate and blood pressure decrease.

B. Sleep

1. **Sleep–wake cycles occur in a circadian rhythm,** with a period of about 24 hours. The circadian periodicity is thought to be driven by the suprachiasmatic nucleus of the **hypothalamus,** which receives input from the retina.
2. **Rapid eye movement (REM) sleep** occurs every 90 minutes.
   - During REM sleep, the **EEG** resembles that of a person who is awake or in stage I non-REM sleep.
   - Most **dreams** occur during REM sleep.
   - REM sleep is characterized by eye movements, **loss of muscle tone,** pupillary constriction, and **penile erection.**
   - Use of **benzodiazepines** and **increasing age** decrease the duration of REM sleep.

C. Language

- Information is transferred between the two hemispheres of the cerebral cortex through the corpus callosum.
- The **right hemisphere** is dominant in facial expression, intonation, body language, and spatial tasks.
- The **left hemisphere** is usually dominant with respect to **language,** even in left-handed people. Lesions of the left hemisphere cause **aphasia.**

1. Damage to **Wernicke area** causes **sensory aphasia,** in which there is difficulty understanding written or spoken language.
2. Damage to **Broca area** causes **motor aphasia,** in which speech and writing are affected, but understanding is intact.

D. Learning and memory

- **Short-term memory** involves synaptic changes.
- **Long-term memory** involves structural changes in the nervous system and is more stable.
- Bilateral lesions of the **hippocampus** block the ability to form new long-term memories.

table 2.9 Comparison of Cerebrospinal Fluid (CSF) and Blood Concentrations

CSF ≈ Blood	CSF < Blood	CSF > Blood
Na^+	K^+	Mg^{2+}
Cl^-	Ca^{2+}	Creatinine
HCO_3^-	Glucose	
Osmolarity	Cholesterol* Protein*	

*Negligible concentration in CSF.

VI. BLOOD–BRAIN BARRIER AND CEREBROSPINAL FLUID (CSF)

A. Anatomy of the blood–brain barrier

- It is the barrier between cerebral capillary blood and the CSF. CSF fills the ventricles and the subarachnoid space.
- It consists of the **endothelial cells of the cerebral capillaries** and the **choroid plexus epithelium.**

B. Formation of CSF by the choroid plexus epithelium

- Lipid-soluble substances (CO_2 and O_2) and H_2O freely cross the blood–brain barrier and equilibrate between blood and CSF.
- Other substances are transported by carriers in the choroid plexus epithelium. They may be secreted from blood into the CSF or absorbed from the CSF into blood.
- **Protein** and **cholesterol** are excluded from the CSF because of their large molecular size.
- The composition of CSF is approximately the same as that of the interstitial fluid of the brain but differs significantly from blood (Table 2.9).
- CSF can be sampled with a **lumbar puncture**.

C. Functions of the blood–brain barrier

1. **It maintains a constant environment** for neurons in the CNS and protects the brain from endogenous or exogenous toxins.
2. **It prevents the escape of neurotransmitters** from their functional sites in the CNS into the general circulation.
3. **Drugs** penetrate the blood–brain barrier to varying degrees. For example, nonionized (lipid-soluble) drugs cross more readily than ionized (water-soluble) drugs.
   - **Inflammation**, **irradiation**, and **tumors** may destroy the blood–brain barrier and permit entry into the brain of substances that are usually excluded (e.g., antibiotics, radiolabeled markers).

VII. TEMPERATURE REGULATION

A. Sources of heat gain and heat loss from the body

1. **Heat-generating mechanisms—response to cold**
   a. **Thyroid hormone** increases metabolic rate and heat production by stimulating Na^+, K^+-adenosine triphosphatase (ATPase).
   b. **Cold temperatures activate the sympathetic nervous system** and, via activation of β receptors in **brown fat**, increase metabolic rate and heat production.
   c. **Shivering** is the most potent mechanism for increasing heat production.
      - Cold temperatures activate the shivering response, which is orchestrated by the *posterior* hypothalamus.

- α-Motoneurons and γ-motoneurons are activated, causing contraction of skeletal muscle and heat production.

2. **Heat-loss mechanisms—response to heat**

   a. Heat loss by **radiation** and **convection** increases when the ambient temperature increases.
   - The response is orchestrated by the *anterior* hypothalamus.
   - Increases in temperature cause a **decrease in sympathetic tone to cutaneous blood vessels**, increasing blood flow through the arterioles and increasing **arteriovenous shunting of blood** to the venous plexus near the surface of the skin. Shunting of warm blood to the surface of the skin increases heat loss by radiation and convection.

   b. Heat loss by **evaporation** depends on the activity of **sweat glands**, which are under **sympathetic muscarinic** control.

B. Hypothalamic set point for body temperature

1. **Temperature sensors on the skin and in the hypothalamus** "read" the core temperature and relay this information to the **anterior hypothalamus**.
2. **The anterior hypothalamus** compares the detected core temperature to the **set-point temperature**.

   a. **If the core temperature is below the set point**, heat-generating mechanisms (e.g., increased metabolism, shivering, vasoconstriction of cutaneous blood vessels) are activated by the *posterior* hypothalamus.

   b. **If the core temperature is above the set point**, mechanisms for heat loss (e.g., vasodilation of the cutaneous blood vessels, increased sympathetic outflow to the sweat glands) are activated by the *anterior* hypothalamus.
3. **Pyrogens increase the set-point temperature.** Core temperature will be recognized as lower than the new set-point temperature by the anterior hypothalamus. As a result, heat-generating mechanisms (e.g., shivering) will be initiated.

C. Fever

1. **Pyrogens** increase the production of **interleukin-1** (IL-1) in phagocytic cells.
   - Macrophages release cytokines into the circulation, which cross the blood–brain barrier.
   - IL-1 acts on the anterior hypothalamus to increase the production of prostaglandin E_2. **Prostaglandins increase the set-point temperature**, setting in motion the heat-generating mechanisms that increase body temperature and produce fever.
2. **Aspirin** reduces fever by **inhibiting cyclooxygenase**, thereby inhibiting the production of prostaglandins. Therefore, aspirin **decreases the set-point temperature**. In response, mechanisms that cause heat loss (e.g., sweating, vasodilation) are activated.
3. **Steroids** reduce fever by blocking the release of arachidonic acid from brain phospholipids, thereby preventing the production of prostaglandins.

D. Heat exhaustion and heat stroke

1. **Heat exhaustion** is caused by excessive sweating. As a result, blood volume and arterial blood pressure decrease and syncope (fainting) occurs.
2. **Heat stroke** occurs when body temperature increases to the point of tissue damage. The normal response to increased ambient temperature (sweating) is impaired, and core temperature increases further.

E. Hypothermia

- results when the ambient temperature is so low that heat-generating mechanisms (e.g., shivering, metabolism) cannot adequately maintain core temperature near the set point.

F. Malignant hyperthermia

- is caused in susceptible individuals by inhalation anesthetics.
- is characterized by a massive increase in oxygen consumption and heat production by skeletal muscle, which causes a rapid rise in body temperature.

Review Test

1. Which autonomic receptor is blocked by hexamethonium at the ganglia, but not at the neuromuscular junction?

(A) Adrenergic α_1 receptors
(B) Adrenergic β_1 receptors
(C) Adrenergic β_2 receptors
(D) Cholinergic muscarinic receptors
(E) Cholinergic nicotinic receptors

2. A 66-year-old man with chronic hypertension is treated with prazosin by his physician. The treatment successfully decreases his blood pressure to within the normal range. What is the mechanism of the drug's action?

(A) Inhibition of β_1 receptors in the sinoatrial (SA) node
(B) Inhibition of β_2 receptors in the SA node
(C) Stimulation of muscarinic receptors in the SA node
(D) Stimulation of nicotinic receptors in the SA node
(E) Inhibition of β_1 receptors in ventricular muscle
(F) Stimulation of β_1 receptors in ventricular muscle
(G) Inhibition of α_1 receptors in ventricular muscle
(H) Stimulation of α_1 receptors in the SA node
(I) Inhibition of α_1 receptors in the SA node
(J) Inhibition of α_1 receptors on vascular smooth muscle
(K) Stimulation of α_1 receptors on vascular smooth muscle
(L) Stimulation of α_2 receptors on vascular smooth muscle

3. Which of the following responses is mediated by parasympathetic muscarinic receptors?

(A) Dilation of bronchiolar smooth muscle
(B) Miosis
(C) Ejaculation
(D) Constriction of gastrointestinal (GI) sphincters
(E) Increased cardiac contractility

4. Which of the following is a property of C fibers?

(A) Have the slowest conduction velocity of any nerve fiber type
(B) Have the largest diameter of any nerve fiber type
(C) Are afferent nerves from muscle spindles
(D) Are afferent nerves from Golgi tendon organs
(E) Are preganglionic autonomic fibers

5. When compared with the cones of the retina, the rods

(A) are more sensitive to low-intensity light
(B) adapt to darkness before the cones
(C) are most highly concentrated on the fovea
(D) are primarily involved in color vision

6. Which of the following statements best describes the basilar membrane of the organ of Corti?

(A) The apex responds better to low frequencies than the base does
(B) The base is wider than the apex
(C) The base is more compliant than the apex
(D) High frequencies produce maximal displacement of the basilar membrane near the helicotrema
(E) The apex is relatively stiff compared to the base

7. Which of the following is a feature of the sympathetic, but not the parasympathetic nervous system?

(A) Ganglia located in the effector organs
(B) Long preganglionic neurons
(C) Preganglionic neurons release norepinephrine
(D) Preganglionic neurons release acetylcholine (ACh)
(E) Preganglionic neurons originate in the thoracolumbar spinal cord
(F) Postganglionic neurons synapse on effector organs
(G) Postganglionic neurons release epinephrine
(H) Postganglionic neurons release ACh

8. Which autonomic receptor mediates an increase in heart rate?

(A) Adrenergic α_1 receptors
(B) Adrenergic β_1 receptors
(C) Adrenergic β_2 receptors
(D) Cholinergic muscarinic receptors
(E) Cholinergic nicotinic receptors

9. Cutting which structure on the left side causes total blindness in the left eye?

(A) Optic nerve
(B) Optic chiasm
(C) Optic tract
(D) Geniculocalcarine tract

10. Which reflex is responsible for monosynaptic excitation of ipsilateral homonymous muscle?

(A) Stretch reflex (myotatic)
(B) Golgi tendon reflex (inverse myotatic)
(C) Flexor withdrawal reflex
(D) Subliminal occlusion reflex

11. Which type of cell in the visual cortex responds best to a moving bar of light?

(A) Simple
(B) Complex
(C) Hypercomplex
(D) Bipolar
(E) Ganglion

12. Administration of which of the following drugs is contraindicated in a 10-year-old child with a history of asthma?

(A) Albuterol
(B) Epinephrine
(C) Isoproterenol
(D) Norepinephrine
(E) Propranolol

13. Which adrenergic receptor produces its stimulatory effects by the formation of inositol 1,4,5-triphosphate (IP3) and an increase in intracellular [Ca^{2+}]?

(A) α_1 Receptors
(B) α_2 Receptors
(C) β_1 Receptors
(D) β_2 Receptors
(E) Muscarinic receptors
(F) Nicotinic receptors

14. The excessive muscle tone produced in decerebrate rigidity can be reversed by

(A) stimulation of group Ia afferents
(B) cutting the dorsal roots
(C) transection of cerebellar connections to the lateral vestibular nucleus
(D) stimulation of α-motoneurons
(E) stimulation of γ-motoneurons

15. Which of the following parts of the body has cortical motoneurons with the largest representation on the primary motor cortex (area 4)?

(A) Shoulder
(B) Ankle
(C) Fingers
(D) Elbow
(E) Knee

16. Which autonomic receptor mediates secretion of epinephrine by the adrenal medulla?

(A) Adrenergic α_1 receptors
(B) Adrenergic β_1 receptors
(C) Adrenergic β_2 receptors
(D) Cholinergic muscarinic receptors
(E) Cholinergic nicotinic receptors

17. Cutting which structure on the right side causes blindness in the temporal field of the left eye and the nasal field of the right eye?

(A) Optic nerve
(B) Optic chiasm
(C) Optic tract
(D) Geniculocalcarine tract

18. A ballet dancer spins to the left. During the spin, her eyes snap quickly to the left. This fast eye movement is

(A) nystagmus
(B) postrotatory nystagmus
(C) ataxia
(D) aphasia

19. Which of the following has a much lower concentration in the cerebrospinal fluid (CSF) than in cerebral capillary blood?

(A) Na^+
(B) K^+
(C) Osmolarity
(D) Protein
(E) Mg^{2+}

20. Which of the following autonomic drugs acts by stimulating adenylate cyclase?

(A) Atropine
(B) Clonidine
(C) Curare
(D) Norepinephrine

(E) Phentolamine
(F) Phenylephrine
(G) Propranolol

21. Which of the following is a step in photoreception in the rods?

(A) Light converts all-*trans* retinal to 11-*cis* retinal
(B) Metarhodopsin II activates transducin
(C) Cyclic guanosine monophosphate (cGMP) levels increase
(D) Rods depolarize
(E) Glutamate release increases

22. Pathogens that produce fever cause

(A) decreased production of interleukin-1 (IL-1)
(B) decreased set-point temperature in the hypothalamus
(C) shivering
(D) vasodilation of blood vessels in the skin

23. Which of the following statements about the olfactory system is true?

(A) The receptor cells are neurons
(B) The receptor cells are sloughed off and are not replaced
(C) Axons of cranial nerve (CN) I are A-delta fibers
(D) Axons from receptor cells synapse in the prepiriform cortex
(E) Fractures of the cribriform plate can cause inability to detect ammonia

24. A lesion of the chorda tympani nerve would most likely result in

(A) impaired olfactory function
(B) impaired vestibular function
(C) impaired auditory function
(D) impaired taste function
(E) nerve deafness

25. Which of the following would produce maximum excitation of the hair cells in the right horizontal semicircular canal?

(A) Hyperpolarization of the hair cells
(B) Bending the stereocilia away from the kinocilia
(C) Rapid ascent in an elevator
(D) Rotating the head to the right

26. The inability to perform rapidly alternating movements (dysdiadochokinesia) is associated with lesions of the

(A) premotor cortex
(B) motor cortex
(C) cerebellum
(D) substantia nigra
(E) medulla

27. Which autonomic receptor is activated by low concentrations of epinephrine released from the adrenal medulla and causes vasodilation?

(A) Adrenergic α_1 receptors
(B) Adrenergic β_1 receptors
(C) Adrenergic β_2 receptors
(D) Cholinergic muscarinic receptors
(E) Cholinergic nicotinic receptors

28. Complete transection of the spinal cord at the level of T1 would most likely result in

(A) temporary loss of stretch reflexes below the lesion
(B) temporary loss of conscious proprioception below the lesion
(C) permanent loss of voluntary control of movement above the lesion
(D) permanent loss of consciousness above the lesion

29. Sensory receptor potentials

(A) are action potentials
(B) always bring the membrane potential of a receptor cell toward threshold
(C) always bring the membrane potential of a receptor cell away from threshold
(D) are graded in size, depending on stimulus intensity
(E) are all or none

30. Cutting which structure causes blindness in the temporal fields of the left and right eyes?

(A) Optic nerve
(B) Optic chiasm
(C) Optic tract
(D) Geniculocalcarine tract

31. Which of the following structures has a primary function to coordinate rate, range, force, and direction of movement?

(A) Primary motor cortex
(B) Premotor cortex and supplementary motor cortex
(C) Prefrontal cortex
(D) Basal ganglia
(E) Cerebellum

32. Which reflex is responsible for polysynaptic excitation of contralateral extensors?

(A) Stretch reflex (myotatic)
(B) Golgi tendon reflex (inverse myotatic)
(C) Flexor withdrawal reflex
(D) Subliminal occlusion reflex

33. Which of the following is a characteristic of nuclear bag fibers?

(A) They are one type of extrafusal muscle fiber
(B) They detect dynamic changes in muscle length
(C) They give rise to group Ib afferents
(D) They are innervated by α-motoneurons

34. Muscle stretch leads to a direct increase in firing rate of which type of nerve?

(A) α-Motoneurons
(B) γ-Motoneurons
(C) Group Ia fibers
(D) Group Ib fibers

35. A 42-year-old woman with elevated blood pressure, visual disturbances, and vomiting has increased urinary excretion of 3-methoxy-4-hydroxymandelic acid (VMA). A computerized tomographic scan shows an adrenal mass that is consistent with a diagnosis of pheochromocytoma. While awaiting surgery to remove the tumor, she is treated with phenoxybenzamine to lower her blood pressure. What is the mechanism of this action of the drug?

(A) Increasing cyclic adenosine monophosphate (cAMP)
(B) Decreasing cAMP
(C) Increasing inositol 1,4,5-triphosphate (IP3)/Ca^{2+}
(D) Decreasing IP3/Ca^{2+}
(E) Opening Na^+/K^+ channels
(F) Closing Na^+/K^+ channels

36. Patients are enrolled in trials of a new atropine analogue. Which of the following would be expected?

(A) Increased AV node conduction velocity
(B) Increased gastric acidity
(C) Pupillary constriction
(D) Sustained erection
(E) Increased sweating

Answers and Explanations

1. **The answer is E** [I C 2 a]. Hexamethonium is a nicotinic blocker, but it acts only at ganglionic (not neuromuscular junction) nicotinic receptors. This pharmacologic distinction emphasizes that nicotinic receptors at these two locations, although similar, are not identical.

2. **The answer is J** [I C 1 a; Table 2.2]. Prazosin is a specific antagonist of α_1 receptors, which are present in vascular smooth muscle, but not in the heart. Inhibition of α_1 receptors results in vasodilation of the cutaneous and splanchnic vascular beds, decreased total peripheral resistance, and decreased blood pressure.

3. **The answer is B** [I C 2 b; Table 2.6]. Miosis is a parasympathetic muscarinic response that involves contraction of the circular muscle of the iris. Dilation of the bronchioles, ejaculation, constriction of the gastrointestinal (GI) sphincters, and increased cardiac contractility are all sympathetic α or β responses.

4. **The answer is A** [II F 1 b; Table 2.5]. C fibers (slow pain) are the smallest nerve fibers and therefore have the slowest conduction velocity.

5. **The answer is A** [II C 2 c (2); Table 2.7]. Of the two types of photoreceptors, the rods are more sensitive to low-intensity light and therefore are more important than the cones for night vision. They adapt to darkness after the cones. Rods are not present in the fovea. The cones are primarily involved in color vision.

6. **The answer is A** [II D 4]. Sound frequencies can be encoded by the organ of Corti because of differences in properties along the basilar membrane. The base of the basilar membrane is narrow and stiff, and hair cells on it are activated by high frequencies. The apex of the basilar membrane is wide and compliant, and hair cells on it are activated by low frequencies.

7. **The answer is E** [I A, B; Table 2.1; Figure 2.1]. Sympathetic preganglionic neurons originate in spinal cord segments T1–L3. Thus, the designation is thoracolumbar. The sympathetic nervous system is further characterized by short preganglionic neurons that synapse in ganglia located in the paravertebral chain (not in the effector organs) and postganglionic neurons that release norepinephrine (not epinephrine). Common features of the sympathetic and parasympathetic nervous systems are preganglionic neurons that release acetylcholine (ACh) and postganglionic neurons that synapse in effector organs.

8. **The answer is B** [I C 1 c]. Heart rate is increased by the stimulatory effect of norepinephrine on β_1 receptors in the sinoatrial (SA) node. There are also sympathetic β_1 receptors in the heart that regulate contractility.

9. **The answer is A** [II C 3 a]. Cutting the optic nerve from the left eye causes blindness in the left eye because the fibers have not yet crossed at the optic chiasm.

10. **The answer is A** [III C 1]. The stretch reflex is the monosynaptic response to stretching of a muscle. The reflex produces contraction and then shortening of the muscle that was originally stretched (homonymous muscle).

11. **The answer is B** [II C 5 b (2)]. Complex cells respond to moving bars or edges with the correct orientation. Simple cells respond to stationary bars, and hypercomplex cells respond to lines, curves, and angles. Bipolar and ganglion cells are found in the retina, not in the visual cortex.

12. **The answer is E** [I C 1 d; Table 2.2]. Asthma, a disease involving increased resistance of the upper airways, is treated by administering drugs that produce bronchiolar dilation (i.e., β_2 agonists). β_2 Agonists include isoproterenol, albuterol, epinephrine, and, to a lesser extent, norepinephrine. β_2 Antagonists, such as propranolol, are strictly contraindicated because they cause constriction of the bronchioles.

13. **The answer is A** [I C 1 a]. Adrenergic α_1 receptors produce physiologic actions by stimulating the formation of inositol 1,4,5-triphosphate (IP_3) and causing a subsequent increase in intracellular [Ca^{2+}]. Both β_1 and β_2 receptors act by stimulating adenylate cyclase and increasing the production of cyclic adenosine monophosphate (cAMP). α_2 Receptors inhibit adenylate cyclase and decrease cAMP levels. Muscarinic and nicotinic receptors are cholinergic.

14. **The answer is B** [III E 3 a, b]. Decerebrate rigidity is caused by increased reflex muscle spindle activity. Stimulation of group Ia afferents would enhance, not diminish, this reflex activity. Cutting the dorsal roots would block the reflexes. Stimulation of α- and γ-motoneurons would stimulate muscles directly.

15. **The answer is C** [II B 4]. Representation on the motor homunculus is greatest for those structures that are involved in the most complicated movements—the fingers, hands, and face.

16. **The answer is E** [I C 2 a; Figure 2.1]. Preganglionic sympathetic fibers synapse on the chromaffin cells of the adrenal medulla at a nicotinic receptor. Epinephrine and, to a lesser extent, norepinephrine are released into the circulation.

17. **The answer is C** [II C 3 c]. Fibers from the left temporal field and the right nasal field ascend together in the right optic tract.

18. **The answer is A** [II E 3]. The fast eye movement that occurs during a spin is nystagmus. It occurs in the same direction as the rotation. After the spin, postrotatory nystagmus occurs in the opposite direction.

19. **The answer is D** [V B; Table 2.9]. Cerebrospinal fluid (CSF) is similar in composition to the interstitial fluid of the brain. Therefore, it is similar to an ultrafiltrate of plasma and has a very low protein concentration because large protein molecules cannot cross the blood–brain barrier. There are other differences in composition between CSF and blood that are created by transporters in the choroid plexus, but the low protein concentration of CSF is the most dramatic difference.

20. **The answer is D** [I C 1 c, d; Table 2.2]. Among the autonomic drugs, only β_1 and β_2 adrenergic agonists act by stimulating adenylate cyclase. Norepinephrine is a $\beta_1$agonist. Atropine is a muscarinic cholinergic antagonist. Clonidine is an α_2 adrenergic agonist. Curare is a nicotinic cholinergic antagonist. Phentolamine is an α_1 adrenergic antagonist. Phenylephrine is an α_1 adrenergic agonist. Propranolol is a β_1 and β_2 adrenergic antagonist.

21. **The answer is B** [II C 4]. Photoreception involves the following steps. Light converts 11-*cis* retinal to all-*trans* retinal, which is converted to such intermediates as metarhodopsin II. Metarhodopsin II activates a stimulatory G protein (transducin), which activates a phosphodiesterase. Phosphodiesterase breaks down cyclic guanosine monophosphate (cGMP), so intracellular cGMP levels decrease, causing closure of Na^+ channels in the photoreceptor cell membrane and hyperpolarization. Hyperpolarization of the photoreceptor cell membrane inhibits release of the neurotransmitter, glutamate. If the decreased release of glutamate interacts with ionotropic receptors on bipolar cells, there will be inhibition (decreased excitation). If the decreased release of glutamate interacts with metabotropic receptors on bipolar cells, there will be excitation (decreased inhibition).

22. **The answer is C** [VI C 1]. Pathogens release interleukin-1 (IL-1) from phagocytic cells. IL-1 then acts to increase the production of prostaglandins, ultimately raising the temperature set point in the anterior hypothalamus. The hypothalamus now "thinks" that the body

temperature is too low (because the core temperature is lower than the new set-point temperature) and initiates mechanisms for generating heat—shivering, vasoconstriction, and shunting of blood away from the venous plexus near the skin surface.

23. **The answer is A** [II F 1 a, b]. Cranial nerve (CN) I innervates the olfactory epithelium. Its axons are C fibers. Fracture of the cribriform plate can tear the delicate olfactory nerves and thereby eliminate the sense of smell (anosmia); however, the ability to detect ammonia is left intact. Olfactory receptor cells are unique in that they are true neurons that are continuously replaced from undifferentiated stem cells.

24. **The answer is D** [II G 1 b]. The chorda tympani (cranial nerve [CN] VII) is involved in taste; it innervates the anterior two-thirds of the tongue.

25. **The answer is D** [II E 1 a, 2 a, b]. The semicircular canals are involved in angular acceleration or rotation. Hair cells of the right semicircular canal are excited (depolarized) when there is rotation to the right. This rotation causes bending of the stereocilia toward the kinocilia, and this bending produces depolarization of the hair cell. Ascent in an elevator would activate the saccules, which detect linear acceleration.

26. **The answer is C** [III F 1 c, 3 c]. Coordination of movement (synergy) is the function of the cerebellum. Lesions of the cerebellum cause ataxia, lack of coordination, poor execution of movement, delay in initiation of movement, and inability to perform rapidly alternating movements. The premotor and motor cortices plan and execute movements. Lesions of the substantia nigra, a component of the basal ganglia, result in tremors, lead-pipe rigidity, and poor muscle tone (Parkinson disease).

27. **The answer is C** [I C 1 d]. β_2 Receptors on vascular smooth muscle produce vasodilation. α_1 Receptors on vascular smooth muscle produce vasoconstriction. Because β_2 receptors are more sensitive to epinephrine than are α receptors, low doses of epinephrine produce vasodilation, and high doses produce vasoconstriction.

28. **The answer is A** [III E 2]. Transection of the spinal cord causes "spinal shock" and loss of all reflexes below the level of the lesion. These reflexes, which are local circuits within the spinal cord, will return with time or become hypersensitive. Proprioception is permanently (rather than temporarily) lost because of the interruption of sensory nerve fibers. Fibers above the lesion are intact.

29. **The answer is D** [II A 4 c]. Receptor potentials are graded potentials that may bring the membrane potential of the receptor cell either toward (depolarizing) or away from (hyperpolarizing) threshold. Receptor potentials are not action potentials, although action potentials (which are all-or-none) may result if the membrane potential reaches threshold.

30. **The answer is B** [II C 3 b]. Optic nerve fibers from both temporal receptor fields cross at the optic chiasm.

31. **The answer is E** [III F 3 b]. Output of Purkinje cells from the cerebellar cortex to deep cerebellar nuclei is inhibitory. This output modulates movement and is responsible for the coordination that allows one to "catch a fly."

32. **The answer is C** [III C 3]. Flexor withdrawal is a polysynaptic reflex that is used when a person touches a hot stove or steps on a tack. On the ipsilateral side of the painful stimulus, there is flexion (withdrawal); on the contralateral side, there is extension to maintain balance.

33. **The answer is B** [III B 3 a (1)]. Nuclear bag fibers are one type of intrafusal muscle fiber that make up muscle spindles. They detect dynamic changes in muscle length, give rise to group Ia afferent fibers, and are innervated by γ-motoneurons. The other type of intrafusal fiber, the nuclear chain fiber, detects static changes in muscle length.

34. **The answer is C** [III B 3 b]. Group Ia afferent fibers innervate intrafusal fibers of the muscle spindle. When the intrafusal fibers are stretched, the group Ia fibers fire and activate the stretch reflex, which causes the muscle to return to its resting length.

35. **The answer is D** [I C; Tables 2.2 and 2.5]. Pheochromocytoma is a tumor of the adrenal medulla that secretes excessive amounts of norepinephrine and epinephrine. Increased blood pressure is due to activation of α_1 receptors on vascular smooth muscle and activation of β_1 receptors in the heart. Phenoxybenzamine decreases blood pressure by acting as an α_1 receptor antagonist, thus decreasing intracellular IP_3/Ca^{2+}.

36. **The answer is A** [I C 3; I D]. An atropine analogue would block muscarinic receptors and thus block actions that are mediated by muscarinic receptors. Muscarinic receptors slow AV node conduction velocity; thus, muscarinic blocking agents would increase AV node conduction velocity. Muscarinic receptors increase gastric acid secretion, constrict the pupils, mediate erection, and cause sweating (via sympathetic cholinergic innervation of sweat glands); thus, blocking muscarinic receptors will inhibit all of those actions.

Chapter 3

Cardiovascular Physiology

I. CIRCUITRY OF THE CARDIOVASCULAR SYSTEM (FIGURE 3.1)

A. Cardiac output of the left heart equals cardiac output of the right heart

- Cardiac output from the left side of the heart is the systemic blood flow.
- Cardiac output from the right side of the heart is the pulmonary blood flow.

B. Direction of blood flow

- Blood flows along the following course:

1. From the lungs to the left atrium via the pulmonary vein
2. From the left atrium to the left ventricle through the mitral valve
3. From the left ventricle to the aorta through the aortic valve
4. From the aorta to the systemic arteries and the systemic tissues (i.e., cerebral, coronary, renal, splanchnic, skeletal muscle, and skin)
5. From the tissues to the systemic veins and vena cava
6. From the vena cava (mixed venous blood) to the right atrium
7. From the right atrium to the right ventricle through the tricuspid valve
8. From the right ventricle to the pulmonary artery through the pulmonic valve
9. From the pulmonary artery to the lungs for oxygenation

II. HEMODYNAMICS

A. Components of the vasculature

1. Arteries

- deliver oxygenated blood to the tissues.
- are thick walled, with extensive **elastic tissue** and **smooth muscle.**
- are under **high pressure.**
- The blood volume contained in the arteries is called the **stressed volume.**

2. Arterioles

- are the smallest branches of the arteries.
- are the **site of highest resistance in the cardiovascular system.**
- have a smooth muscle wall that is extensively innervated by autonomic nerve fibers.
- Arteriolar resistance is regulated by the autonomic nervous system (ANS).

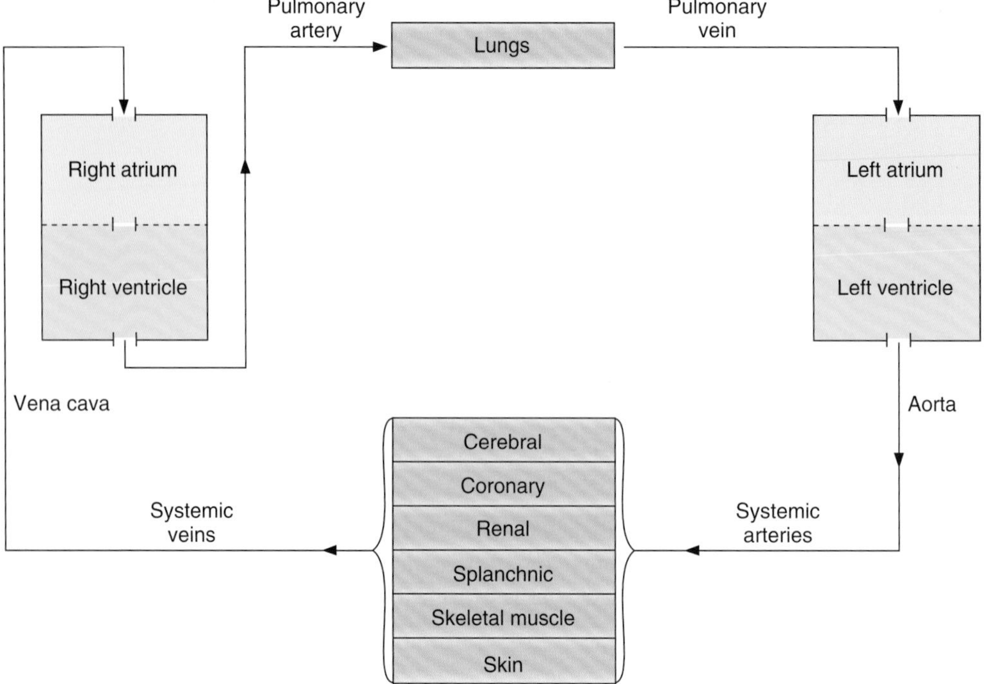

FIGURE 3.1. Circuitry of the cardiovascular system.

- α_1-Adrenergic receptors are found on the arterioles of the skin, splanchnic, and renal circulations.
- β_2-Adrenergic receptors are found on arterioles of skeletal muscle.

3. Capillaries

- have the **largest total cross-sectional and surface area.**
- consist of a single layer of endothelial cells surrounded by basal lamina.
- are thin walled.
- are the site of exchange of nutrients, water, and gases.

4. Venules

- are formed from merged capillaries.

5. Veins

- progressively merge to form larger veins. The largest vein, the vena cava, returns blood to the heart.
- are thin walled.
- are under **low pressure.**
- contain the **highest proportion of the blood** in the cardiovascular system.
- The blood volume contained in the veins is called the **unstressed volume.**
- have α_1-adrenergic receptors.

B. Velocity of blood flow

- can be expressed by the following equation:

$$v = Q/A$$

where:
v = velocity (cm/sec)
Q = blood flow (mL/min)
A = cross-sectional area (cm^2)

- Velocity is directly proportional to blood flow and inversely proportional to the cross-sectional area at any level of the cardiovascular system.
- **For example**, blood velocity is higher in the aorta (small cross-sectional area) than in the sum of all of the capillaries (large cross-sectional area). The lower velocity of blood in the capillaries optimizes conditions for exchange of substances across the capillary wall.

C. Blood flow

- can be expressed by the following equation:

$$\mathbf{Q = \Delta P/R}$$

or

$$\textbf{Cardiac output} = \frac{\textbf{Mean arterial pressure} - \textbf{Right atrial pressure}}{\textbf{Total peripheral resitance (TPR)}}$$

where:
Q = flow or cardiac output (mL/min)
ΔP = pressure gradient (mm Hg)
R = resistance or total peripheral resistance (mm Hg/mL/min)

- The equation for blood flow (or cardiac output) is analogous to **Ohm law** for electrical circuits ($I = V/R$), where flow is analogous to current, and pressure is analogous to voltage.
- The pressure gradient (ΔP) drives blood flow.
- Thus, the direction of blood flow is from high pressure to low pressure.
- Blood flow is inversely proportional to the resistance of the blood vessels.

D. Resistance

- Poiseuille equation gives factors that change the resistance of blood vessels.

$$\mathbf{R = \frac{8\eta l}{\pi r^4}}$$

where:
R = resistance
η = viscosity of blood
l = length of blood vessel
r^4 = radius of blood vessel to the fourth power

- Resistance is directly proportional to the viscosity of the blood. For example, increasing viscosity by increasing hematocrit will increase resistance and decrease blood flow.
- Resistance is directly proportional to the length of the vessel.
- Resistance is inversely proportional to the **fourth power of the vessel** radius. This relationship is powerful. **For example,** if blood vessel radius decreases by a factor of 2, then resistance increases by a factor of 16 (2^4), and blood flow accordingly decreases by a factor of 16.

1. Resistances in parallel or series

a. Parallel resistance is illustrated by the systemic circulation. Each organ is supplied by an artery that branches off the aorta. The total resistance of this parallel arrangement is expressed by the following equation:

$$\frac{1}{R_{total}} = \frac{1}{R_a} + \frac{1}{R_b} + \cdots \frac{1}{R_n}$$

R_a, R_b, and R_n are the resistances of the renal, hepatic, and other circulations, respectively.

- Each artery in parallel receives a fraction of the total blood flow.
- The total resistance is less than the resistance of any of the individual arteries.
- When an **artery is added in parallel, the total resistance decreases.**
- In each parallel artery, the pressure is the same.

b. Series resistance is illustrated by the arrangement of blood vessels within a given organ. Each organ is supplied by a large artery, smaller arteries, arterioles, capillaries, and veins arranged in series. The total resistance is the sum of the individual resistances, as expressed by the following equation:

$$R_{total} = R_{artery} + R_{arterioles} + R_{capillaries}$$

- The largest proportion of resistance in this series is contributed by the **arterioles.**
- Each blood vessel (e.g., the largest artery) or set of blood vessels (e.g., all of the capillaries) in series receives the same total blood flow. Thus, blood flow through the largest artery is the same as the total blood flow through all of the capillaries.
- As blood flows through the series of blood vessels, the pressure decreases.

2. Laminar flow versus turbulent flow

- Laminar flow is streamlined (in a straight line); turbulent flow is not.
- The **Korotkoff sounds** used in the auscultatory measurement of blood pressure are caused by turbulent blood flow.
- **Reynolds number** predicts whether blood flow will be laminar or turbulent.
- When Reynolds number is increased, there is a greater tendency for **turbulence,** which causes audible vibrations called **bruits.** Reynolds number (and therefore turbulence) is increased by the following factors:

a. ↓ blood viscosity (e.g., ↓ hematocrit, **anemia**).

b. ↑ blood velocity (e.g., **narrowing of a vessel**).

3. Shear

- is a consequence of the fact that adjacent layers of blood travel at different velocities within a blood vessel.
- Velocity of blood is zero at the wall and highest at the center of the vessel.
- Shear is therefore highest at the wall, where the *difference* in blood velocity of adjacent layers is greatest; shear is lowest at the center of the vessel, where blood velocity is constant.

E. Capacitance (compliance)

- describes the **distensibility** of blood vessels.
- is **inversely related to elastance,** or stiffness. The greater the amount of elastic tissue there is in a blood vessel, the higher the elastance is, and the lower the compliance is.
- is expressed by the following equation:

$$C = \frac{V}{P}$$

where:
C = capacitance or compliance (mL/mm Hg)
V = volume (mL)
P = pressure (mm Hg)

- is directly proportional to volume and inversely proportional to pressure.
- describes how volume changes in response to a change in pressure.
- is much **greater for veins than for arteries.** As a result, more blood volume is contained in the veins **(unstressed volume)** than in the arteries **(stressed volume)**.
- Changes in the capacitance of the veins produce changes in unstressed volume. For example, a decrease in venous capacitance decreases unstressed volume and increases stressed volume by shifting blood from the veins to the arteries.
- Capacitance of the arteries **decreases with age;** as a person ages, the arteries become stiffer and less distensible.

F. Pressure profile in blood vessels

- As blood flows through the systemic circulation, pressure decreases progressively because of the resistance to blood flow.

- Thus, pressure is highest in the aorta and large arteries and lowest in the venae cavae.
- The **largest decrease in pressure occurs across the arterioles** because they are the site of highest resistance.
- Mean pressures in the systemic circulation are as follows:

1. Aorta, 100 mm Hg
2. Arterioles, 50 mm Hg
3. Capillaries, 20 mm Hg
4. Vena cava, 4 mm Hg

G. Arterial pressure (Figure 3.2)

- is pulsatile.
- is not constant during a cardiac cycle.

1. Systolic pressure

- is the highest arterial pressure during a cardiac cycle.
- is measured after the heart contracts (systole) and blood is ejected into the arterial system.

2. Diastolic pressure

- is the lowest arterial pressure during a cardiac cycle.
- is measured when the heart is relaxed (diastole) and blood is returned to the heart via the veins.

3. Pulse pressure

- is the difference between the systolic and diastolic pressures.
- The most important determinant of pulse pressure is **stroke volume.** As blood is ejected from the left ventricle into the arterial system, arterial pressure increases because of the relatively low capacitance of the arteries. Because diastolic pressure remains unchanged during ventricular systole, the pulse pressure increases to the same extent as the systolic pressure.
- Decreases in capacitance, such as those that occur with the **aging** process, cause **increases in pulse pressure**.

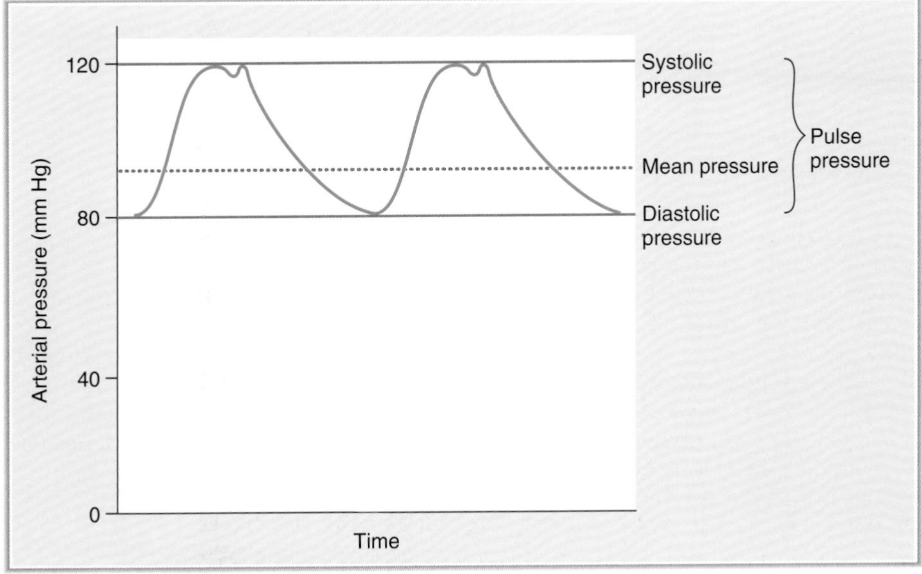

FIGURE 3.2. Arterial pressure during the cardiac cycle.

4. Mean arterial pressure

- is the average arterial pressure with respect to time.
- is *not* the simple average of diastolic and systolic pressures (because a greater fraction of the cardiac cycle is spent in diastole).
- can be calculated approximately as **diastolic pressure plus one-third of pulse pressure.**

H. Venous pressure

- is very low.
- The veins have a high capacitance and, therefore, can hold large volumes of blood at low pressure.

I. Atrial pressure

- is slightly lower than venous pressure.
- Left atrial pressure is estimated by the **pulmonary wedge pressure.** A catheter, inserted into the smallest branches of the pulmonary artery, makes almost direct contact with the pulmonary capillaries. The measured pulmonary capillary pressure is approximately equal to the left atrial pressure.

III. CARDIAC ELECTROPHYSIOLOGY

A. Electrocardiogram (ECG) (Figure 3.3)

1. P wave

- represents atrial depolarization.
- does not include atrial repolarization, which is "buried" in the QRS complex.

2. PR interval

- is the interval from the beginning of the P wave to the beginning of the Q wave (initial depolarization of the ventricle).

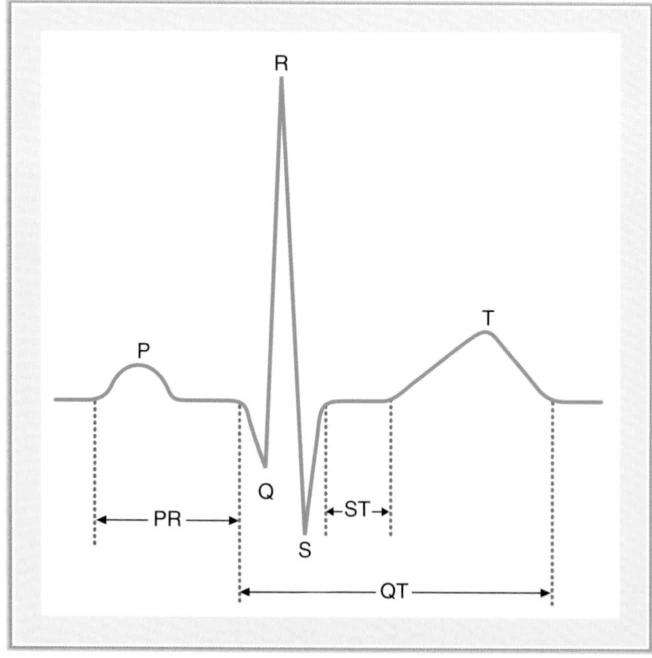

FIGURE 3.3. Normal electrocardiogram measured from lead II.

- depends on **conduction velocity through the atrioventricular (AV) node.** For example, if AV nodal conduction decreases (as in **heart block**), the PR interval increases.
- is decreased (i.e., increased conduction velocity through AV node) by stimulation of the sympathetic nervous system.
- is increased (i.e., decreased conduction velocity through AV node) by stimulation of the parasympathetic nervous system.

3. QRS complex

- represents depolarization of the ventricles.

4. QT interval

- is the interval from the beginning of the Q wave to the end of the T wave.
- represents the entire period of depolarization and repolarization of the ventricles.

5. ST segment

- is the segment from the end of the S wave to the beginning of the T wave.
- is isoelectric.
- represents the period when the ventricles are depolarized.

6. T wave

- represents ventricular repolarization.

B. Cardiac action potentials (see Table 1.3)

- The **resting membrane potential** is determined by the conductance to K^+ and approaches the K^+ equilibrium potential.
- **Inward current** brings positive charge into the cell and **depolarizes** the membrane potential. **Outward current** takes positive charge out of the cell and **hyperpolarizes** the membrane potential.
- The role of Na^+, K^+-adenosine triphosphatase (ATPase) is to maintain ionic gradients across cell membranes.

1. Ventricles, atria, and the Purkinje system (Figure 3.4)

- have stableresting membrane potentials of about −90 millivolts (mV). This value approaches the K^+ equilibrium potential.
- Action potentials are of long duration, especially in Purkinje fibers, where they last 300 milliseconds (msec).

a. Phase 0

- is the **upstroke** of the action potential.
- is caused by a transient increase in **Na^+ conductance.** This increase results in an inward Na^+ current that depolarizes the membrane.

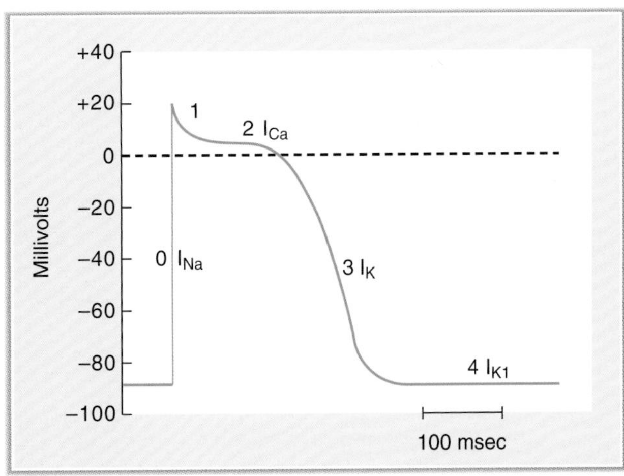

FIGURE 3.4. Ventricular action potential.

- At the peak of the action potential, the membrane potential approaches the Na^+ equilibrium potential.

b. Phase 1

- is a brief period of initial repolarization.
- **Initial repolarization** is caused by an outward current, in part because of the movement of K^+ ions (favored by both chemical and electrical gradients) out of the cell and in part because of a decrease in Na^+ conductance.

c. Phase 2

- is the **plateau** of the action potential.
- is caused by a **transient increase in Ca^{2+} conductance,** which results in an **inward Ca^{2+} current,** and by an increase in K^+ conductance.
- During phase 2, outward and inward currents are approximately equal, so the membrane potential is stable at the plateau level.

d. Phase 3

- is **repolarization.**
- During phase 3, Ca^{2+} conductance decreases, and K^+ conductance increases and therefore predominates.
- The high K^+ conductance results in a large **outward K^+ current (I_K),** which hyperpolarizes the membrane back toward the K^+ equilibrium potential.

e. Phase 4

- is the **resting membrane potential.**
- is a period during which inward and outward currents (I_{K1}) are equal and the membrane potential approaches the K^+ equilibrium potential.

2. Sinoatrial (SA) node (Figure 3.5)

- is normally the **pacemaker** of the heart.
- has an **unstable resting potential.**
- exhibits phase 4 depolarization, or automaticity.
- The AV node and the His–Purkinje systems are **latent pacemakers** that may exhibit automaticity and override the SA node if it is suppressed.
- The intrinsic rate of phase 4 depolarization (and heart rate) is fastest in the SA node and slowest in the His–Purkinje system:

SA node > AV node > His–Purkinje

a. Phase 0

- is the **upstroke** of the action potential.
- is caused by an increase in Ca^{2+} conductance. This increase causes an **inward Ca^{2+} current** that drives the membrane potential toward the Ca^{2+} equilibrium potential.

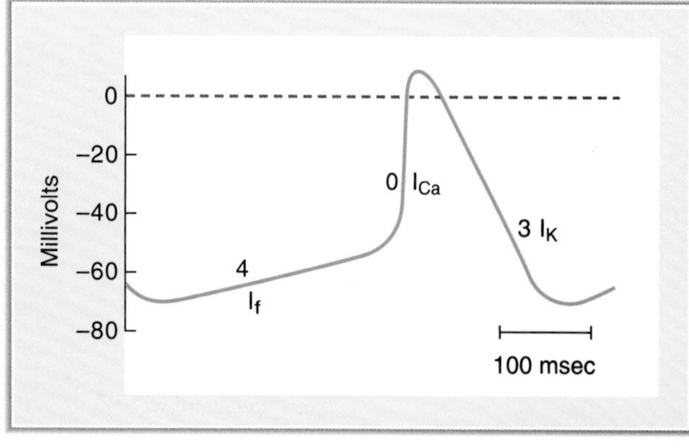

FIGURE 3.5. Sinoatrial nodal action potential.

- The ionic basis for phase 0 in the SA node is different from that in the ventricles, atria, and Purkinje fibers (where it is the result of an inward Na^+ current).

b. Phase 3

- is **repolarization.**
- is caused by an increase in K^+ conductance. This increase results in an **outward K^+ current** that causes repolarization of the membrane potential.

c. Phase 4

- is **slow depolarization.**
- accounts for the pacemaker activity of the SA node (automaticity).
- is caused by an increase in Na^+ conductance, which results in an **inward Na^+ current** called I_f.
- **I_f is turned on by repolarization** of the membrane potential during the preceding action potential.

d. Phases 1 and 2

- are not present in the SA node action potential.

3. AV node

- Upstroke of the action potential in the AV node is the result of an **inward Ca^+ current** (as in the SA node).

C. Conduction velocity

- reflects the time required for excitation to spread throughout cardiac tissue.
- depends on the **size of the inward current during the upstroke** of the action potential. The larger the inward current, the higher the conduction velocity.
- is **fastest in the Purkinje system.**
- is **slowest in the AV node** (seen as the PR interval on the ECG), allowing time for **ventricular filling** before ventricular contraction. If conduction velocity through the AV node is increased, ventricular filling may be compromised.

D. Excitability

- is the ability of cardiac cells to initiate action potentials in response to inward, depolarizing current.
- reflects the recovery of channels that carry the inward currents for the upstroke of the action potential.
- changes over the course of the action potential. These changes in excitability are described by **refractory periods** (Figure 3.6).

1. Absolute refractory period (ARP)

- begins with the upstroke of the action potential and ends after the plateau.

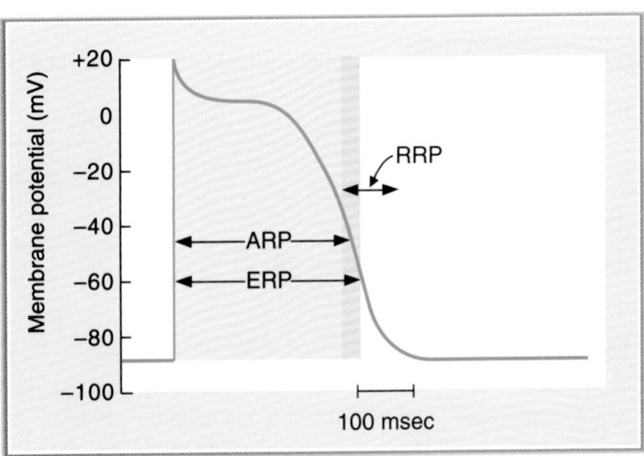

FIGURE 3.6. Absolute refractory period (ARP), effective refractory period (ERP), and relative refractory period (RRP) in the ventricle.

- occurs because, during this period, most channels carrying inward current for the upstroke (Na^+ or Ca^{2+}) are closed and unavailable.
- reflects the period during which **no action potential can be initiated,** regardless of how much inward current is supplied.

2. Effective refractory period (ERP)

- is slightly longer than the ARP.
- is the period during which a **conducted action potential cannot be elicited.**

3. Relative refractory period (RRP)

- is the period immediately after the ARP when repolarization is almost complete.
- is the period during which an **action potential can be elicited, but more than the usual inward current is required.**

E. Autonomic effects on heart rate and conduction velocity (Table 3.1)

- See IV C for a discussion of inotropic effects.

1. Definitions of chronotropic and dromotropic effects

a. Chronotropic effects

- produce changes in heart rate.
- A **negative chronotropic effect** decreases heart rate by decreasing the firing rate of the SA node.
- A **positive chronotropic effect** increases heart rate by increasing the firing rate of the SA node.

b. Dromotropic effects

- produce changes in conduction velocity, primarily in the AV node.
- A **negative dromotropic effect** decreases conduction velocity through the AV node, slowing the conduction of action potentials from the atria to the ventricles and increasing the PR interval.
- A **positive dromotropic effect** increases conduction velocity through the AV node, speeding the conduction of action potentials from the atria to the ventricles and decreasing the PR interval.

2. Parasympathetic effects on heart rate and conduction velocity

- The SA node, atria, and AV node have parasympathetic vagal innervation, but the ventricles do not. The neurotransmitter is **acetylcholine** (ACh), which acts at **muscarinic receptors.**

a. Negative chronotropic effect

- **decreases heart rate** by decreasing the rate of phase 4 depolarization.
- Fewer action potentials occur per unit time because the threshold potential is reached more slowly and, therefore, less frequently.
- The mechanism of the negative chronotropic effect is **decreased I_f**, the inward Na^+ current that is responsible for phase 4 depolarization in the SA node.

table 3.1 Autonomic Effects on the Heart and Blood Vessels

	Sympathetic		Parasympathetic	
	Effect	Receptor	Effect	Receptor
Heart rate	↑	β_1	↓	Muscarinic
Conduction velocity (AV node)	↑	β_1	↓	Muscarinic
Contractility	↑	β_1	↓ (atria only)	Muscarinic
Vascular smooth muscle				
Skin, splanchnic	Constriction	α_1		
Skeletal muscle	Constriction Relaxation	α_1 β_2		

AV = atrioventricular.

b. **Negative dromotropic effect**

- **decreases conduction velocity through the AV node.**
- Action potentials are conducted more slowly from the atria to the ventricles.
- **increases the PR interval.**
- The mechanism of the negative dromotropic effect is **decreased inward Ca^{2+} current** and increased outward K^+ current.

3. **Sympathetic effects on heart rate and conduction velocity**

- **Norepinephrine** is the neurotransmitter, acting at **β_1 receptors.**

a. **Positive chronotropic effect**

- **increases heart rate** by increasing the rate of phase 4 depolarization.
- More action potentials occur per unit time because the threshold potential is reached more quickly and, therefore, more frequently.
- The mechanism of the positive chronotropic effect is **increased I_f**, the inward Na^+ current that is responsible for phase 4 depolarization in the SA node.

b. **Positive dromotropic effect**

- **increases conduction velocity through the AV node.**
- Action potentials are conducted more rapidly from the atria to the ventricles, and ventricular filling may be compromised.
- **decreases the PR interval.**
- The mechanism of the positive dromotropic effect is **increased inward Ca^{2+} current.**

IV. CARDIAC MUSCLE AND CARDIAC OUTPUT

A. Myocardial cell structure

1. **Sarcomere**

- is the contractile unit of the myocardial cell.
- is similar to the contractile unit in skeletal muscle.
- runs from Z line to Z line.
- contains thick filaments (myosin) and thin filaments (actin, troponin, tropomyosin).
- As in skeletal muscle, shortening occurs according to a sliding filament model, which states that thin filaments slide along adjacent thick filaments by forming and breaking crossbridges between actin and myosin.

2. **Intercalated disks**

- occur at the ends of the cells.
- maintain cell-to-cell cohesion.

3. **Gap junctions**

- are present at the intercalated disks.
- are **low-resistance paths** between cells that allow for rapid electrical spread of action potentials.
- account for the observation that the heart behaves as an **electrical syncytium.**

4. **Mitochondria**

- are more numerous in cardiac muscle than in skeletal muscle.

5. **T tubules**

- are continuous with the cell membrane.
- invaginate the cells at the Z lines and **carry action potentials into the cell interior.**
- are well developed in the ventricles, but poorly developed in the atria.
- form **dyads** with the sarcoplasmic reticulum.

6. Sarcoplasmic reticulum (SR)

- are small-diameter tubules in close proximity to the contractile elements.
- are the site of **storage and release of Ca^{2+} for excitation–contraction coupling.**

B. Steps in excitation–contraction coupling

1. The action potential spreads from the cell membrane into the T tubules.
2. During the **plateau** of the action potential, Ca^{2+} conductance is increased and Ca^{2+} enters the cell from the extracellular fluid **(inward Ca^{2+} current)** through L-type Ca^{2+} channels **(dihydropyridine receptors).**
3. This Ca^{2+} entry triggers the release of even more Ca^{2+} from the SR **(Ca^{2+}-induced Ca^{2+} release)** through Ca^{2+} release channels **(ryanodine receptors).**
   - The amount of Ca^{2+} released from the SR depends on the:

   a. amount of Ca^{2+} previously stored in the SR.
   b. size of the inward Ca^{2+} current during the plateau of the action potential.
4. As a result of this Ca^{2+} release, **intracellular [Ca^{2+}] increases.**
5. Ca^{2+} binds to troponin C, and tropomyosin is moved out of the way, removing the inhibition of actin and myosin binding.
6. Actin and myosin bind, the thick and thin filaments slide past each other, and the myocardial cell contracts. **The magnitude of the tension that develops is proportional to the intracellular [Ca^{2+}].**
7. **Relaxation** occurs when Ca^{2+} is reaccumulated by the SR by an active Ca^{2+}-ATPase pump, intracellular Ca^{2+} concentration decreases, and Ca^{2+} dissociates from troponin C.

C. Contractibility

- is the **intrinsic ability of cardiac muscle to develop force at a given muscle length.**
- is also called **inotropism.**
- is related to the **intracellular Ca^{2+} concentration.**
- can be estimated by the **ejection fraction** (stroke volume/end-diastolic volume), which is normally 0.55 (55%).
- **Positive inotropic agents** produce an increase in contractility.
- **Negative inotropic agents** produce a decrease in contractility.

1. Factors that increase contractility (positive inotropism) [see Table 3.1]

a. Increased heart rate

- When more action potentials occur per unit time, more Ca^{2+} enters the myocardial cells during the action potential plateaus, more Ca^{2+} is stored in the SR, more Ca^{2+} is released from the SR, and greater tension is produced during contraction.
- Examples of the effect of increased heart rate are

(1) *positive staircase* or Bowditch staircase (or treppe). Increased heart rate increases the force of contraction in a stepwise fashion as the intracellular [Ca^{2+}] increases cumulatively over several beats.

(2) *postextrasystolic potentiation.* The beat that occurs after an extrasystolic beat has increased force of contraction because "extra" Ca^{2+} entered the cells during the extrasystole.

b. Sympathetic stimulation (catecholamines) via β_1 receptors (see Table 3.1)

- increases the force of contraction by two mechanisms:

(1) It increases the **inward Ca^{2+} current** during the plateau of each cardiac action potential.

(2) It increases the activity of the Ca^{2+} pump of the SR (by phosphorylation of **phospholamban**); as a result, more Ca^{2+} is accumulated by the SR and thus, more Ca^{2+} is available for release in subsequent beats.

c. Cardiac glycosides (digitalis)

- increase the force of contraction by inhibiting Na^+, K^+-ATPase in the myocardial cell membrane (Figure 3.7).

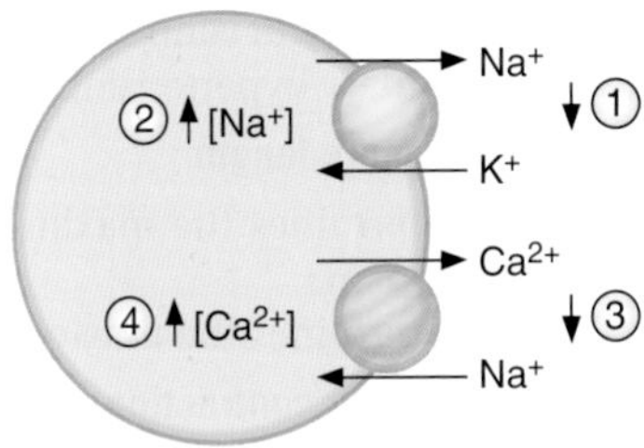

FIGURE 3.7. Stepwise explanation of how ouabain (digitalis) causes an increase in intracellular [Ca^{2+}] and myocardial contractility. The *circled numbers* show the sequence of events.

- As a result of this inhibition, the intracellular [Na^+] increases, diminishing the Na^+ gradient across the cell membrane.
- Na^+–Ca^{2+} exchange (a mechanism that extrudes Ca^{2+} from the cell) depends on the size of the Na^+ gradient and thus is diminished, producing an increase in intracellular [Ca^{2+}].

2. Factors that decrease contractility (negative inotropism) [see Table 3.1]

- **Parasympathetic stimulation (ACh) via muscarinic receptors** decreases the force of contraction in the **atria** by decreasing the inward Ca^{2+} current during the plateau of the cardiac action potential.

D. Length–tension relationship in the ventricles (Figure 3.8)

- describes the effect of ventricular muscle cell length on the force of contraction.
- is analogous to the relationship in skeletal muscle.

1. Preload

- is **end-diastolic volume,** which is related to **right atrial pressure**.
- When venous return increases, end-diastolic volume increases and stretches or lengthens the ventricular muscle fibers (see Frank-Starling relationship, IV D 5).

2. Afterload

- for the left ventricle is **aortic pressure.** Increases in aortic pressure (i.e., systemic hypertension) cause an increase in afterload on the left ventricle.

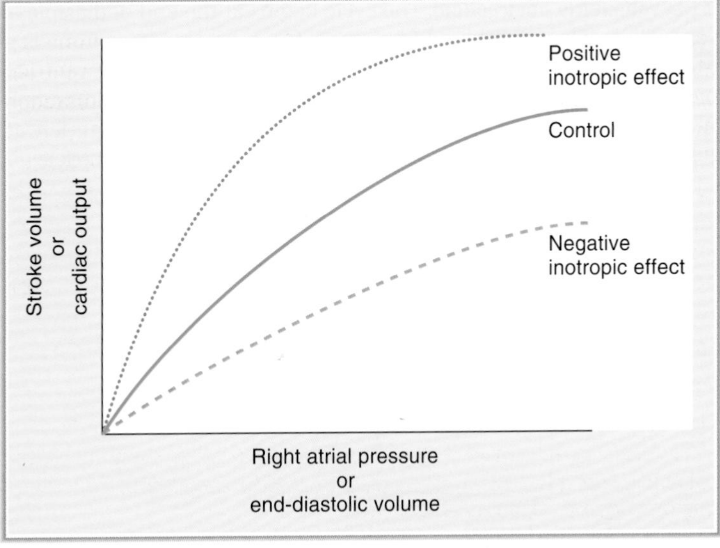

FIGURE 3.8. Frank-Starling relationship and the effect of positive and negative inotropic agents.

- for the right ventricle is **pulmonary artery pressure.** Increases in pulmonary artery (i.e., pulmonary hypertension) pressure cause an increase in afterload on the right ventricle.

3. Sarcomere length

- determines the maximum number of crossbridges that can form between actin and myosin.
- determines the maximum tension, or force of contraction.

4. Velocity of contraction at a fixed muscle length

- is maximal when the afterload is zero.
- is decreased by increases in afterload.

5. Frank-Starling relationship

- describes the increases in stroke volume and cardiac output that occur in response to an increase in venous return or end-diastolic volume (see Figure 3.8).
- is based on the length–tension relationship in the ventricle. **Increases in end-diastolic volume cause an increase in ventricular fiber length, which produces an increase in developed tension.**
- is the mechanism that **matches cardiac output to venous return.** The greater the venous return, the greater the cardiac output.
- Changes in contractility shift the Frank-Starling curve upward (increased contractility) or downward (decreased contractility).

a. Increases in contractility cause an increase in cardiac output for any level of right atrial pressure or end-diastolic volume.

b. Decreases in contractility cause a decrease in cardiac output for any level of right atrial pressure or end-diastolic volume.

E. Ventricular pressure–volume loops (Figure 3.9)

- are constructed by combining systolic and diastolic pressure curves.
- The diastolic pressure curve is the relationship between diastolic pressure and diastolic volume in the ventricle.
- The systolic pressure curve is the corresponding relationship between systolic pressure and systolic volume in the ventricle.
- **A single left ventricular cycle of contraction, ejection, relaxation, and refilling** can be visualized by combining the two curves into a pressure–volume loop.

1. Steps in the cycle

a. 1 → 2 (isovolumetric contraction). The cycle begins at the end of diastole at point 1. The left ventricle is filled with blood from the left atrium and its volume is about 140 mL (end-diastolic volume). Ventricular pressure is low because the ventricular muscle is relaxed. On excitation, the ventricle contracts and ventricular pressure increases. The mitral valve closes when left ventricular pressure is greater than left atrial pressure. Because all valves are closed, no blood can be ejected from the ventricle (isovolumetric).

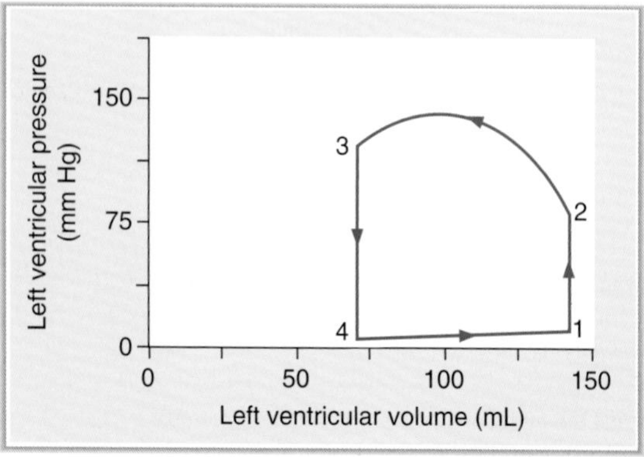

FIGURE 3.9. Left ventricular pressure–volume loop.

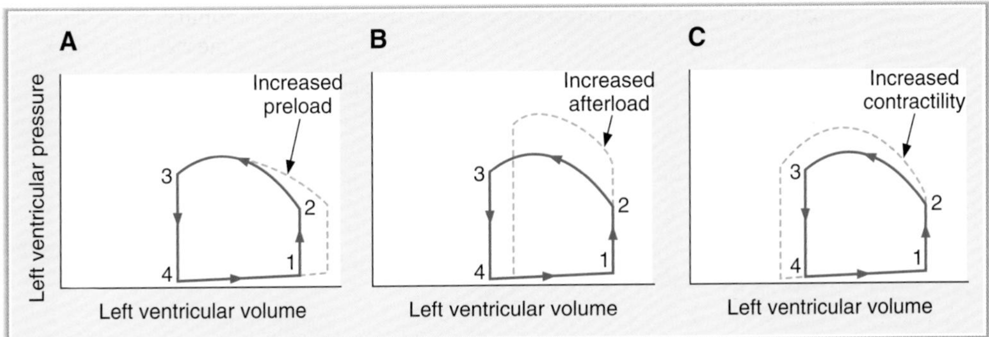

FIGURE 3.10. Effects of changes in **(A)** preload, **(B)** afterload, and **(C)** contractility on the ventricular pressure–volume loop.

b. **2 → 3 (ventricular ejection).** The aortic valve opens at point 2 when pressure in the left ventricle exceeds pressure in the aorta. Blood is ejected into the aorta, and ventricular volume decreases. The volume that is ejected in this phase is the **stroke volume.** Thus, stroke volume can be measured graphically by the **width of the pressure–volume loop.** The volume remaining in the left ventricle at point 3 is end-systolic volume.

c. **3 → 4 (isovolumetric relaxation).** At point 3, the ventricle relaxes. When ventricular pressure decreases to less than aortic pressure, the aortic valve closes. Because all of the valves are closed again, ventricular volume is constant (isovolumetric) during this phase.

d. **4 → 1 (ventricular filling).** Once left ventricular pressure decreases to less than left atrial pressure, the mitral valve opens and filling of the ventricle begins. During this phase, ventricular volume increases to about 140 mL (the end-diastolic volume).

2. **Changes in the ventricular pressure–volume loop** are caused by several factors (Figure 3.10).

a. **Increased preload** (see Figure 3.10A)

- refers to an increase in end-diastolic volume and is the result of increased venous return (e.g., increased blood volume or decreased venous capacitance)
- causes an **increase in stroke volume** based on the Frank-Starling relationship.
- The increase in stroke volume is reflected in increased width of the pressure–volume loop.

b. **Increased afterload** (see Figure 3.10B)

- refers to an increase in aortic pressure.
- The ventricle must eject blood against a higher pressure, resulting in a **decrease in stroke volume.**
- The decrease in stroke volume is reflected in decreased width of the pressure–volume loop.
- The decrease in stroke volume results in an increase in end-systolic volume.

c. **Increased contractility** (see Figure 3.10C)

- The ventricle develops greater tension than usual during systole, causing an **increase in stroke volume.**
- The increase in stroke volume results in a decrease in end-systolic volume.

F. Cardiac and vascular function curves (Figure 3.11)

- are simultaneous plots of cardiac output and venous return as a function of right atrial pressure or end-diastolic volume.

1. **The cardiac function (cardiac output) curve**

- depicts the Frank-Starling relationship for the ventricle.
- shows that cardiac output is a function of end-diastolic volume.

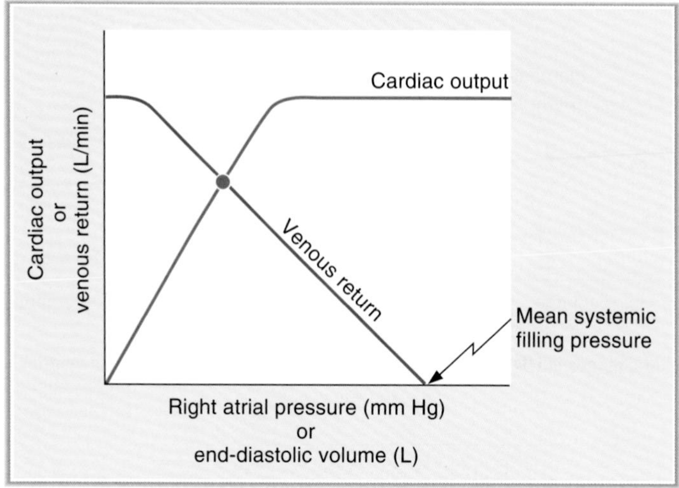

FIGURE 3.11. Simultaneous plots of the cardiac and vascular function curves. The curves cross at the equilibrium point for the cardiovascular system.

2. The vascular function (venous return) curve

- depicts the relationship between blood flow through the vascular system (or venous return) and right atrial pressure.

a. Mean systemic filling pressure

- is the point at which the vascular function curve intersects the x-axis.
- equals right atrial pressure when there is "no flow" in the cardiovascular system.
- is measured when the heart is stopped experimentally. Under these conditions, cardiac output and venous return are zero, and pressure is equal throughout the cardiovascular system.

(1) Mean systemic filling pressure is increased by an **increase in blood volume** or by a **decrease in venous capacitance** (where blood is shifted from the veins to the arteries). An increase in mean systemic filling pressure is reflected in a **shift of the vascular function curve to the right** (Figure 3.12).

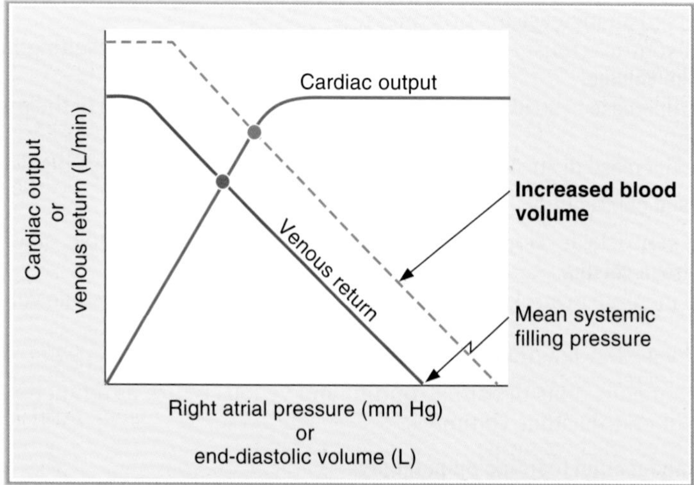

FIGURE 3.12. Effect of increased blood volume on the mean systemic filling pressure, vascular function curve, cardiac output, and right atrial pressure.

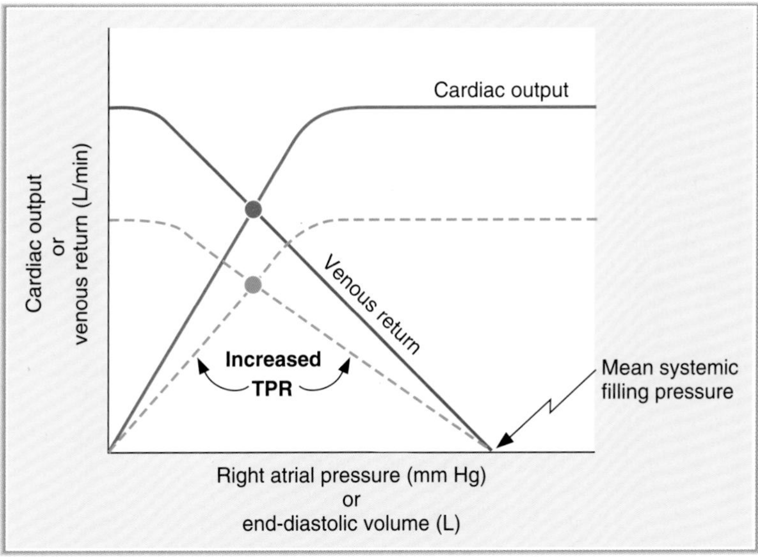

FIGURE 3.13. Effect of increased total peripheral resistance (TPR) on the cardiac and vascular function curves and on cardiac output.

(2) Mean systemic filling pressure is decreased by a **decrease in blood volume** or by an **increase in venous capacitance** (where blood is shifted from the arteries to the veins). A decrease in mean systemic filling pressure is reflected in a **shift of the vascular function curve to the left.**

b. Slope of the venous return curve

- is determined by the **resistance of the arterioles.**

(1) A clockwise rotation (not illustrated) of the venous return curve indicates a **decrease in total peripheral resistance (TPR).** When TPR is decreased for a given right atrial pressure, there is an increase in venous return (i.e., vasodilation of the arterioles "allows" more blood to flow from the arteries to the veins and back to the heart).

(2) A counterclockwise rotation of the venous return curve indicates an **increase in TPR** (Figure 3.13). When TPR is increased for a given right atrial pressure, there is a decrease in venous return to the heart (i.e., vasoconstriction of the arterioles decreases blood flow from the arteries to the veins and back to the heart).

3. Combining cardiac output and venous return curves

- When cardiac output and venous return are simultaneously plotted as a function of right atrial pressure, they intersect at a single value of right atrial pressure.
- The point at which the two curves intersect is the **equilibrium, or steady-state, point** (see Figure 3.11). Equilibrium occurs when cardiac output equals venous return.
- Cardiac output can be changed by altering the cardiac output curve, the venous return curve, or both curves simultaneously.
- The superimposed curves can be used to predict the direction and magnitude of changes in cardiac output and the corresponding values of right atrial pressure.

a. Inotropic agents change the cardiac output curve.

(1) *Positive inotropic agents* (e.g., **cardiac glycosides**) produce increased contractility and increased cardiac output (Figure 3.14).

- The equilibrium, or intersection, point shifts to a higher cardiac output and a correspondingly lower right atrial pressure.
- Right atrial pressure decreases because more blood is ejected from the heart on each beat (increased stroke volume).

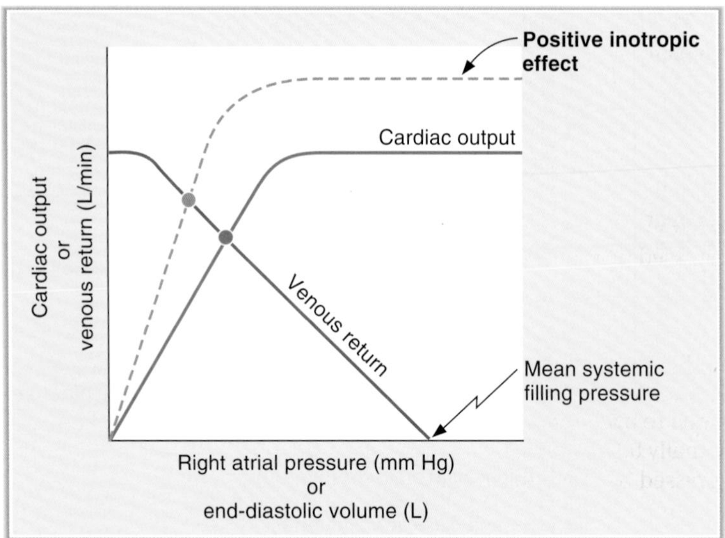

FIGURE 3.14. Effect of a positive inotropic agent on the cardiac function curve, cardiac output, and right atrial pressure.

(2) *Negative inotropic agents* produce decreased contractility and decreased cardiac output (not illustrated).

b. Changes in blood volume or venous capacitance change the venous return curve.

(1) *Increases in blood volume or decreases in venous capacitance* increase mean systemic filling pressure, shifting the venous return curve to the right in a parallel fashion (see Figure 3.12). A new equilibrium, or intersection, point is established at which **both cardiac output and right atrial pressure are increased.**

(2) *Decreases in blood volume* (e.g., hemorrhage) *or increases in venous capacitance* have the opposite effect—decreased mean systemic filling pressure and a shift of the venous return curve to the left in a parallel fashion. A new equilibrium point is established at which **both cardiac output and right atrial pressure are decreased** (not illustrated).

c. Changes in TPR change both the cardiac output and the venous return curves.

- Changes in TPR alter both curves simultaneously; therefore, the responses are more complicated than those noted in the previous examples.

(1) *Increasing TPR causes a decrease in both cardiac output and venous return* (see Figure 3.13).

(a) A **counterclockwise rotation of the venous return curve** occurs. Increased TPR results in decreased venous return as blood is retained on the arterial side.

(b) A **downward shift of the cardiac output curve** is caused by the increased aortic pressure (increased afterload) as the heart pumps against a higher pressure.

(c) As a result of these simultaneous changes, a new equilibrium point is established at which **both cardiac output and venous return are decreased,** but right atrial pressure is unchanged.

(2) *Decreasing TPR causes an increase in both cardiac output and venous return* (not illustrated).

(a) A **clockwise rotation of the venous return curve** occurs. Decreased TPR results in increased venous return as more blood is allowed to flow back to the heart from the arterial side.

(b) **An upward shift of the cardiac output curve** is caused by the decreased aortic pressure (decreased afterload) as the heart pumps against a lower pressure.

(c) As a result of these simultaneous changes, a new equilibrium point is established at which **both cardiac output and venous return are increased,** but right atrial pressure is unchanged.

G. Stroke volume, cardiac output, and ejection fraction

1. Stroke volume

- is the volume ejected from the ventricle on each beat.
- is expressed by the following equation:

$$\text{Stroke volume} = \text{End-diastolic volume} - \text{End-systolic volume}$$

2. Cardiac output

- is expressed by the following equation:

$$\text{Cardiac output} = \text{stroke volume} \times \text{Heart rate}$$

3. Ejection fraction

- is the fraction of the end-diastolic volume ejected in each stroke volume.
- is related to **contractility.**
- is normally 0.55, or **55%.**
- is expressed by the following equation:

$$\text{Ejection fraction} = \frac{\text{Stroke volume}}{\text{End-diastolic volume}}$$

H. Stroke work

- is the work the heart performs on each beat.
- is equal to **pressure** × **volume.** For the left ventricle, pressure is aortic pressure and volume is stroke volume.
- is expressed by the following equation:

$$\text{Stroke work} = \text{Aortic pressure} \times \text{Stroke volume}$$

- Fatty acids are the primary energy source for stroke work.

I. Cardiac oxygen (O_2) consumption

- is directly related to the amount of tension developed by the ventricles.
- is increased by

1. increased **afterload** (increased aortic pressure)
2. increased **size of the heart** (Laplace law states that tension is proportional to the radius of a sphere)
3. increased **contractility**
4. increased **heart rate**

J. Measurement of cardiac output by the Fick principle

- The Fick principle for measuring cardiac output is expressed by the following equation:

$$\text{Cardiac output} = \frac{O_2 \text{ consumption}}{[O_2]_{\text{pulmonary vein}} - [O_2]_{\text{pulmonary artery}}}$$

- The equation is solved as follows:

1. O_2 consumption for the whole body is measured.
2. Pulmonary vein $[O_2]$ is measured in systemic arterial blood.
3. Pulmonary artery $[O_2]$ is measured in systemic mixed venous blood.
   - **For example,** a 70-kg man has a resting O_2 consumption of 250 mL/min, a systemic arterial O_2 content of 0.20 mL O_2/mL of blood, a systemic mixed venous O_2 content of 0.15 mL O_2/mL of blood, and a heart rate of 72 beats/min. What is his cardiac output? What is his stroke volume?

$$\text{Cardiac output} = \frac{250\ \text{mL/min}}{0.20\ \text{mL}\ O_2/\text{mL} - 0.15\ \text{mL}\ O_2/\text{mL}}$$
$$= 5000\ \text{mL/min, or } 5.0\ \text{L/min}$$

$$\text{Stroke volume} = \frac{\text{Cardiac output}}{\text{Heart rate}}$$
$$= \frac{5000\ \text{mL/min}}{72\ \text{beats/min}}$$
$$= 69.4\ \text{mL/beat}$$

V. CARDIAC CYCLE

- Figure 3.15 shows the mechanical and electrical events of a single cardiac cycle. The seven phases are separated by vertical lines.
- Use the **ECG** as an event marker.
- Opening and closing of valves cause the physiologic **heart sounds.**
- When all valves are closed, ventricular volume is constant, and the phase is called **isovolumetric.**

A. Atrial systole

- is preceded by the P wave, which represents electrical activation of the atria.
- contributes to, but is not essential for, ventricular filling.
- The increase in atrial pressure (venous pressure) caused by atrial systole is the **a wave** on the venous pulse curve.
- In ventricular hypertrophy, filling of the ventricle by atrial systole causes the **fourth heart sound**, which is not audible in normal adults.

B. Isovolumetric ventricular contraction

- begins during the QRS complex, which represents electrical activation of the ventricles.
- When ventricular pressure becomes greater than atrial pressure, the AV valves close. Their closure corresponds to the **first heart sound.** Because the mitral valve closes before the tricuspid valve, the first heart sound may be split.
- Ventricular pressure increases isovolumetrically as a result of ventricular contraction. However, no blood leaves the ventricle during this phase because the **aortic valve is closed.**
- Ventricular volume is constant (isovolumetric) because all valves are closed.

C. Rapid ventricular ejection

- Ventricular pressure reaches its maximum value during this phase.
- C wave on venous pulse curve occurs because of bulging of tricuspid value into right atrium during right ventricular contraction.
- When ventricular pressure becomes greater than aortic pressure, the **aortic valve opens.**
- Rapid ejection of blood into the aorta occurs because of the pressure gradient between the ventricle and the aorta.
- Ventricular volume decreases dramatically because **most of the stroke volume is ejected** during this phase.
- Atrial filling begins.
- The onset of the T wave, which represents repolarization of the ventricles, marks the end of both ventricular contraction and rapid ventricular ejection.

D. Reduced ventricular ejection

- Ejection of blood from the ventricle continues but is slower.
- Ventricular pressure begins to decrease.
- Aortic pressure also decreases because of the runoff of blood from large arteries into smaller arteries.

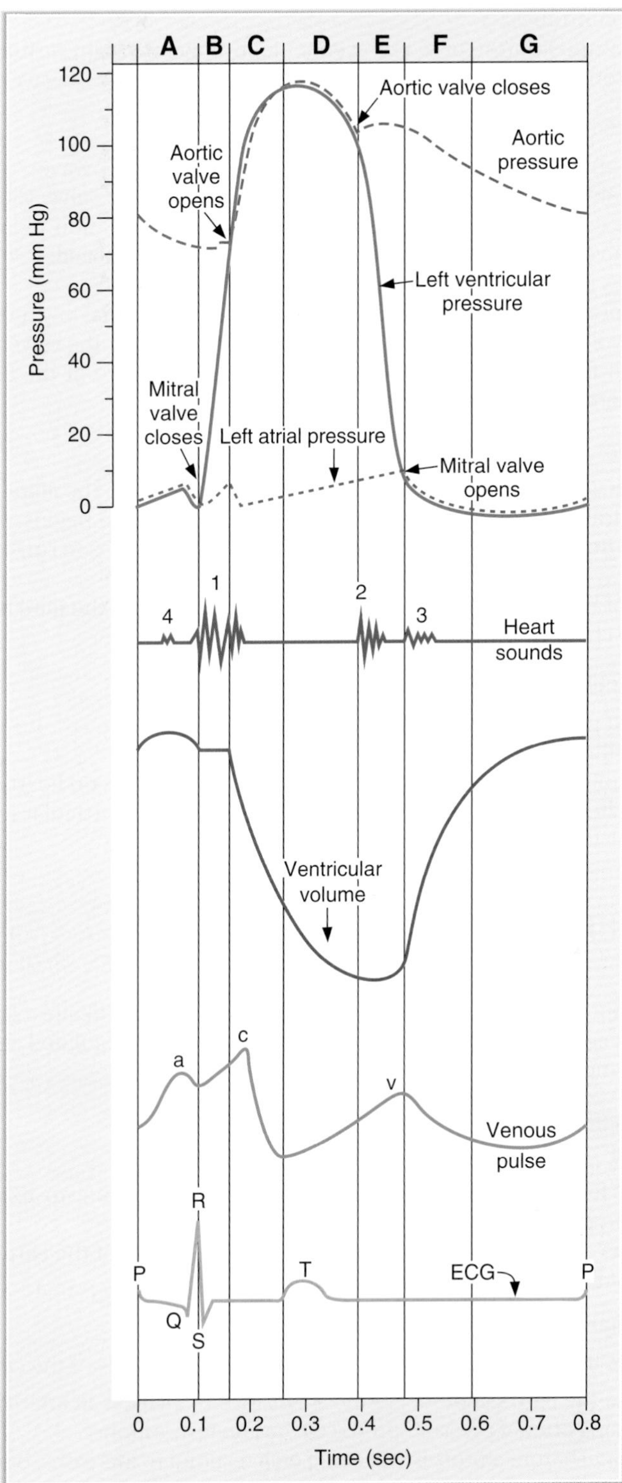

FIGURE 3.15. The cardiac cycle. ECG = electrocardiogram; A = atrial systole; B = isovolumetric ventricular contraction; C = rapid ventricular ejection; D = reduced ventricular ejection; E = isovolumetric ventricular relaxation; F = rapid ventricular filling; G = reduced ventricular filling.

- Atrial filling continues.
- V wave on venous pulse curve represents blood flow into right atrium (rising phase of wave) and from right atrium into right ventricle (falling phase of wave).

E. Isovolumetric ventricular relaxation

- Repolarization of the ventricles is now complete (end of the T wave).
- The aortic valve closes, followed by closure of the pulmonic valve. Closure of the semilunar valves corresponds to the **second heart sound.** Inspiration delays closure of the pulmonic valve and thus causes **splitting of the second heart sound.**
- The AV valves remain closed during most of this phase.
- Ventricular pressure decreases rapidly because the ventricle is now relaxed.
- Ventricular volume is constant (isovolumetric) because all of the valves are closed.
- The “blip” in the aortic pressure tracing occurs after closure of the aortic valve and is called the **dicrotic notch, or incisura.**

F. Rapid ventricular filling

- When ventricular pressure becomes less than atrial pressure, the **mitral valve opens.**
- With the mitral valve open, ventricular filling from the atrium begins.
- Aortic pressure continues to decrease because blood continues to run off into the smaller arteries.
- Rapid flow of blood from the atria into the ventricles causes the **third heart sound,** which is normal in children but, in adults, is associated with disease.

G. Reduced ventricular filling (diastasis)

- is the longest phase of the cardiac cycle.
- Ventricular filling continues, but at a slower rate.
- The time required for diastasis and ventricular filling depends on heart rate. For example, increases in heart rate cause decreased time available for ventricular refilling, decreased end-diastolic volume, and decreased stroke volume.

VI. REGULATION OF ARTERIAL PRESSURE

- The most important mechanisms for regulating arterial pressure are a fast, neurally mediated baroreceptor mechanism and a slower, hormonally regulated renin–angiotensin–aldosterone mechanism.

A. Baroreceptor reflex

- includes **fast, neural mechanisms.**
- is a negative feedback system that is responsible for the minute-to-minute regulation of arterial blood pressure.
- **Baroreceptors** are stretch receptors located within the walls of the carotid sinus near the **bifurcation of the common carotid arteries.**

1. Steps in the baroreceptor reflex (Figure 3.16)

a. A **decrease in arterial pressure** decreases stretch on the walls of the carotid sinus.

- Because the baroreceptors are most sensitive to **changes in arterial pressure,** rapidly decreasing arterial pressure produces the greatest response.
- Additional baroreceptors in the **aortic arch** respond to increases, but not to decreases, in arterial pressure.

b. Decreased stretch **decreases the firing rate of the carotid sinus nerve** [Hering nerve, cranial nerve (CN) IX], which carries information to the vasomotor center in the brainstem.

c. The set point for mean arterial pressure in the vasomotor center is about 100 mm Hg. Therefore, if mean arterial pressure is less than 100 mm Hg, a series of autonomic responses is coordinated by the vasomotor center. These changes will attempt to increase blood pressure toward normal.

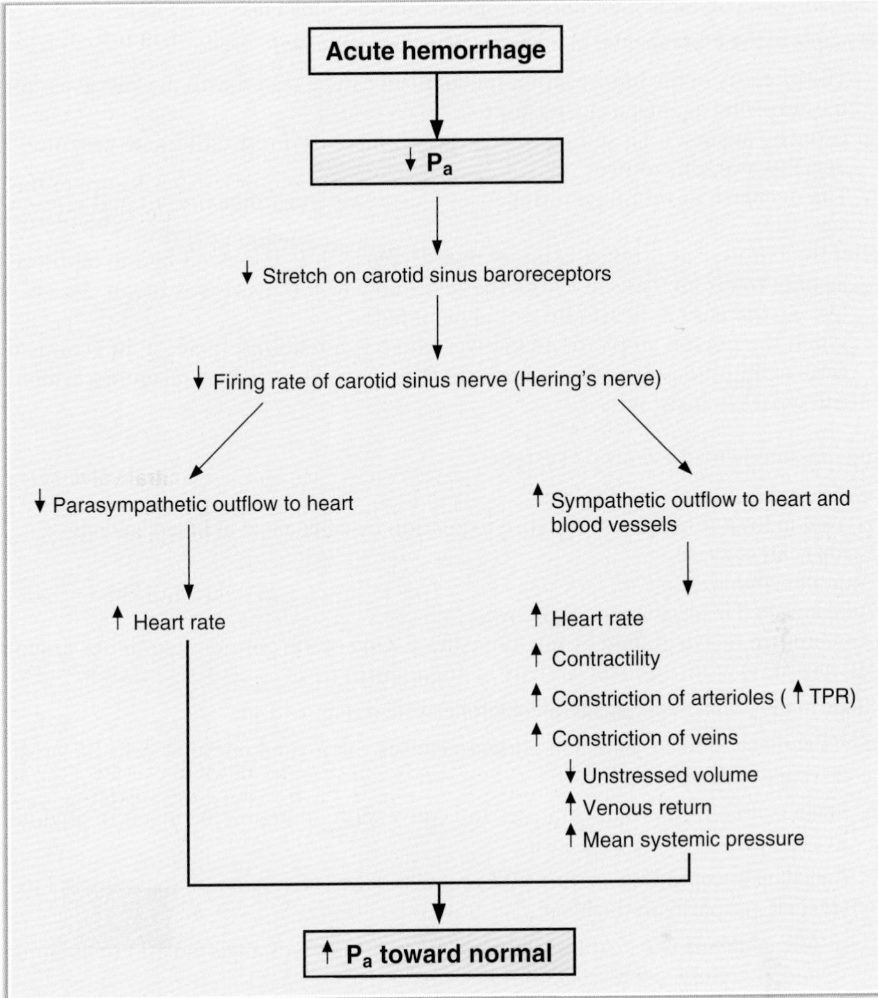

FIGURE 3.16. Role of the baroreceptor reflex in the cardiovascular response to hemorrhage. P_a = mean arterial pressure; TPR = total peripheral resistance.

d. The responses of the vasomotor center to a decrease in mean arterial blood pressure are coordinated to increase the arterial pressure back to 100 mm Hg. The responses are **decreased parasympathetic (vagal) outflow to the heart** and **increased sympathetic outflow to the heart and blood vessels.**

- The following four effects attempt to increase the arterial pressure back toward normal:

(1) ↑ *heart rate,* resulting from decreased parasympathetic tone and increased sympathetic tone to the SA node of the heart.

(2) ↑ *contractility and stroke volume,* resulting from increased sympathetic tone to the heart. Together with the increase in heart rate, the increases in contractility and stroke volume produce an increase in cardiac output that increases arterial pressure.

(3) ↑ *vasoconstriction of arterioles,* resulting from the increased sympathetic outflow. As a result, TPR and arterial pressure will increase.

(4) ↑ *vasoconstriction of veins* (venoconstriction), resulting from the increased sympathetic outflow. Constriction of the veins causes a decrease in unstressed volume and an increase in venous return to the heart. The increase in venous return causes an increase in cardiac output by the Frank-Starling mechanism.

2. Example of the baroreceptor reflex: response to acute blood loss (see Figure 3.16)

3. Example of the baroreceptor mechanism: Valsalva maneuver

- The integrity of the baroreceptor mechanism can be tested with the Valsalva maneuver (i.e., expiring against a closed glottis).
- Expiring against a closed glottis causes an increase in intrathoracic pressure, which decreases venous return.
- The decrease in venous return causes a decrease in cardiac output and arterial pressure (P_a).
- If the baroreceptor reflex is intact, the decrease in P_a is sensed by the baroreceptors, leading to an increase in sympathetic outflow to the heart and blood vessels. In the test, an increase in heart rate would be noted.
- When the person stops the maneuver, there is a rebound increase in venous return, cardiac output, and P_a. The increase in P_a is sensed by the baroreceptors, which direct a decrease in heart rate.

B. Renin–angiotensin–aldosterone system

- is a slow, hormonal mechanism.
- is used in long-term blood pressure regulation by **adjustment of blood volume.**
- **Renin** is an enzyme.
- Angiotensin I is inactive.
- **Angiotensin II is physiologically active.**
- Angiotensin II is degraded by angiotensinase. One of the peptide fragments, angiotensin III, has some of the biologic activity of angiotensin II.

1. Steps in the renin–angiotensin–aldosterone system (Figure 3.17)

a. A **decrease in renal perfusion pressure** causes the juxtaglomerular cells of the afferent arteriole to secrete renin.

b. Renin is an enzyme that catalyzes the conversion of angiotensinogen to angiotensin I in plasma.

c. Angiotensin-converting enzyme (ACE) catalyzes the conversion of angiotensin I to **angiotensin II,** primarily in the **lungs.**

- **ACE inhibitors** (e.g., captopril) block the conversion of angiotensin I to angiotensin II and, therefore, decrease blood pressure.
- **Angiotensin receptor (AT_1) antagonists** (e.g., losartan) block the action of angiotensin II at its receptor and decrease blood pressure.

d. Angiotensin II has four effects:

(1) It stimulates the synthesis and **secretion of aldosterone** by the adrenal cortex.

- Aldosterone increases **Na^+ reabsorption** by the renal distal tubule, thereby increasing extracellular fluid (ECF) volume, blood volume, and arterial pressure.
- This action of aldosterone is **slow** because it requires new protein synthesis.

(2) It increases **Na^+–H^+ exchange** in the proximal convoluted tubule.

- This action of angiotensin II directly increases Na^+ reabsorption, complementing the indirect stimulation of Na^+ reabsorption via aldosterone.
- This action of angiotensin II leads to contraction alkalosis.

(3) It increases **thirst** and therefore water intake.

(4) It causes **vasoconstriction of the arterioles,** thereby increasing TPR and arterial pressure.

2. Example: response of the renin–angiotensin–aldosterone system to acute blood loss (see Figure 3.17)

C. Other regulation of arterial blood pressure

1. Cerebral ischemia

a. When the brain is ischemic, the partial pressure of carbon dioxide (P_{CO_2}) in brain tissue increases.

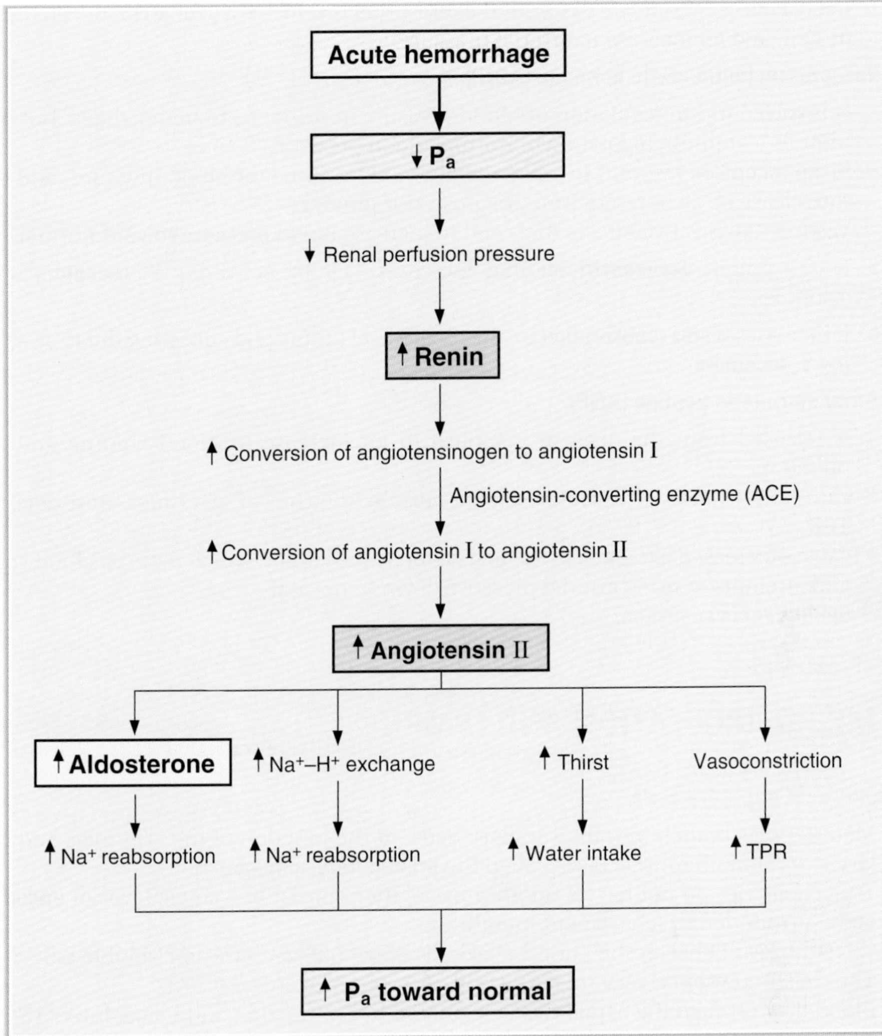

FIGURE 3.17. Role of the renin–angiotensin–aldosterone system in the cardiovascular response to hemorrhage. P_a = mean arterial pressure; TPR = total peripheral resistance.

b. Chemoreceptors in the vasomotor center respond by **increasing sympathetic outflow** to the heart and blood vessels.

- Constriction of arterioles causes intense **peripheral vasoconstriction** and increased TPR. Blood flow to other organs (e.g., kidneys) is significantly reduced in an attempt to preserve blood flow to the brain.
- **Mean arterial pressure can increase to life-threatening levels.**

c. The **Cushing reaction** is an example of the response to cerebral ischemia. Increases in intracranial pressure cause compression of the cerebral blood vessels, leading to cerebral ischemia and increased cerebral P_{CO_2}. The vasomotor center directs an increase in sympathetic outflow to the heart and blood vessels, which causes a profound increase in arterial pressure.

2. Chemoreceptors in the carotid and aortic bodies

- are located near the bifurcation of the common carotid arteries and along the aortic arch.
- have very high rates of O_2 consumption and are very sensitive to decreases in the partial pressure of oxygen (P_{O_2}).

- **Decreases in Po_2** activate vasomotor centers that produce vasoconstriction, an increase in TPR, and an increase in arterial pressure.

3. Vasopressin [antidiuretic hormone (ADH)]

- is involved in the regulation of blood pressure in response to hemorrhage, but not in minute-to-minute regulation of normal blood pressure.
- Atrial receptors respond to a decrease in blood volume (or blood pressure) and cause the release of vasopressin from the posterior pituitary.
- Vasopressin has two effects that tend to increase blood pressure toward normal:

a. It is a potent **vasoconstrictor** that increases TPR by activating **V_1 receptors** on the arterioles.

b. It increases **water reabsorption** by the renal distal tubule and collecting ducts by activating **V_2 receptors.**

4. Atrial natriuretic peptide (ANP)

- is released from the atria in response to an increase in blood volume and atrial pressure.
- causes **relaxation of vascular smooth muscle,** dilation of arterioles, and decreased TPR.
- causes increased **excretion of Na^+ and water** by the kidney, which reduces blood volume and attempts to bring arterial pressure down to normal.
- **inhibits renin secretion.**

VII. MICROCIRCULATION AND LYMPH

A. Structure of capillary beds

- Metarterioles branch into the capillary beds. At the junction of the arterioles and capillaries is a smooth muscle band called the **precapillary sphincter.**
- True capillaries do not have smooth muscle; they consist of a single layer of **endothelial cells** surrounded by a basement membrane.
- Clefts (pores) between the endothelial cells allow passage of water-soluble substances. The clefts represent a very small fraction of the surface area (<0.1%).
- Blood flow through the capillaries is regulated by contraction and relaxation of the arterioles and the precapillary sphincters.

B. Passage of substances across the capillary wall

1. Lipid-soluble substances

- cross the membranes of the capillary endothelial cells by **simple diffusion.**
- include O_2 and CO_2.

2. Small water-soluble substances

- cross via the water-filled clefts between the endothelial cells.
- include **water, glucose,** and **amino acids.**
- Generally, protein molecules are too large to pass freely through the clefts.
- In the brain, the clefts between endothelial cells are exceptionally tight **(blood–brain barrier).**
- In the liver and intestine, the clefts are exceptionally wide and allow passage of protein. These capillaries are called **sinusoids.**

3. Large water-soluble substances

- can cross by **pinocytosis.**

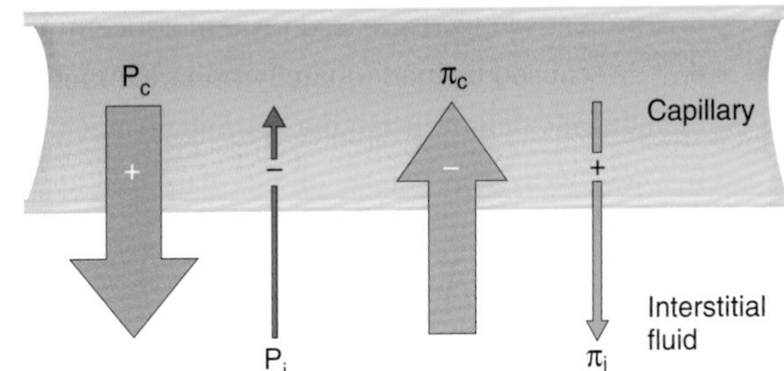

FIGURE 3.18. Starling forces across the capillary wall. + *sign* = favors filtration; − *sign* = opposes filtration; P_c = capillary hydrostatic pressure; P_i = interstitial hydrostatic pressure; π_c = capillary oncotic pressure; π_i = interstitial oncotic pressure.

C. Fluid exchange across capillaries

1. The Starling equation (Figure 3.18)

$$J_v = K_f\left[(P_c - P_i) - (\pi_c - \pi_i)\right]$$

where:
J_v = fluid movement (mL/min)
K_f = hydraulic conductance (mL/min·mm Hg)
P_c = capillary hydrostatic pressure (mm Hg)
P_i = interstitial hydrostatic pressure (mm Hg)
π_c = capillary oncotic pressure (mm Hg)
π_i = interstitial oncotic pressure (mm Hg)

a. J_v is fluid flow.

- When **J_v is positive,** there is net fluid movement out of the capillary **(filtration).**
- When **J_v is negative,** there is net fluid movement into the capillary **(absorption).**

b. K_f is the filtration coefficient.

- It is the hydraulic conductance (water permeability) of the capillary wall.

c. P_c is capillary hydrostatic pressure.

- An increase in P_c **favors filtration** out of the capillary.
- P_c is determined by arterial and venous pressures and resistances.
- An increase in either arterial or venous pressure produces an increase in P_c; increases in venous pressure have a greater effect on P_c.
- P_c is higher at the arteriolar end of the capillary than at the venous end (except in glomerular capillaries, where it is nearly constant).

d. P_i is interstitial fluid hydrostatic pressure.

- An increase in P_i **opposes filtration** out of the capillary.
- It is normally close to 0 mm Hg (or it is slightly negative).

e. π_c is capillary oncotic, or colloidosmotic, pressure.

- An increase in π_c **opposes filtration** out of the capillary.
- π_c is increased by increases in the protein concentration in the blood (e.g., dehydration).
- π_c is decreased by decreases in the protein concentration in the blood (e.g., nephrotic syndrome, protein malnutrition, liver failure).
- Small solutes do not contribute to π_c.

f. π_i is interstitial fluid oncotic pressure.

- An increase in π_i **favors filtration** out of the capillary.
- π_i is dependent on the protein concentration of the interstitial fluid, which is normally quite low because very little protein is filtered.

2. Factors that increase filtration

a. ↑ P_c—caused by increased arterial pressure, increased venous pressure, arteriolar dilation, and venous constriction

b. ↓ P_i

c. ↓ π_c—caused by decreased protein concentration in the blood

d. ↑ π_i—caused by inadequate lymphatic function

3. Sample calculations using the Starling equation

a. Example 1: At the arteriolar end of a capillary, P_c is 30 mm Hg, π_c is 28 mm Hg, P_i is 0 mm Hg, and π_i is 4 mm Hg. Will filtration or absorption occur?

$$\text{Net pressure} = (30-0)-(28-4)\ \text{mm Hg}$$
$$= +6\ \text{mm Hg}$$

Because the net pressure is positive, **filtration will occur.**

b. Example 2: At the venous end of the same capillary, P_c has decreased to 16 mm Hg, π_c remains at 28 mm Hg, P_i is 0 mm Hg, and π_i is 4 mm Hg. Will filtration or absorption occur?

$$\text{Net pressure} = (16-0)-(28-4)\ \text{mm Hg}$$
$$= -8\ \text{mm Hg}$$

Because the net pressure is negative, **absorption will occur.**

4. Lymph

a. Function of lymph

- Normally, filtration of fluid out of the capillaries is slightly greater than absorption of fluid into the capillaries. The **excess filtered fluid is returned to the circulation via the lymph.**
- Lymph also returns any filtered protein to the circulation.

b. Unidirectional flow of lymph

- **One-way flap valves** permit interstitial fluid to enter, but not leave, the lymph vessels.
- Flow through larger lymphatic vessels is also unidirectional and is aided by one-way valves and skeletal muscle contraction.

c. Edema (Table 3.2)

- occurs when the volume of interstitial fluid exceeds the capacity of the lymphatics to return it to the circulation.
- can be caused by excess filtration or blocked lymphatics.
- **Histamine** causes both arteriolar dilation and venous constriction, which together produce a large increase in P_c and local edema.

D. Nitric oxide (NO)

- is produced in the endothelial cells.
- causes local **relaxation of vascular smooth muscle.**

table 3.2 Causes and Examples of Edema

Cause	Examples
↑ P_c	Arteriolar dilation Venous constriction Increased venous pressure Heart failure Extracellular volume expansion Standing (edema in the dependent limbs)
↓ π_c	Decreased plasma protein concentration Severe liver disease (failure to synthesize proteins) Protein malnutrition Nephrotic syndrome (loss of protein in urine)
↑ K_f	Burn Inflammation (release of histamine; cytokines)

- Mechanism of action involves the activation of guanylate cyclase and production of **cyclic guanosine monophosphate (cGMP).**
- is one form of endothelial-derived relaxing factor (EDRF).
- Circulating ACh causes vasodilation by stimulating the production of NO in vascular smooth muscle.

VIII. SPECIAL CIRCULATIONS (TABLE 3.3)

- Blood flow varies from one organ to another.
- Blood flow to an organ is regulated by altering arteriolar resistance and can be varied, depending on the organ's metabolic demands.
- Pulmonary and renal blood flows are discussed in Chapters 4 and 5, respectively.

A. Local (intrinsic) control of blood flow

1. Examples of local control

a. Autoregulation

- Blood flow to an organ remains constant over a wide range of perfusion pressures.
- Organs that exhibit autoregulation are the heart, brain, and kidney.
- **For example,** if perfusion pressure to the heart is suddenly decreased, compensatory vasodilation of the arterioles will occur to maintain a constant flow.

b. Active hyperemia

- Blood flow to an organ is proportional to its metabolic activity.
- **For example,** if metabolic activity in skeletal muscle increases as a result of strenuous exercise, blood flow to the muscle will increase proportionately to meet metabolic demands.

table 3.3 Summary of Control of Special Circulations

Circulation* (% of Resting Cardiac Output)	Local Metabolic Control	Vasoactive Metabolites	Sympathetic Control	Mechanical Effects
Coronary (5%)	Most important mechanism	Hypoxia Adenosine	Least important mechanism	Mechanical compression during systole
Cerebral (15%)	Most important mechanism	CO_2 H^+	Least important mechanism	Increases in intracranial pressure decrease cerebral blood flow
Muscle (20%)	Most important mechanism during exercise	Lactate K^+ Adenosine	Most important mechanism at rest (α_1 receptor causes vasoconstriction; β_2 receptor causes vasodilation)	Muscular activity causes temporary decrease in blood flow
Skin (5%)	Least important mechanism		Most important mechanism (temperature regulation)	
Pulmonary† (100%)	Most important mechanism	Hypoxia vasoconstricts	Least important mechanism	Lung inflation

*Renal blood flow (25% of resting cardiac output) is discussed in Chapter 5.
†Pulmonary blood flow is discussed in Chapter 4.

c. **Reactive hyperemia**

- is an increase in blood flow to an organ that occurs after a period of occlusion of flow.
- The longer the period of occlusion is, the greater the increase in blood flow is above preocclusion levels.

2. **Mechanisms that explain local control of blood flow**

a. **Myogenic hypothesis**

- explains autoregulation, but not active or reactive hyperemia.
- is based on the observation that **vascular smooth muscle contracts when it is stretched.**
- **For example,** if perfusion pressure to an organ suddenly increases, the arteriolar smooth muscle will be stretched and will contract. The resulting vasoconstriction will maintain a constant flow. (Without vasoconstriction, blood flow would increase as a result of the increased pressure.)

b. **Metabolic hypothesis**

- is based on the observation that the **tissue supply of O_2 is matched to the tissue demand for O_2.**
- **Vasodilator metabolites** are produced as a result of metabolic activity in tissue. These vasodilators are **CO_2, H^+, K^+, lactate, and adenosine.**
- Examples of **active hyperemia:**

(1) If the metabolic activity of a tissue increases (e.g., strenuous exercise), both the demand for O_2 and the production of vasodilator metabolites increase. These metabolites cause arteriolar vasodilation, increased blood flow, and increased O_2 delivery to the tissue to meet demand.

(2) If blood flow to an organ suddenly increases as a result of a spontaneous increase in arterial pressure, then more O_2 is provided for metabolic activity. At the same time, the increased flow "washes out" vasodilator metabolites. As a result of this "washout," arteriolar vasoconstriction occurs, resistance increases, and blood flow is decreased to normal.

B. Hormonal (extrinsic) control of blood flow

1. **Sympathetic innervation of vascular smooth muscle**

- Increases in sympathetic tone cause vasoconstriction.
- Decreases in sympathetic tone cause vasodilation.
- The density of sympathetic innervation varies widely among tissues. Skin has the greatest innervation, whereas coronary, pulmonary, and cerebral vessels have little innervation.

2. **Other vasoactive hormones**

a. **Histamine**

- causes **arteriolar dilation and venous constriction.** The combined effects of arteriolar dilation and venous constriction cause **increased P_c** and **increased filtration** out of the capillaries, resulting in local **edema.**
- is released in response to tissue trauma.

b. **Bradykinin**

- causes **arteriolar dilation and venous constriction.**
- produces increased filtration out of the capillaries (similar to histamine) and causes local edema.

c. **Serotonin (5-hydroxytryptamine)**

- causes arteriolar constriction and is released in response to blood vessel damage to help prevent blood loss.
- has been implicated in the vascular spasms of **migraine headaches.**

d. **Prostaglandins**

- **Prostacyclin** is a vasodilator in several vascular beds.
- **E-series prostaglandins** are vasodilators.

- **F-series prostaglandins** are vasoconstrictors.
- **Thromboxane A_2** is a vasoconstrictor.

C. Coronary circulation

- is controlled almost entirely by **local metabolic factors.**
- exhibits autoregulation.
- exhibits active and reactive hyperemia.
- The most important local metabolic factors are **hypoxia** and **adenosine.**
- For example, **increases in myocardial contractility** are accompanied by an increased demand for O_2. To meet this demand, compensatory vasodilation of coronary vessels occurs and, accordingly, both blood flow and O_2 delivery to the contracting heart muscle increase (active hyperemia).
- During **systole,** mechanical compression of the coronary vessels reduces blood flow. After the period of occlusion, blood flow increases to repay the O_2 debt (reactive hyperemia).
- Sympathetic nerves play a minor role.

D. Cerebral circulation

- is controlled almost entirely by **local metabolic factors.**
- exhibits autoregulation.
- exhibits active and reactive hyperemia.
- The **most important local vasodilator for the cerebral circulation is CO_2.** Increases in P_{CO_2} cause vasodilation of the cerebral arterioles and increased blood flow to the brain. Decreases in P_{CO_2} cause vasoconstriction of cerebral arterioles and decreased blood flow to the brain.
- Sympathetic nerves play a minor role.
- Vasoactive substances in the systemic circulation have little or no effect on cerebral circulation because such substances are excluded by the blood–brain barrier.

E. Skeletal muscle

- is controlled by the extrinsic **sympathetic innervation** of blood vessels in skeletal muscle and by **local metabolic factors.**

1. Sympathetic innervation

- is the primary regulator of blood flow to the skeletal muscle at **rest.**
- The arterioles of skeletal muscle are densely innervated by sympathetic fibers. The veins also are innervated, but less densely.
- There are both α_1 and β_2 receptors on the blood vessels of skeletal muscle.
- Stimulation of **α_1 receptors** causes **vasoconstriction.**
- Stimulation of **β_2 receptors** causes **vasodilation.**
- The state of constriction of skeletal muscle arterioles is a major contributor to the TPR (because of the large mass of skeletal muscle).

2. Local metabolic control

- Blood flow in skeletal muscle exhibits autoregulation and active and reactive hyperemia.
- Demand for O_2 in skeletal muscle varies with metabolic activity level, and blood flow is regulated to meet demand.
- During **exercise,** when demand is high, these local metabolic mechanisms are dominant.
- The local vasodilator substances are **lactate, adenosine, and K^+.**
- Mechanical effects during exercise temporarily compress the arteries and decrease blood flow. During the postocclusion period, reactive hyperemia increases blood flow to repay the O_2 debt.

F. Skin

- has extensive **sympathetic innervation.** Cutaneous blood flow is under extrinsic control.
- **Temperature regulation** is the principal function of the cutaneous sympathetic nerves. Increased ambient temperature leads to cutaneous vasodilation, allowing dissipation of excess body heat.

- **Trauma** produces the "triple response" in skin—a red line, a red flare, and a wheal. A **wheal** is local **edema** that results from the local release of **histamine,** which increases capillary filtration.

IX. INTEGRATIVE FUNCTIONS OF THE CARDIOVASCULAR SYSTEM: GRAVITY, EXERCISE, AND HEMORRHAGE

- The responses to changes in gravitational force, exercise, and hemorrhage demonstrate the integrative functions of the cardiovascular system.

A. Changes in gravitational forces (Table 3.4 and Figure 3.19)

- The following changes occur when an individual **moves from a supine position to a standing position:**

1. **When a person stands,** a significant volume of blood pools in the lower extremities because of the high compliance of the veins. (Muscular activity would prevent this pooling.)
2. **As a result of venous pooling** and increased local venous pressure, P_c in the legs increases and fluid is filtered into the interstitium. If net filtration of fluid exceeds the ability of the lymphatics to return it to the circulation, **edema** will occur.
3. **Venous return decreases.** As a result of the decrease in venous return, both **stroke volume and cardiac output decrease** (Frank-Starling relationship, IV D 5).
4. **Arterial pressure decreases** because of the reduction in cardiac output. If cerebral blood pressure becomes low enough, fainting may occur.
5. **Compensatory mechanisms** will attempt to increase blood pressure to normal (see Figure 3.19). The **carotid sinus baroreceptors** respond to the decrease in arterial pressure by decreasing the firing rate of the carotid sinus nerves. A coordinated response from the vasomotor center then increases sympathetic outflow to the heart and blood vessels and decreases parasympathetic outflow to the heart. As a result, heart rate, contractility, TPR, and venous return increase, and blood pressure increases toward normal.
6. **Orthostatic hypotension** (fainting or light-headedness on standing) may occur in individuals whose baroreceptor reflex mechanism is impaired (e.g., individuals treated with sympatholytic agents) or who are volume depleted.

B. Exercise (Table 3.5 and Figure 3.20)

1. **The central command (anticipation of exercise)**
   - originates in the motor cortex or from reflexes initiated in muscle proprioceptors when exercise is anticipated.
   - initiates the following changes:

table 3.4 Summary of Responses to Standing

Parameter	Initial Response to Standing	Compensatory Response
Arterial blood pressure	↓	↑ (toward normal)
Heart rate	—	↑
Cardiac output	↓	↑ (toward normal)
Stroke volume	↓	↑ (toward normal)
TPR	—	↑
Central venous pressure	↓	↑ (toward normal)

TPR = total peripheral resistance.

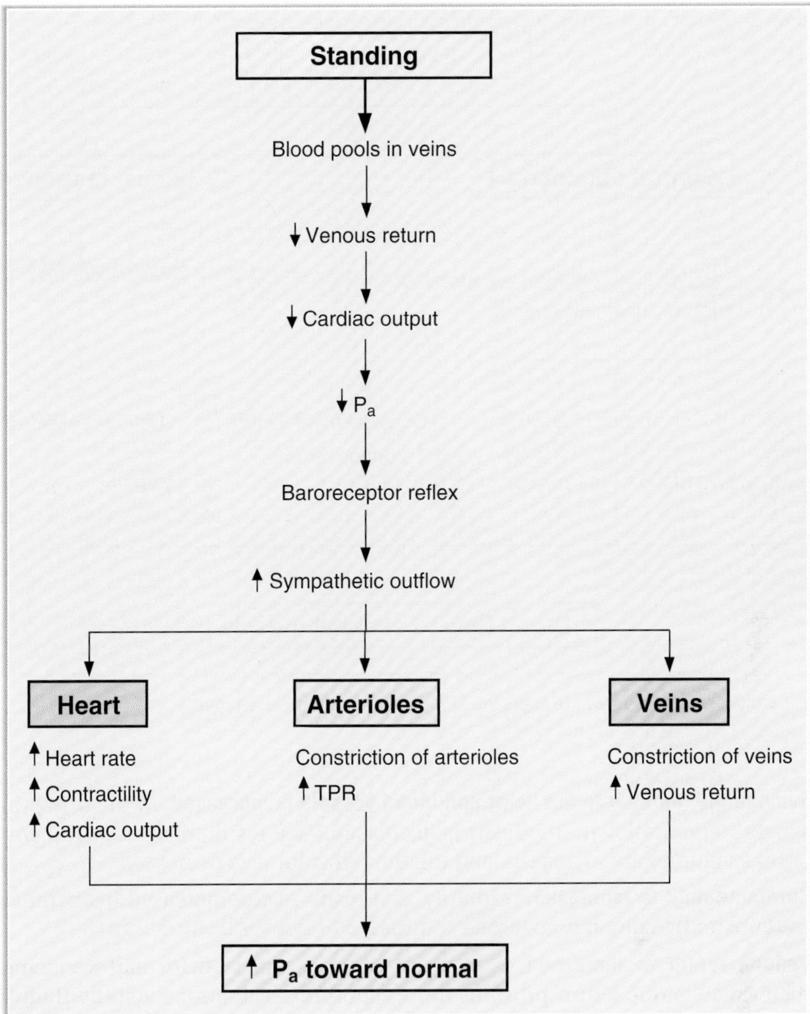

FIGURE 3.19. Cardiovascular responses to standing. P_a = arterial pressure; TPR = total peripheral resistance.

table 3.5 Summary of Effects of Exercise

Parameter	Effect
Heart rate	↑↑
Stroke volume	↑
Cardiac output	↑↑
Arterial pressure	↑ (slight)
Pulse pressure	↑ (due to increased stroke volume)
TPR	↓↓ (due to vasodilation of skeletal muscle beds)
AV O_2 difference	↑↑ (due to increased O_2 consumption)

AV = arteriovenous; TPR = total peripheral resistance.

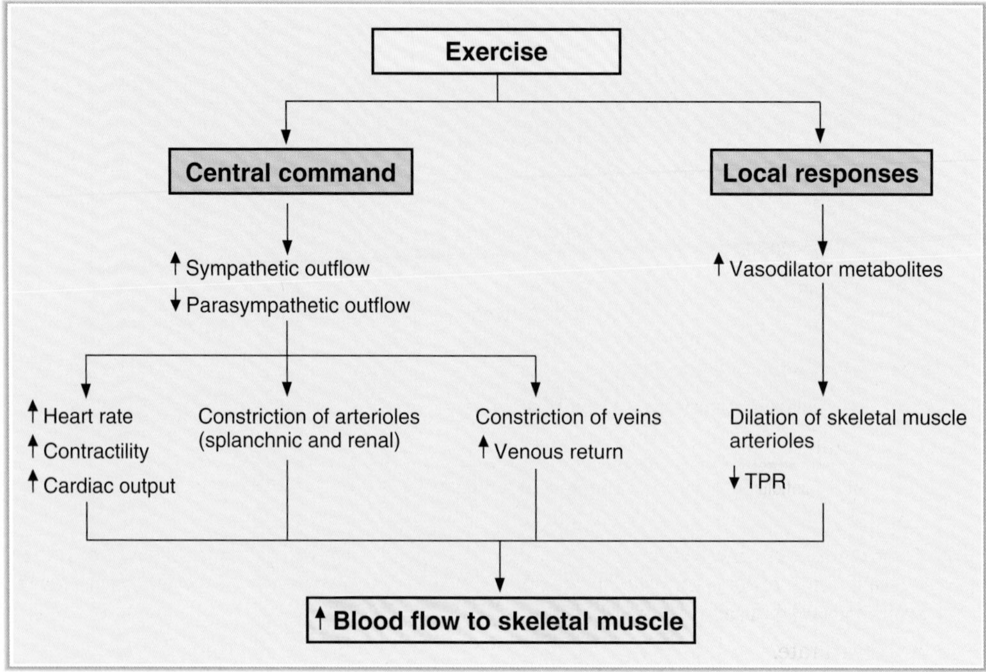

FIGURE 3.20. Cardiovascular responses to exercise. TPR = total peripheral resistance.

a. **Sympathetic outflow to the heart and blood vessels is increased.** At the same time, parasympathetic outflow to the heart is decreased. As a result, heart rate and contractility (stroke volume) are increased, and unstressed volume is decreased.

b. **Cardiac output is increased,** primarily as a result of the increased heart rate and, to a lesser extent, the increased stroke volume.

c. **Venous return is increased** as a result of muscular activity and venoconstriction. Increased venous return provides more blood for each stroke volume (Frank-Starling relationship, IV D 5).

d. **Arteriolar resistance in the skin, splanchnic regions, kidneys, and inactive muscles is increased.** Accordingly, blood flow to these organs is decreased.

2. **Increased metabolic activity of skeletal muscle**

- **Vasodilator metabolites (lactate, K^+,** and **adenosine)** accumulate because of increased metabolism of the exercising muscle.
- These metabolites cause arteriolar dilation in the active skeletal muscle, thus increasing skeletal muscle blood flow (active hyperemia).
- As a result of the increased blood flow, O_2 delivery to the muscle is increased. The number of perfused capillaries is increased so that the diffusion distance for O_2 is decreased.
- This vasodilation accounts for the **overall decrease in TPR** that occurs with exercise. Note that activation of the sympathetic nervous system alone (by the central command) would cause an increase in TPR.

C. Hemorrhage (Table 3.6 and Figure 3.21)

- The **compensatory responses** to acute blood loss are as follows:

1. **A decrease in blood volume** produces a decrease in venous return. As a result, there is a **decrease in both cardiac output** and **arterial pressure.**

table 3.6 Summary of Compensatory Responses to Hemorrhage

Parameter	Compensatory Response
Heart rate	↑
Contractility	↑
TPR	↑
Venoconstriction	↑
Renin	↑
Angiotensin II	↑
Aldosterone	↑
Circulating epinephrine and norepinephrine	↑
ADH	↑

ADH = antidiuretic hormone; TPR = total peripheral resistance.

2. **The carotid sinus baroreceptors** detect the decrease in arterial pressure. As a result of the baroreceptor reflex, there is **increased sympathetic outflow to the heart and blood vessels** and **decreased parasympathetic outflow to the heart, producing:**
   a. ↑ heart rate.
   b. ↑ contractility.
   c. ↑ TPR (due to arteriolar constriction).
   d. venoconstriction, which increases venous return.

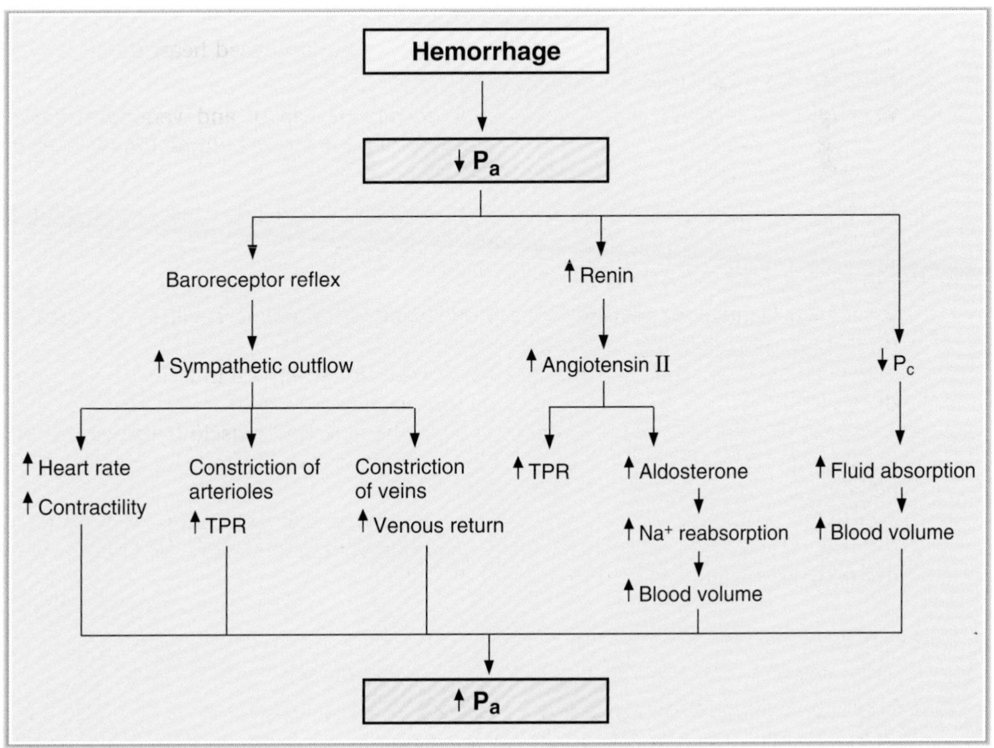

FIGURE 3.21. Cardiovascular responses to hemorrhage. P_a = arterial pressure; P_c = capillary hydrostatic pressure; TPR = total peripheral resistance.

   e. constriction of arterioles in skeletal, splanchnic, and cutaneous vascular beds. However, it does not occur in coronary or cerebral vascular beds, ensuring that adequate blood flow will be maintained to the heart and brain.

   f. These responses attempt to restore normal arterial blood pressure.

3. **Chemoreceptors in the carotid and aortic bodies** are very sensitive to hypoxia. They supplement the baroreceptor mechanism by increasing sympathetic outflow to the heart and blood vessels.
4. **Cerebral ischemia** (if present) causes an **increase in P_{CO_2}**, which activates chemoreceptors in the vasomotor center to increase sympathetic outflow.
5. **Arteriolar vasoconstriction** causes **a decrease in P_c**. As a result, capillary absorption is favored, which helps to restore circulating blood volume.
6. **The adrenal medulla releases epinephrine and norepinephrine,** which supplement the actions of the sympathetic nervous system on the heart and blood vessels.
7. **The renin–angiotensin–aldosterone system** is activated by the decrease in renal perfusion pressure. Because **angiotensin II** is a potent vasoconstrictor, it reinforces the stimulatory effect of the sympathetic nervous system on TPR. **Aldosterone** increases NaCl reabsorption in the kidney, increasing the circulating blood volume.
8. **ADH** is released when atrial receptors detect the decrease in blood volume. ADH causes both vasoconstriction and increased water reabsorption, both of which tend to increase blood pressure.

Review Test

1. A 53-year-old woman is found, by arteriography, to have 50% narrowing of her left renal artery. What is the expected change in blood flow through the stenotic artery?

(A) Decrease to ½
(B) Decrease to ¼
(C) Decrease to ⅛
(D) Decrease to $\frac{1}{16}$
(E) No change

2. When a person moves from a supine position to a standing position, which of the following compensatory changes occurs?

(A) Decreased heart rate
(B) Increased contractility
(C) Decreased total peripheral resistance (TPR)
(D) Decreased cardiac output
(E) Increased PR intervals

3. At which site is systolic blood pressure the highest?

(A) Aorta
(B) Central vein
(C) Pulmonary artery
(D) Right atrium
(E) Renal artery
(F) Renal vein

4. A person's electrocardiogram (ECG) has no P wave but has a normal QRS complex and a normal T wave. Therefore, his pacemaker is located in the

(A) sinoatrial (SA) node
(B) atrioventricular (AV) node
(C) bundle of His
(D) Purkinje system
(E) ventricular muscle

5. If the ejection fraction increases, there will be a decrease in

(A) cardiac output
(B) end-systolic volume
(C) heart rate
(D) pulse pressure
(E) stroke volume
(F) systolic pressure

QUESTIONS 6 AND 7

An electrocardiogram (ECG) on a person shows ventricular extrasystoles.

6. The extrasystolic beat would produce

(A) increased pulse pressure because contractility is increased
(B) increased pulse pressure because heart rate is increased
(C) decreased pulse pressure because ventricular filling time is increased
(D) decreased pulse pressure because stroke volume is decreased
(E) decreased pulse pressure because the PR interval is increased

7. After an extrasystole, the next "normal" ventricular contraction produces

(A) increased pulse pressure because the contractility of the ventricle is increased
(B) increased pulse pressure because total peripheral resistance (TPR) is decreased
(C) increased pulse pressure because compliance of the veins is decreased
(D) decreased pulse pressure because the contractility of the ventricle is increased
(E) decreased pulse pressure because TPR is decreased

8. An increase in contractility is demonstrated on a Frank-Starling diagram by

(A) increased cardiac output for a given end-diastolic volume
(B) increased cardiac output for a given end-systolic volume
(C) decreased cardiac output for a given end-diastolic volume
(D) decreased cardiac output for a given end-systolic volume

QUESTIONS 9–12

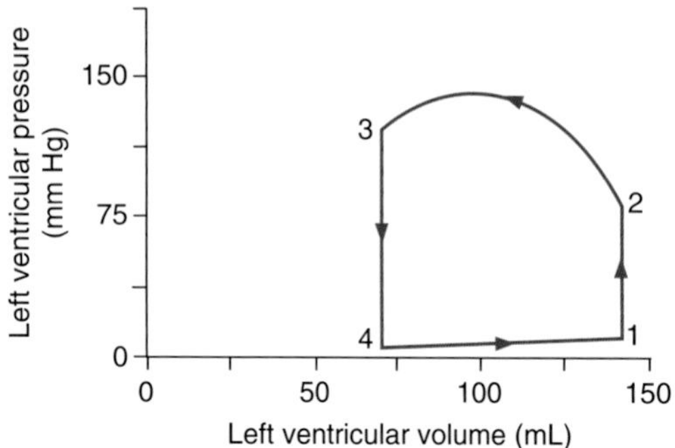

9. On the graph showing left ventricular volume and pressure, isovolumetric contraction occurs between points

(A) 4 → 1
(B) 1 → 2
(C) 2 → 3
(D) 3 → 4

10. The aortic valve closes at point

(A) 1
(B) 2
(C) 3
(D) 4

11. The first heart sound corresponds to point

(A) 1
(B) 2
(C) 3
(D) 4

12. If the heart rate is 70 beats/min, then the cardiac output of this ventricle is closest to

(A) 3.45 L/min
(B) 4.55 L/min
(C) 5.25 L/min
(D) 8.00 L/min
(E) 9.85 L/min

QUESTIONS 13 AND 14

In a capillary, P_c is 30 mm Hg, P_i is −2 mm Hg, π_c is 25 mm Hg, and π_i is 2 mm Hg.

13. What is the direction of fluid movement and the net driving force?

(A) Absorption; 6 mm Hg
(B) Absorption; 9 mm Hg
(C) Filtration; 6 mm Hg
(D) Filtration; 9 mm Hg
(E) There is no net fluid movement

14. If K_f is 0.5 mL/min/mm Hg, what is the rate of water flow across the capillary wall?

(A) 0.06 mL/min
(B) 0.45 mL/min
(C) 4.50 mL/min
(D) 9.00 mL/min
(E) 18.00 mL/min

15. The tendency for blood flow to be turbulent is increased by

(A) increased viscosity
(B) increased hematocrit
(C) partial occlusion of a blood vessel
(D) decreased velocity of blood flow

16. A 66-year-old man, who has had a sympathectomy, experiences a greater-than-normal fall in arterial pressure upon standing up. The explanation for this occurrence is

(A) an exaggerated response of the renin–angiotensin–aldosterone system
(B) a suppressed response of the renin–angiotensin–aldosterone system
(C) an exaggerated response of the baroreceptor mechanism
(D) a suppressed response of the baroreceptor mechanism

17. The ventricles are completely depolarized during which isoelectric portion of the electrocardiogram (ECG)?

(A) PR interval
(B) QRS complex
(C) QT interval
(D) ST segment
(E) T wave

18. In which of the following situations is pulmonary blood flow greater than aortic blood flow?

(A) Normal adult
(B) Fetus
(C) Left-to-right ventricular shunt
(D) Right-to-left ventricular shunt
(E) Right ventricular failure
(F) Administration of a positive inotropic agent

19. The change indicated by the dashed lines on the cardiac output/venous return curves shows

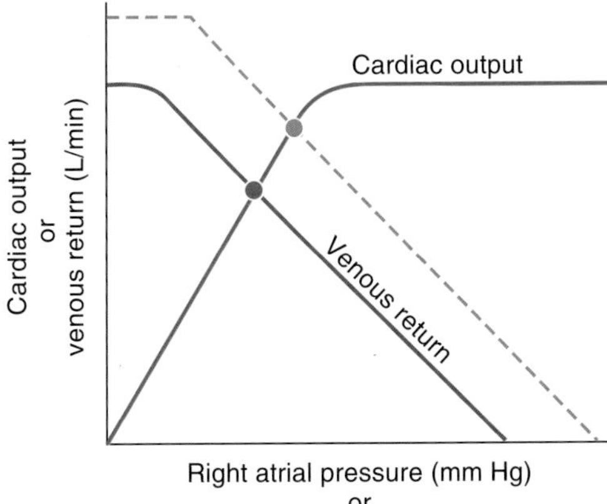

(A) decreased cardiac output in the "new" steady state
(B) decreased venous return in the "new" steady state
(C) increased mean systemic filling pressure
(D) decreased blood volume
(E) increased myocardial contractility

20. A 30-year-old female patient's electrocardiogram (ECG) shows two P waves preceding each QRS complex. The interpretation of this pattern is

(A) decreased firing rate of the pacemaker in the sinoatrial (SA) node
(B) decreased firing rate of the pacemaker in the atrioventricular (AV) node
(C) increased firing rate of the pacemaker in the SA node
(D) decreased conduction through the AV node
(E) increased conduction through the His–Purkinje system

21. An acute decrease in arterial blood pressure elicits which of the following compensatory changes?

(A) Decreased firing rate of the carotid sinus nerve
(B) Increased parasympathetic outflow to the heart
(C) Decreased heart rate
(D) Decreased contractility
(E) Decreased mean systemic filling pressure

22. The tendency for edema to occur will be increased by

(A) arteriolar constriction
(B) increased venous pressure
(C) increased plasma protein concentration
(D) muscular activity

23. Inspiration "splits" the second heart sound because

(A) the aortic valve closes before the pulmonic valve
(B) the pulmonic valve closes before the aortic valve
(C) the mitral valve closes before the tricuspid valve
(D) the tricuspid valve closes before the mitral valve
(E) filling of the ventricles has fast and slow components

24. During exercise, total peripheral resistance (TPR) decreases because of the effect of

(A) the sympathetic nervous system on splanchnic arterioles
(B) the parasympathetic nervous system on skeletal muscle arterioles
(C) local metabolites on skeletal muscle arterioles
(D) local metabolites on cerebral arterioles
(E) histamine on skeletal muscle arterioles

QUESTIONS 25 AND 26

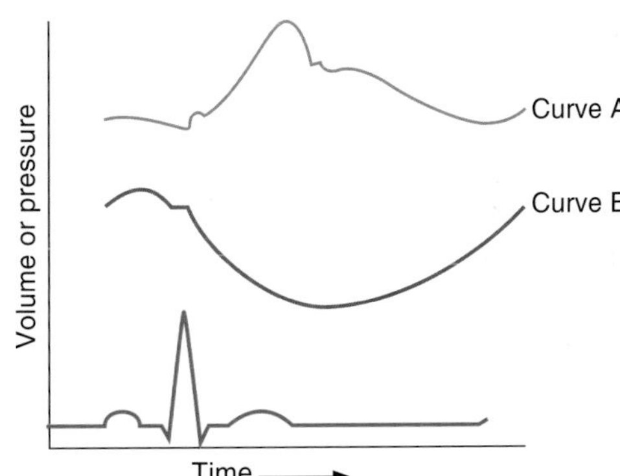

25. Curve A in the figure represents

(A) aortic pressure
(B) ventricular pressure
(C) atrial pressure
(D) ventricular volume

26. Curve B in the figure represents

(A) left atrial pressure
(B) ventricular pressure
(C) atrial pressure
(D) ventricular volume

27. An increase in arteriolar resistance, without a change in any other component of the cardiovascular system, will produce

(A) a decrease in total peripheral resistance (TPR)
(B) an increase in capillary filtration
(C) an increase in arterial pressure
(D) a decrease in afterload

28. The following measurements were obtained in a male patient:

Central venous pressure: 10 mm Hg
Heart rate: 70 beats/min
Systemic arterial $[O_2] = 0.24$ mL O_2/mL
Mixed venous $[O_2] = 0.16$ mL O_2/mL
Whole body O_2 consumption: 500 mL/min

What is this patient's cardiac output?

(A) 1.65 L/min
(B) 4.55 L/min
(C) 5.00 L/min
(D) 6.25 L/min
(E) 8.00 L/min

29. Which of the following is the result of an inward Na^+ current?

(A) Upstroke of the action potential in the sinoatrial (SA) node
(B) Upstroke of the action potential in Purkinje fibers
(C) Plateau of the action potential in ventricular muscle
(D) Repolarization of the action potential in ventricular muscle
(E) Repolarization of the action potential in the SA node

QUESTIONS 30 AND 31

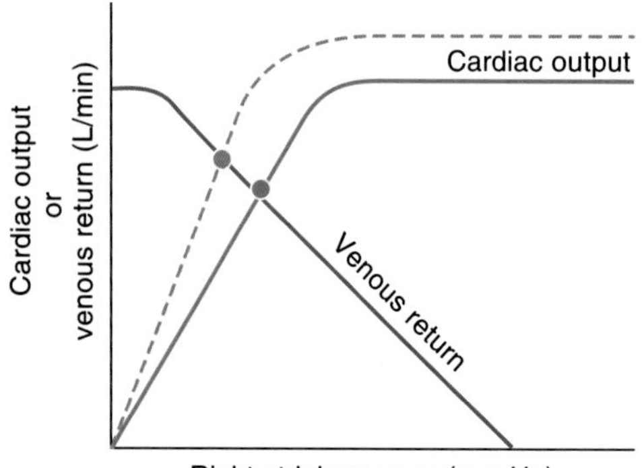

30. The dashed line in the figure illustrates the effect of

(A) increased total peripheral resistance (TPR)
(B) increased blood volume
(C) increased contractility
(D) a negative inotropic agent
(E) increased mean systemic filling pressure

31. The x-axis in the figure could have been labeled

(A) end-systolic volume
(B) end-diastolic volume
(C) pulse pressure
(D) mean systemic filling pressure
(E) heart rate

32. The greatest pressure decrease in the circulation occurs across the arterioles because

(A) they have the greatest surface area
(B) they have the greatest cross-sectional area
(C) the velocity of blood flow through them is the highest
(D) the velocity of blood flow through them is the lowest
(E) they have the greatest resistance

33. Pulse pressure is

(A) the highest pressure measured in the arteries
(B) the lowest pressure measured in the arteries
(C) measured only during diastole
(D) determined by stroke volume
(E) decreased when the capacitance of the arteries decreases
(F) the difference between mean arterial pressure and central venous pressure

34. In the sinoatrial (SA) node, phase 4 depolarization (pacemaker potential) is attributable to

(A) an increase in K^+ conductance
(B) an increase in Na^+ conductance
(C) a decrease in Cl^- conductance
(D) a decrease in Ca^{2+} conductance
(E) simultaneous increases in K^+ and Cl^- conductances

35. A healthy 35-year-old man is running a marathon. During the run, there is an increase in his splanchnic vascular resistance. Which receptor is responsible for the increased resistance?

(A) α_1 Receptors
(B) β_1 Receptors
(C) β_2 Receptors
(D) Muscarinic receptors

36. During which phase of the cardiac cycle is aortic pressure highest?

(A) Atrial systole
(B) Isovolumetric ventricular contraction
(C) Rapid ventricular ejection
(D) Reduced ventricular ejection
(E) Isovolumetric ventricular relaxation
(F) Rapid ventricular filling
(G) Reduced ventricular filling (diastasis)

37. Myocardial contractility is best correlated with the intracellular concentration of

(A) Na^+
(B) K^+
(C) Ca^{2+}
(D) Cl^-
(E) Mg^{2+}

38. Which of the following is an effect of histamine?

(A) Decreased capillary filtration
(B) Vasodilation of the arterioles
(C) Vasodilation of the veins
(D) Decreased P_c
(E) Interaction with the muscarinic receptors on the blood vessels

39. Carbon dioxide (CO_2) regulates blood flow to which one of the following organs?

(A) Heart
(B) Skin
(C) Brain
(D) Skeletal muscle at rest
(E) Skeletal muscle during exercise

40. Cardiac output of the right side of the heart is what percentage of the cardiac output of the left side of the heart?

(A) 25%
(B) 50%
(C) 75%
(D) 100%
(E) 125%

41. The physiologic function of the relatively slow conduction through the atrioventricular (AV) node is to allow sufficient time for

(A) runoff of blood from the aorta to the arteries
(B) venous return to the atria
(C) filling of the ventricles
(D) contraction of the ventricles
(E) repolarization of the ventricles

42. Blood flow to which organ is controlled primarily by the sympathetic nervous system rather than by local metabolites?

(A) Skin
(B) Heart
(C) Brain
(D) Skeletal muscle during exercise

43. Which of the following parameters is decreased during moderate exercise?

(A) Arteriovenous O_2 difference
(B) Heart rate
(C) Cardiac output
(D) Pulse pressure
(E) Total peripheral resistance (TPR)

44. A 72-year-old woman, who is being treated with propranolol, finds that she cannot maintain her previous exercise routine. Her physician explains that the drug has reduced her cardiac output. Blockade of which receptor is responsible for the decrease in cardiac output?

(A) α_1 Receptors
(B) β_1 Receptors
(C) β_2 Receptors
(D) Muscarinic receptors
(E) Nicotinic receptors

45. During which phase of the cardiac cycle is ventricular volume lowest?

(A) Atrial systole
(B) Isovolumetric ventricular contraction
(C) Rapid ventricular ejection
(D) Reduced ventricular ejection
(E) Isovolumetric ventricular relaxation
(F) Rapid ventricular filling
(G) Reduced ventricular filling (diastasis)

46. Which of the following changes will cause an increase in myocardial O_2 consumption?

(A) Decreased aortic pressure
(B) Decreased heart rate
(C) Decreased contractility
(D) Increased size of the heart
(E) Increased influx of Na^+ during the upstroke of the action potential

47. Which of the following substances crosses capillary walls primarily through water-filled clefts between the endothelial cells?

(A) O_2
(B) CO_2
(C) CO
(D) Glucose

48. A 24-year-old woman presents to the emergency department with severe diarrhea. When she is supine (lying down), her blood pressure is 90/60 mm Hg (decreased) and her heart rate is 100 beats/min (increased). When she is moved to a standing position, her heart rate further increases to 120 beats/min. Which of the following accounts for the further increase in heart rate upon standing?

(A) Decreased total peripheral resistance
(B) Increased venoconstriction
(C) Increased contractility
(D) Increased afterload
(E) Decreased venous return

49. A 60-year-old businessman is evaluated by his physician, who determines that his blood pressure is significantly elevated at 185/130 mm Hg. Laboratory tests reveal an increase in plasma renin activity, plasma aldosterone level, and left renal vein renin level. His right renal vein renin level is decreased. What is the most likely cause of the patient's hypertension?

(A) Aldosterone-secreting tumor
(B) Adrenal adenoma secreting aldosterone and cortisol
(C) Pheochromocytoma
(D) Left renal artery stenosis
(E) Right renal artery stenosis

QUESTIONS 50–52

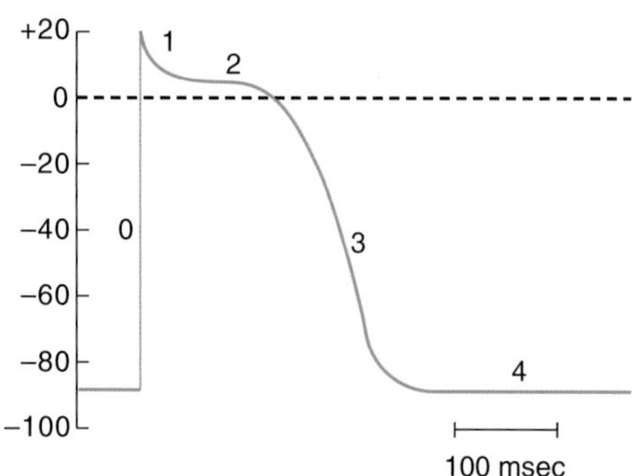

50. During which phase of the ventricular action potential is the membrane potential closest to the K^+ equilibrium potential?

(A) Phase 0
(B) Phase 1
(C) Phase 2
(D) Phase 3
(E) Phase 4

51. During which phase of the ventricular action potential is the conductance to Ca^{2+} highest?

(A) Phase 0
(B) Phase 1
(C) Phase 2
(D) Phase 3
(E) Phase 4

52. Which phase of the ventricular action potential coincides with diastole?

(A) Phase 0
(B) Phase 1
(C) Phase 2
(D) Phase 3
(E) Phase 4

53. Propranolol has which of the following effects?

(A) Decreases heart rate
(B) Increases left ventricular ejection fraction
(C) Increases stroke volume
(D) Decreases splanchnic vascular resistance
(E) Decreases cutaneous vascular resistance

54. Which receptor mediates slowing of the heart?

(A) α_1 Receptors
(B) β_1 Receptors
(C) β_2 Receptors
(D) Muscarinic receptors

55. Which of the following agents or changes has a negative inotropic effect on the heart?

(A) Increased heart rate
(B) Sympathetic stimulation
(C) Norepinephrine
(D) Acetylcholine (ACh)
(E) Cardiac glycosides

56. The low-resistance pathways between myocardial cells that allow for the spread of action potentials are the

(A) gap junctions
(B) T tubules
(C) sarcoplasmic reticulum (SR)
(D) intercalated disks
(E) mitochondria

57. Which agent is released or secreted after a hemorrhage and causes an increase in renal Na^+ reabsorption?

(A) Aldosterone
(B) Angiotensin I
(C) Angiotensinogen
(D) Antidiuretic hormone (ADH)
(E) Atrial natriuretic peptide

58. During which phase of the cardiac cycle does the mitral valve open?

(A) Atrial systole
(B) Isovolumetric ventricular contraction
(C) Rapid ventricular ejection
(D) Reduced ventricular ejection
(E) Isovolumetric ventricular relaxation
(F) Rapid ventricular filling
(G) Reduced ventricular filling (diastasis)

59. A hospitalized patient has an ejection fraction of 0.4, a heart rate of 95 beats/min, and a cardiac output of 3.5 L/min. What is the patient's end-diastolic volume?

(A) 14 mL
(B) 37 mL
(C) 55 mL
(D) 92 mL
(E) 140 mL

60. A 38-year-old woman has a bout of "intestinal flu," with vomiting and diarrhea for several days. Although she is feeling better, when she stands up quickly, she feels faint and light-headed. Which of the following explains why she is light-headed?

(A) Decreased blood volume, decreased preload, decreased cardiac output
(B) Increased heart rate, increased cardiac output
(C) Increased sympathetic output, increased total peripheral resistance, increased arterial pressure
(D) Increased renin levels, increased angiotensin II levels, increased aldosterone levels
(E) Decreased atrial natriuretic peptide levels, decreased Na^+ reabsorption

Answers and Explanations

1. **The answer is D** [II C, D]. If the radius of the artery decreased by 50% (1/2), then resistance would increase by 2^4, or 16 ($R = 8\eta l/\pi r^4$). Because blood flow is inversely proportional to resistance ($Q = \Delta P/R$), flow will decrease to 1/16 of the original value.
2. **The answer is B** [IX A; Table 3.4]. When a person moves to a standing position, blood pools in the leg veins, causing decreased venous return to the heart, decreased cardiac output, and decreased arterial pressure. The baroreceptors detect the decrease in arterial pressure, and the vasomotor center is activated to increase sympathetic outflow and decrease parasympathetic outflow. There is an increase in heart rate (resulting in a decreased PR interval), contractility, and total peripheral resistance (TPR). Because both heart rate and contractility are increased, cardiac output will increase toward normal.
3. **The answer is E** [II G, H, I]. Pressures on the venous side of the circulation (e.g., central vein, right atrium, renal vein) are lower than pressures on the arterial side. Pressure in the pulmonary artery (and all pressures on the right side of the heart) is much lower than their counterparts on the left side of the heart. In the systemic circulation, systolic pressure is actually slightly higher in the downstream arteries (e.g., renal artery) than in the aorta because of the reflection of pressure waves at branch points.
4. **The answer is B** [III A]. The absent P wave indicates that the atrium is not depolarizing and, therefore, the pacemaker cannot be in the sinoatrial (SA) node. Because the QRS and T waves are normal, depolarization and repolarization of the ventricle must be proceeding in the normal sequence. This situation can occur if the pacemaker is located in the atrioventricular (AV) node. If the pacemaker were located in the bundle of His or in the Purkinje system, the ventricles would activate in an abnormal sequence (depending on the exact location of the pacemaker) and the QRS wave would have an abnormal configuration. Ventricular muscle does not have pacemaker properties.
5. **The answer is B** [IV G 3]. An increase in ejection fraction means that a higher fraction of the end-diastolic volume is ejected in the stroke volume (e.g., because of the administration of a positive inotropic agent). When this situation occurs, the volume remaining in the ventricle after systole, the end-systolic volume, will be reduced. Cardiac output, pulse pressure, stroke volume, and systolic pressure will be increased.
6. **The answer is D** [V G]. On the extrasystolic beat, pulse pressure decreases because there is inadequate ventricular filling time—the ventricle beats "too soon." As a result, stroke volume decreases.
7. **The answer is A** [IV C I a (2)]. The postextrasystolic contraction produces increased pulse pressure because contractility is increased. Extra Ca^{2+} enters the cell during the extrasystolic beat. Contractility is directly related to the amount of intracellular Ca^{2+} available for binding to troponin C.
8. **The answer is A** [IV D 5 a]. An increase in contractility produces an increase in cardiac output for a given end-diastolic volume, or pressure. The Frank-Starling relationship demonstrates the matching of cardiac output (what leaves the heart) with venous return (what returns to the heart). An increase in contractility (positive inotropic effect) will shift the curve upward.
9. **The answer is B** [IV E 1 a]. Isovolumetric contraction occurs during ventricular systole, before the aortic valve opens. Ventricular pressure increases, but volume remains constant because blood cannot be ejected into the aorta against a closed valve.
10. **The answer is C** [IV 1 c]. Closure of the aortic valve occurs once ejection of blood from the ventricle has occurred and the left ventricular pressure has decreased to less than the aortic pressure.

11. The answer is A [V B]. The first heart sound corresponds to closure of the atrial–ventricular valves. Before this closure occurs, the ventricle fills (phase 4 → 1). After the valves close, isovolumetric contraction begins and ventricular pressure increases (phase 1 → 2).

12. The answer is C [IV E 1, G 1, 2]. Stroke volume is the volume ejected from the ventricle and is represented on the pressure–volume loop as phase 2 → 3; end-diastolic volume is about 140 mL and end-systolic volume is about 65 mL; the difference, or stroke volume, is 75 mL. Cardiac output is calculated as stroke volume × heart rate or 75 mL × 70 beats/min = 5250 mL/min or 5.25 L/min.

13. The answer is D [VII C 1]. The net driving force can be calculated with the Starling equation

$$\begin{aligned}\text{Net pressure} &= (P_c - P_i) - (\pi_c - \pi_i)\\ &= [(30-(-2)) - (25-2)]\,\text{mm Hg}\\ &= 32\text{ mm Hg} - 23\text{ mm Hg}\\ &= +9\text{ mm Hg}\end{aligned}$$

Because the net pressure is positive, filtration out of the capillary will occur.

14. The answer is C [VII C 1]. K_f is the filtration coefficient for the capillary and describes the intrinsic water permeability.

$$\begin{aligned}\text{Water flow} &= K_f \times \text{Net pressure}\\ &= 0.5\text{ mL/min/mm Hg} \times 9\text{ mm Hg}\\ &= 4.5\text{ mL/min}\end{aligned}$$

15. The answer is C [II D 2 a, b]. Turbulent flow is predicted when the Reynolds number is increased. Factors that increase the Reynolds number and produce turbulent flow are decreased viscosity (hematocrit) and increased velocity. Partial occlusion of a blood vessel increases the Reynolds number (and turbulence) because the decrease in cross-sectional area results in increased blood velocity (v = Q/A).

16. The answer is D [IX A]. Orthostatic hypotension is a decrease in arterial pressure that occurs when a person moves from a supine to a standing position. A person with a normal baroreceptor mechanism responds to a decrease in arterial pressure through the vasomotor center by increasing sympathetic outflow and decreasing parasympathetic outflow. The sympathetic component helps to restore blood pressure by increasing heart rate, contractility, total peripheral resistance (TPR), and mean systemic filling pressure. In a patient who has undergone a sympathectomy, the sympathetic component of the baroreceptor mechanism is absent.

17. The answer is D [III A]. The PR segment (part of the PR interval) and the ST segment are the only portions of the electrocardiogram (ECG) that are isoelectric. The PR interval includes the P wave (atrial depolarization) and the PR segment, which represents conduction through the atrioventricular (AV) node; during this phase, the ventricles are not yet depolarized. The ST segment is the only isoelectric period when the entire ventricle is depolarized.

18. The answer is C [I A]. In a left-to-right ventricular shunt, a defect in the ventricular septum allows blood to flow from the left ventricle to the right ventricle instead of being ejected into the aorta. The "shunted" fraction of the left ventricular output is therefore added to the output of the right ventricle, making pulmonary blood flow (the cardiac output of the right ventricle) higher than systemic blood flow (the cardiac output of the left ventricle). In normal adults, the outputs of both ventricles are equal in the steady state. In the fetus, pulmonary blood flow is near zero. Right ventricular failure results in decreased pulmonary blood flow. Administration of a positive inotropic agent should have the same effect on contractility and cardiac output in both ventricles.

19. The answer is C [IV F 2 a]. The shift in the venous return curve to the right is consistent with an increase in blood volume and, as a consequence, mean systemic filling pressure.

Both cardiac output and venous return are increased in the new steady state (and are equal to each other). Contractility is unaffected.

20. The answer is D [III E 1 b]. A pattern of two P waves preceding each QRS complex indicates that only every other P wave is conducted through the atrioventricular (AV) node to the ventricle. Thus, conduction velocity through the AV node must be decreased.

21. The answer is A [VI A 1 a to d]. A decrease in blood pressure causes decreased stretch of the carotid sinus baroreceptors and decreased firing of the carotid sinus nerve. In an attempt to restore blood pressure, the parasympathetic outflow to the heart is decreased and sympathetic outflow is increased. As a result, heart rate and contractility will be increased. Mean systemic filling pressure will increase because of increased sympathetic tone of the veins (and a shift of blood to the arteries).

22. The answer is B [VII C 4 c; Table 3.2]. Edema occurs when more fluid is filtered out of the capillaries than can be returned to the circulation by the lymphatics. Filtration is increased by changes that increase P_c or decrease π_c. Arteriolar constriction would decrease P_c and decrease filtration. Dehydration would increase plasma protein concentration (by hemoconcentration) and thereby increase π_c and decrease filtration. Increased venous pressure would increase P_c and filtration.

23. The answer is A [V E]. The second heart sound is associated with closure of the aortic and pulmonic valves. Because the aortic valve closes before the pulmonic valve, the sound can be split by inspiration.

24. The answer is C [IX B 2]. During exercise, local metabolites accumulate in the exercising muscle and cause local vasodilation and decreased arteriolar resistance of the skeletal muscle. Because muscle mass is large, it contributes a large fraction of the total peripheral resistance (TPR). Therefore, the skeletal muscle vasodilation results in an overall decrease in TPR, even though there is sympathetic vasoconstriction in other vascular beds.

25. The answer is A [V A to G]. The electrocardiogram (ECG) tracing serves as a reference. The QRS complex marks ventricular depolarization, followed immediately by ventricular contraction. Aortic pressure increases steeply after QRS, as blood is ejected from the ventricles. After reaching peak pressure, aortic pressure decreases as blood runs off into the arteries. The characteristic dicrotic notch ("blip" in the aortic pressure curve) appears when the aortic valve closes. Aortic pressure continues to decrease as blood flows out of the aorta.

26. The answer is D [V A to G]. Ventricular volume increases slightly with atrial systole (P wave), is constant during isovolumetric contraction (QRS), and then decreases dramatically after the QRS, when blood is ejected from the ventricle.

27. The answer is C [II C]. An increase in arteriolar resistance will increase total peripheral resistance (TPR). Arterial pressure = cardiac output × TPR, so arterial pressure will also increase. Capillary filtration decreases when there is arteriolar constriction because P_c decreases. Afterload of the heart would be increased by an increase in TPR.

28. The answer is D [IV J]. Cardiac output is calculated by the Fick principle if whole body oxygen (O_2) consumption and $[O_2]$ in the pulmonary artery and pulmonary vein are measured. Mixed venous blood could substitute for a pulmonary artery sample, and peripheral arterial blood could substitute for a pulmonary vein sample. Central venous pressure and heart rate are not needed for this calculation.

$$\text{Cardiac output} = \frac{500 \text{ mL/min}}{0.24 \text{ mL } O_2/\text{mL} - 0.16 \text{ mL } O_2/\text{mL}}$$
$$= 6250 \text{ mL/min, or } 6.25 \text{ L/min}$$

29. The answer is B [III B 1 a, c, d, 2 a]. The upstroke of the action potential in the atria, ventricles, and Purkinje fibers is the result of a fast inward Na^+ current. The upstroke of the action potential in the sinoatrial (SA) node is the result of an inward Ca^{2+} current.

The plateau of the ventricular action potential is the result of a slow inward Ca^{2+} current. Repolarization in all cardiac tissues is the result of an outward K^+ current.

30. **The answer is C** [IV F 3 a (1)]. An upward shift of the cardiac output curve is consistent with an increase in myocardial contractility; for any right atrial pressure (sarcomere length), the force of contraction is increased. Such a change causes an increase in stroke volume and cardiac output. Increased blood volume and increased mean systemic filling pressure are related and would cause a rightward shift in the venous return curve. A negative inotropic agent would cause a decrease in contractility and a downward shift of the cardiac output curve.

31. **The answer is B** [IV F 3]. End-diastolic volume and right atrial pressure are related and can be used interchangeably.

32. **The answer is E** [II A 2, 3, F]. The decrease in pressure at any level of the cardiovascular system is caused by the resistance of the blood vessels ($\Delta P = Q \times R$). The greater the resistance is, the greater the decrease in pressure is. The arterioles are the site of highest resistance in the vasculature. The arterioles do not have the greatest surface area or cross-sectional area (the capillaries do). Velocity of blood flow is lowest in the capillaries, riot in the arterioles.

33. **The answer is D** [II G 3]. Pulse pressure is the difference between the highest (systolic) and lowest (diastolic) arterial pressures. It reflects the volume ejected by the left ventricle (stroke volume). Pulse pressure increases when the capacitance of the arteries decreases, such as with aging.

34. **The answer is B** [III B 2 c]. Phase 4 depolarization is responsible for the pacemaker property of sinoatrial (SA) nodal cells. It is caused by an increase in Na^+ conductance and an inward Na^+ current (I_f), which depolarizes the cell membrane.

35. **The answer is A** [VIII E 1; Table 3.1]. During exercise, the sympathetic nervous system is activated. The observed increase in splanchnic vascular resistance is due to sympathetic activation of α_1 receptors on splanchnic arterioles.

36. **The answer is D** [V A to G]. Aortic pressure reaches its highest level immediately after the rapid ejection of blood during left ventricular systole. This highest level actually coincides with the beginning of the reduced ventricular ejection phase.

37. **The answer is C** [IV B 6]. Contractility of myocardial cells depends on the intracellular [Ca^{2+}], which is regulated by Ca^{2+} entry across the cell membrane during the plateau of the action potential and by Ca^{2+} uptake into and release from the sarcoplasmic reticulum (SR). Ca^{2+} binds to troponin C and removes the inhibition of actin–myosin interaction, allowing contraction (shortening) to occur.

38. **The answer is B** [VIII B 2 a]. Histamine causes vasodilation of the arterioles, which increases P_c and capillary filtration. It also causes constriction of the veins, which contributes to the increase in P_c. Acetylcholine (ACh) interacts with muscarinic receptors (although these are not present on vascular smooth muscle).

39. **The answer is C** [VIII C, D, E 2, F]. Blood flow to the brain is autoregulated by the P_{CO_2}. If metabolism increases (or arterial pressure decreases), the P_{CO_2} will increase and cause cerebral vasodilation. Blood flow to the heart and to skeletal muscle during exercise is also regulated metabolically, but adenosine and hypoxia are the most important vasodilators for the heart. Adenosine, lactate, and K^+ are the most important vasodilators for exercising skeletal muscle. Blood flow to the skin is regulated by the sympathetic nervous system rather than by local metabolites.

40. **The answer is D** [I A]. Cardiac output of the left and right sides of the heart is equal. Blood ejected from the left side of the heart to the systemic circulation must be oxygenated by passage through the pulmonary circulation.

41. **The answer is C** [III C]. The atrioventricular (AV) delay (which corresponds to the PR interval) allows time for filling of the ventricles from the atria. If the ventricles contracted before they were filled, stroke volume would decrease.

42. The answer is A [VIII C to F]. Circulation of the skin is controlled primarily by the sympathetic nerves. The coronary and cerebral circulations are primarily regulated by local metabolic factors. Skeletal muscle circulation is regulated by metabolic factors (local metabolites) during exercise, although at rest it is controlled by the sympathetic nerves.

43. The answer is E [IX B]. In anticipation of exercise, the central command increases sympathetic outflow to the heart and blood vessels, causing an increase in heart rate and contractility. Venous return is increased by muscular activity and contributes to an increase in cardiac output by the Frank-Starling mechanism. Pulse pressure is increased because stroke volume is increased. Although increased sympathetic outflow to the blood vessels might be expected to increase total peripheral resistance (TPR), it does not because there is an overriding vasodilation of the skeletal muscle arterioles as a result of the buildup of vasodilator metabolites (lactate, K^+ adenosine). Because this vasodilation improves the delivery of O_2, more O_2 can be extracted and used by the contracting muscle.

44. The answer is B [III 3; Table 3.1]. Propranolol is an adrenergic antagonist that blocks both β_1 and β_2 receptors. When propranolol is administered to reduce cardiac output, it inhibits β_1 receptors in the sinoatrial (SA) node (heart rate) and in ventricular muscle (contractility).

45. The answer is E [V E]. Ventricular volume is at its lowest value while the ventricle is relaxed (diastole), just before ventricular filling begins.

46. The answer is D [IV I]. Myocardial O_2 consumption is determined by the amount of tension developed by the heart. It increases when there are increases in aortic pressure (increased afterload), when there is increased heart rate or stroke volume (which increases cardiac output), or when the size (radius) of the heart is increased ($T = P \times r$). Influx of Na^+ ions during an action potential is a purely passive process, driven by the electrochemical driving forces on Na^+ ions. Of course, maintenance of the inwardly directed Na^+ gradient over the long term requires the Na^+–K^+ pump, which is energized by adenosine triphosphate (ATP).

47. The answer is D [VII B 1, 2]. Because O_2, CO_2, and CO are lipophilic, they cross capillary walls primarily by diffusion through the endothelial cell membranes. Glucose is water soluble; it cannot cross through the lipid component of the cell membrane and is restricted to the water-filled clefts, or pores, between the cells.

48. The answer is E [VI A]. Diarrhea causes a loss of extracellular fluid volume, which produces a decrease in arterial pressure. The decrease in arterial pressure activates the baroreceptor mechanism, which produces an increase in heart rate when the patient is supine. When she stands up, blood pools in her leg veins and produces a decrease in venous return, a decrease in cardiac output (by the Frank-Starling mechanism), and a further decrease in arterial pressure. The *further* decrease in arterial pressure causes *further* activation of the baroreceptor mechanism and a *further* increase in heart rate.

49. The answer is D [VI B]. In this patient, hypertension is most likely caused by left renal artery stenosis, which led to increased renin secretion by the left kidney. The increased plasma renin activity causes an increased secretion of aldosterone, which increases Na^+ reabsorption by the renal distal tubule. The increased Na^+ reabsorption leads to increased blood volume and blood pressure. The right kidney responds to the increase in blood pressure by decreasing its renin secretion. Right renal artery stenosis causes a similar pattern of results, except that renin secretion from the right kidney, not the left kidney, is increased. Aldosterone-secreting tumors cause increased levels of aldosterone but decreased plasma renin activity (as a result of decreased renin secretion by both kidneys). Pheochromocytoma is associated with increased circulating levels of catecholamines, which increase blood pressure by their effects on the heart (increased heart rate and contractility) and blood vessels (vasoconstriction); the increase in blood pressure is sensed by the kidneys and results in decreased plasma renin activity and aldosterone levels.

50. The answer is E [III B 1 e]. Phase 4 is the resting membrane potential. Because the conductance K^+ is highest, the membrane potential approaches the equilibrium potential for K^+.

51. **The answer is C** [III B 1 c]. Phase 2 is the plateau of the ventricular action potential. During this phase, the conductance to Ca^{2+} increases transiently. Ca^{2+} that enters the cell during the plateau is the trigger that releases more Ca^{2+} from the sarcoplasmic reticulum (SR) for the contraction.

52. **The answer is E** [III B 1 e]. Phase 4 is electrical diastole.

53. **The answer is A** [III E 2, 3; Table 3.1]. Propranolol, a β-adrenergic antagonist, blocks all sympathetic effects that are mediated by a β_1 or β_2 receptor. The sympathetic effect on the sinoatrial (SA) node is to increase heart rate via a β_1 receptor; therefore, propranolol decreases heart rate. Ejection fraction reflects ventricular contractility, which is another effect of β_1 receptors; thus, propranolol decreases contractility, ejection fraction, and stroke volume. Splanchnic and cutaneous resistance are mediated by α_1 receptors.

54. **The answer is D** [III E 2 a; Table 3.1]. Acetylcholine (ACh) causes slowing of the heart via muscarinic receptors in the sinoatrial (SA) node.

55. **The answer is D** [IV C]. A negative inotropic effect is one that decreases myocardial contractility. Contractility is the ability to develop tension at a fixed muscle length. Factors that decrease contractility are those that decrease the intracellular [Ca^{2+}]. Increasing heart rate increases intracellular [Ca^{2+}] because more Ca^{2+} ions enter the cell during the plateau of each action potential. Sympathetic stimulation and norepinephrine increase intracellular [Ca^{2+}] by increasing entry during the plateau and increasing the storage of Ca^{2+} by the sarcoplasmic reticulum (SR) [for later release]. Cardiac glycosides increase intracellular [Ca^{2+}] by inhibiting the Na^+–K^+ pump, thereby inhibiting Na^+–Ca^{2+} exchange (a mechanism that pumps Ca^{2+} out of the cell). Acetylcholine (ACh) has a negative inotropic effect on the atria.

56. **The answer is A** [IVA 3]. The gap junctions occur at the intercalated disks between cells and are low-resistance sites of current spread.

57. **The answer is A** [VI C 4; IX C]. Angiotensin I and aldosterone are increased in response to a decrease in renal perfusion pressure. Angiotensinogen is the precursor for angiotensin I. Antidiuretic hormone (ADH) is released when atrial receptors detect a decrease in blood volume. Of these, only aldosterone increases Na^+ reabsorption. Atrial natriuretic peptide is released in response to an increase in atrial pressure, and an increase in its secretion would not be anticipated after blood loss.

58. **The answer is E** [V E]. The mitral [atrioventricular (AV)] valve opens when left atrial pressure becomes higher than left ventricular pressure. This situation occurs when the left ventricular pressure is at its lowest level—when the ventricle is relaxed, blood has been ejected from the previous cycle, and before refilling has occurred.

59. **The answer is D** [IV G]. First, calculate stroke volume from the cardiac output and heart rate: Cardiac output = stroke volume × heart rate; thus, stroke volume = cardiac output/heart rate = 3500 mL/95 beats/min = 36.8 mL. Then, calculate end-diastolic volume from stroke volume and ejection fraction: Ejection fraction = stroke volume/end-diastolic volume; thus, end-diastolic volume = stroke volume/ejection fraction = 36.8 mL/0.4 = 92 mL.

60. **The answer is A** (IX A). The woman has significant loss of extracellular fluid (and blood) volume due to vomiting and diarrhea. Decreased blood volume leads to decreased venous return, decreased preload, and decreased cardiac output by the Frank-Starling mechanism; the decrease in cardiac output causes decreased arterial pressure (P_a) and decreased cerebral blood flow, which is responsible for the feeling of light-headedness. The decrease in P_a will activate both the baroreceptor mechanism and the renin–angiotensin II–aldosterone system, but the results of turning on these mechanisms (increased sympathetic output, increased total peripheral resistance, increased heart rate, and compensatory increase in P_a toward normal) are secondary to the decrease in P_a; they are not causes of the light-headedness. Likewise, decreased atrial natriuretic peptide levels and decreased Na^+ reabsorption can occur secondary to the decrease in P_a but are not causes of the light-headedness.

Chapter 4 Respiratory Physiology

I. LUNG VOLUMES AND CAPACITIES

A. Lung volumes (Figure 4.1)

1. Tidal volume (V_T)

- is the volume inspired or expired with each normal breath.

2. Inspiratory reserve volume (IRV)

- is the volume that can be inspired over and above the tidal volume.
- is used during exercise.

3. Expiratory reserve volume (ERV)

- is the volume that can be expired after the expiration of a tidal volume.

4. Residual volume (RV)

- is the volume that remains in the lungs after a maximal expiration.
- cannot be measured by spirometry.

5. Dead space

a. Anatomic dead space

- is the volume of the conducting airways.
- is normally approximately 150 mL.

b. Physiologic dead space

- is a functional measurement.
- is defined as the volume of the lungs that does not participate in gas exchange.
- is approximately equal to the anatomic dead space in normal lungs.
- may be greater than the anatomic dead space in lung diseases in which there are ventilation/perfusion (V/Q) defects.
- is calculated by the following equation:

$$V_D = V_T \times \frac{P_{A_{CO_2}} - P_{E_{CO_2}}}{P_{A_{CO_2}}}$$

where:

V_D = physiologic dead space (mL)

V_T = tidal volume (mL)

$P_{A_{CO_2}}$ = P_{CO_2} of alveolar gas (mm Hg) = Pa_{CO_2} of arterial blood

$P_{E_{CO_2}}$ = P_{CO_2} of expired air (mm Hg)

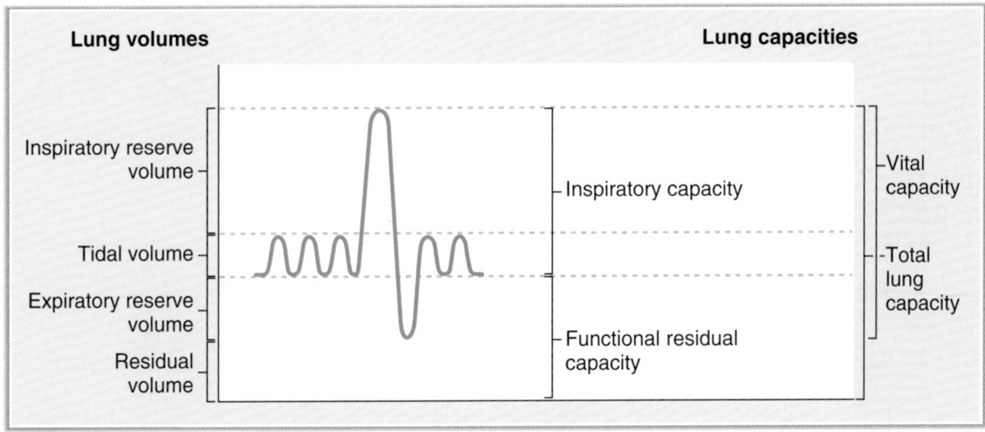

FIGURE 4.1. Lung volumes and capacities.

- In words, the equation states that physiologic dead space is tidal volume multiplied by a fraction. The fraction represents the dilution of alveolar P_{CO_2} by dead space air, which does not participate in gas exchange and does not therefore contribute CO_2 to expired air.

6. **Ventilation rate**

   a. **Minute ventilation** is expressed as follows:

$$\text{Minute ventilation} = V_T \times \text{breaths/min}$$

   b. **Alveolar ventilation (V_A)** is expressed as follows:

$$V_A = (V_T - V_D) \times \text{breaths/min}$$

- **Sample problem:** A person with a tidal volume (V_T) of 0.5 L is breathing at a rate of 15 breaths/min. The P_{CO_2} of his arterial blood is 40 mm Hg, and the P_{CO_2} of his expired air is 36 mm Hg. What is his rate of alveolar ventilation?

$$\begin{aligned}
\text{Dead space} &= V_T \times \frac{P_{A_{CO_2}} - P_{E_{CO_2}}}{P_{A_{CO_2}}} \\
&= 0.5\ \text{L} \times \frac{40\ \text{mm Hg} - 36\ \text{mm Hg}}{40\ \text{mm Hg}} \\
&= 0.05\ \text{L} \\
V_A &= (V_T - V_D) \times \text{breaths/min} \\
&= (0.5\ \text{L} - 0.05\ \text{L}) \times 15\ \text{breaths/min} \\
&= 6.75\ \text{L/min}
\end{aligned}$$

B. Lung capacities (see Figure 4.1)

1. **Inspiratory capacity**
   - is the sum of tidal volume and IRV.
2. **Functional residual capacity (FRC)**
   - is the sum of ERV and RV.
   - is the volume remaining in the lungs after a tidal volume is expired.
   - includes the RV, so it **cannot be measured by spirometry.**
3. **Vital capacity (VC), or forced vital capacity (FVC)**
   - is the sum of tidal volume, IRV, and ERV.
   - is the volume of air that can be forcibly expired after a maximal inspiration.

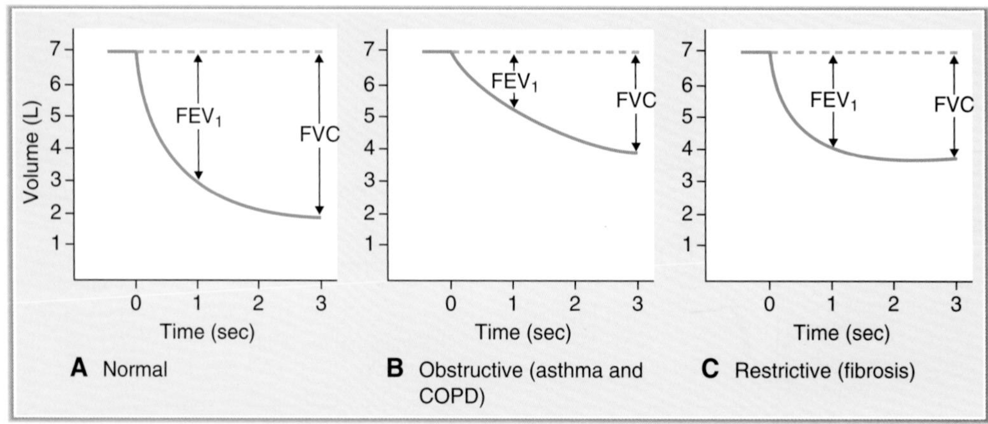

FIGURE 4.2. Forced vital capacity (FVC) and FEV_1 in normal subjects and in patients with lung disease. FEV_1 = volume expired in first second of forced maximal expiration; COPD = chronic obstructive pulmonary disease.

4. Total lung capacity (TLC)

- is the sum of all four lung volumes.
- is the volume in the lungs after a maximal inspiration.
- includes RV, so it **cannot be measured by spirometry.**

C. Forced expiratory volume (FEV_1) (Figure 4.2)

- FEV_1 is the volume of air that can be expired in the first second of a forced maximal expiration.
- FEV_1 is **normally 80% of the forced vital capacity,** which is expressed as

$$FEV_1 / FVC = 0.8$$

- In **obstructive** lung disease, such as asthma and chronic obstructive pulmonary disease (COPD), both FEV_1 and FVC are reduced, but FEV_1 is reduced more than FVC is; thus, **FEV_1/FVC is decreased.**
- In **restrictive** lung disease, such as fibrosis, both FEV_1 and FVC are reduced, but FEV_1 is reduced less than FVC is; thus, **FEV_1/FVC is increased.**

II. MECHANICS OF BREATHING

A. Muscles of inspiration

1. Diaphragm

- is the **most important** muscle for inspiration.
- When the diaphragm contracts, the abdominal contents are pushed downward, and the ribs are lifted upward and outward, increasing the volume of the thoracic cavity.

2. External intercostals and accessory muscles

- are not used for inspiration during normal quiet breathing.
- are used during **exercise** and in **respiratory distress.**

B. Muscles of expiration

- Expiration is **normally passive.**
- Because the lung–chest wall system is elastic, it returns to its resting position after inspiration.
- Expiratory muscles are used **during exercise** or when airway resistance is increased because of disease (e.g., **asthma**).

1. **Abdominal muscles**
   - compress the abdominal cavity, push the diaphragm up, and push air out of the lungs.
2. **Internal intercostal muscles**
   - pull the ribs downward and inward.

C. Compliance of the respiratory system

- is analogous to capacitance in the cardiovascular system.
- is described by the following equation:

$$C = V/P$$

where:
C = compliance (mL/mm Hg)
V = volume (mL)
P = pressure (mm Hg)

- describes the **distensibility** of the lungs and chest wall.
- is **inversely related to elastance**, which depends on the amount of elastic tissue.
- is inversely related to stiffness.
- is the **slope of the pressure–volume curve.**
- is the change in volume for a given change in pressure. Pressure can refer to the pressure inside the lungs and airways or to transpulmonary pressure (i.e., the pressure difference across pulmonary structures).

1. **Compliance of the lungs** (Figure 4.3)
   - Transmural pressure is alveolar pressure minus intrapleural pressure.
   - When the pressure outside of the lungs (i.e., intrapleural pressure) is negative, the lungs expand and lung volume increases.
   - When the pressure outside of the lungs is positive, the lungs collapse and lung volume decreases.
   - In the air-filled lung, inflation (inspiration) follows a different curve than deflation (expiration); this difference is called **hysteresis** and is due to the need to overcome surface tension forces at the air–liquid interface when inflating the lungs.
   - In the middle range of pressures, compliance is greatest and the lungs are most distensible.
   - At high expanding pressures, compliance is lowest, the lungs are least distensible, and the curve flattens.

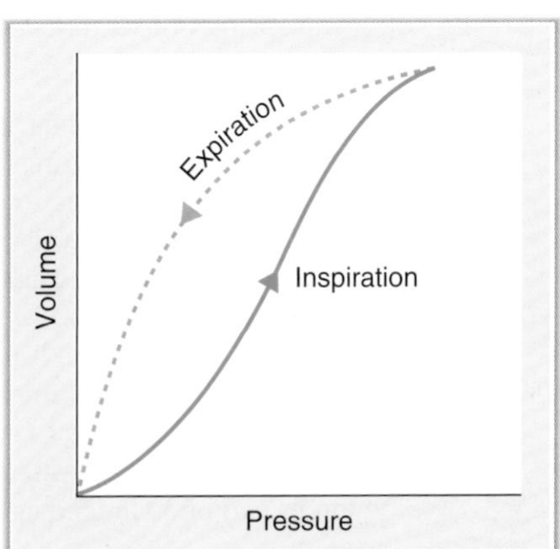

FIGURE 4.3. Compliance of the air-filled lung. Different curves are followed during inspiration and expiration (hysteresis).

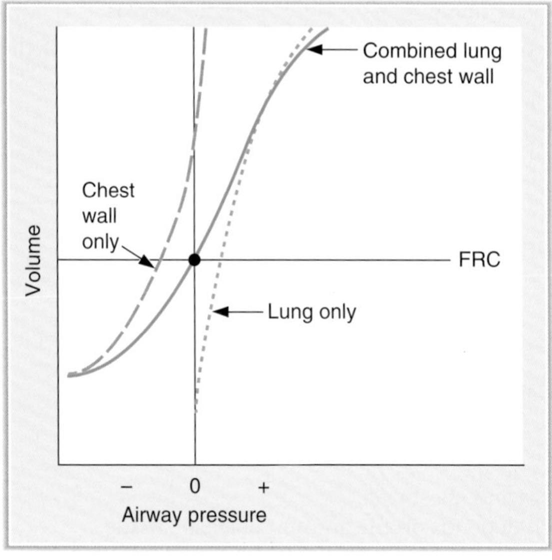

FIGURE 4.4. Compliance of the lungs and chest wall separately and together. FRC = functional residual capacity.

2. **Compliance of the combined lung–chest wall system** (Figure 4.4)
   a. Figure 4.4 shows the pressure–volume relationships for the lungs alone (hysteresis has been eliminated for simplicity), the chest wall alone, and the lungs and chest wall together.
      - **Compliance of the lung–chest wall system** is less than that of the lungs alone or the chest wall alone (the slope is flatter).
   b. At rest (identified by the filled circle in the center of Figure 4.4), lung volume is at FRC and the pressure in the airways and lungs is equal to atmospheric pressure (i.e., zero). Under these equilibrium conditions, there is a collapsing force on the lungs and an expanding force on the chest wall. At **FRC,** these two forces are **equal and opposite** and, therefore, the combined lung–chest wall system neither wants to collapse nor wants to expand (i.e., equilibrium).
   c. As a result of these two opposing forces, **intrapleural pressure is negative** (subatmospheric).
      - If air is introduced into the intrapleural space **(pneumothorax),** the intrapleural pressure becomes equal to atmospheric pressure. Without the normal negative intrapleural pressure, the lungs will collapse (their natural tendency) and the chest wall will spring outward (its natural tendency).
   d. **Changes in lung compliance**
      - In a patient with **emphysema,** lung compliance is increased, elastic recoil is decreased, and the tendency of the lungs to collapse is decreased. Therefore, at the original FRC, the tendency of the lungs to collapse is less than the tendency of the chest wall to expand. The lung–chest wall system will seek a **new, higher FRC** so that the two opposing forces can be balanced again; the patient's chest becomes **barrel-shaped,** reflecting this higher volume.
      - In a patient with **fibrosis,** lung compliance is decreased, elastic recoil is increased, and the tendency of the lungs to collapse is increased. Therefore, at the original FRC, the tendency of the lungs to collapse is greater than the tendency of the chest wall to expand. The lung–chest wall system will seek a **new, lower FRC** so that the two opposing forces can be balanced again.

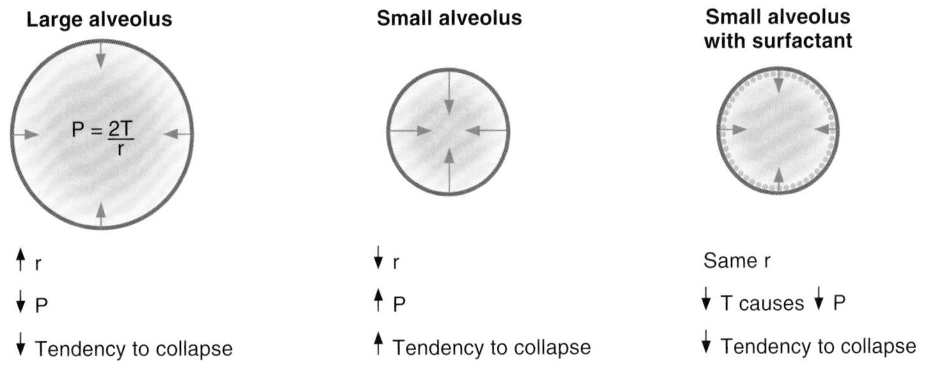

FIGURE 4.5. Effect of alveolar size and surfactant on the pressure that tends to collapse the alveoli. P = pressure; r = radius; T = surface tension.

D. Surface tension of alveoli and surfactant

1. Surface tension of the alveoli (Figure 4.5)

- results from the attractive forces between liquid molecules lining the alveoli at the air–liquid interface.
- creates a collapsing pressure that is directly proportional to surface tension and inversely proportional to alveolar radius **(Laplace law)**, as shown in the following equation:

$$P = \frac{2T}{r}$$

where:
P = collapsing pressure on alveolus (or pressure required to keep alveolus open) [dynes/cm^2]
T = surface tension (dynes/cm)
r = radius of alveolus (cm)

a. Large alveoli (large radii) have low collapsing pressures and are easy to keep open.

b. Small alveoli (small radii) have high collapsing pressures and are more difficult to keep open.

- In the **absence of surfactant,** the small alveoli have a tendency to collapse **(atelectasis).**

2. Surfactant (see Figure 4.5)

- lines the alveoli.
- **reduces surface tension** by disrupting the intermolecular forces between liquid molecules. This reduction in surface tension prevents small alveoli from collapsing and **increases compliance.**
- is synthesized by **type II alveolar cells** and consists primarily of the phospholipid **dipalmitoylphosphatidylcholine (DPPC).**
- In the **fetus,** surfactant synthesis is variable. Surfactant may be present as early as gestational week 24 and is almost always present by gestational week 35.
- Generally, a lecithin:sphingomyelin ratio greater than 2:1 in amniotic fluid reflects mature levels of surfactant.
- **Neonatal respiratory distress syndrome** can occur in premature infants because of the lack of surfactant. The infant exhibits atelectasis (lungs collapse), difficulty reinflating the lungs (as a result of decreased compliance), decreased V/Q and right-to-left shunt, and hypoxemia (as a result of decreased V/Q and right-to-left shunt).

E. Relationships between pressure, airflow, and resistance

- are analogous to the relationships between blood pressure, blood flow, and resistance in the cardiovascular system.

1. **Airflow**

- is driven by, and is directly proportional to, the **pressure difference** between the mouth (or nose) and the alveoli.
- is **inversely proportional to airway resistance;** thus, the higher the airway resistance, the lower the airflow. This inverse relationship is shown in the following equation:

$$Q = \frac{\Delta P}{R}$$

where:
Q = airflow (L/min)
ΔP = pressure gradient (cm H_2O)
R = airway resistance (cm H_2O/L/min)

2. **Resistance of the airways**

- is described by **Poiseuille law,** as shown in the following equation:

$$R = \frac{8\,\eta l}{\pi r^4}$$

where:
R = resistance
η = viscosity of the inspired gas
l = length of the airway
r = radius of the airway

- Notice the powerful inverse fourth-power relationship between resistance and the size (radius) of the airway.
- **For example,** if airway radius decreases by a factor of 4, then resistance will increase by a factor of 256 (4^4), and airflow will decrease by a factor of 256.

3. **Factors that change airway resistance**

- The major site of airway resistance is the **medium-sized bronchi.**
- The smallest airways would seem to offer the highest resistance, but they do not because of their parallel arrangement.

a. **Contraction or relaxation of bronchial smooth muscle**

- changes airway resistance by altering the radius of the airways.

(1) *Parasympathetic stimulation,* irritants, and the slow-reacting substance of anaphylaxis **(asthma)** constrict the airways, decrease the radius, and increase the resistance to airflow.

(2) *Sympathetic stimulation* and sympathetic agonists **(isoproterenol)** dilate the airways via **β_2 receptors,** increase the radius, and decrease the resistance to airflow.

b. **Lung volume**

- alters airway resistance because of the radial traction exerted on the airways by surrounding lung tissue.

(1) *High lung volumes* are associated with greater traction on airways and decreased airway resistance. Patients with increased airway resistance (e.g., asthma) "learn" to breathe at higher lung volumes to offset the high airway resistance associated with their disease.

(2) *Low lung volumes* are associated with less traction and increased airway resistance, even to the point of airway collapse.

c. **Viscosity or density of inspired gas**

- changes the resistance to airflow.
- During a deep-sea dive, both air density and resistance to airflow are increased.
- Breathing a low-density gas, such as helium, reduces the resistance to airflow.

F. Breathing cycle: description of pressures and airflow (Figure 4.6)

1. At rest (before inspiration begins)

a. Alveolar pressure equals atmospheric pressure.

- Because lung pressures are expressed relative to atmospheric pressure, **alveolar pressure is said to be zero.**

b. Intrapleural pressure is negative.

- At FRC, the opposing forces of the lungs trying to collapse and the chest wall trying to expand create a negative pressure in the intrapleural space between them.
- Intrapleural pressure can be measured by a **balloon catheter in the esophagus.**

c. Lung volume is the FRC.

2. During inspiration

a. The inspiratory muscles contract and cause the volume of the thorax to increase.

- As lung volume increases, alveolar pressure decreases to less than atmospheric pressure (i.e., becomes negative).
- The **pressure gradient** between the atmosphere and the alveoli now causes air to flow into the lungs; airflow will continue until the pressure gradient dissipates.

b. Intrapleural pressure becomes more negative.

- Because lung volume increases during inspiration, the elastic recoil strength of the lungs also increases. As a result, intrapleural pressure becomes even more negative than it was at rest.
- Changes in intrapleural pressure during inspiration are used to measure the **dynamic compliance** of the lungs.

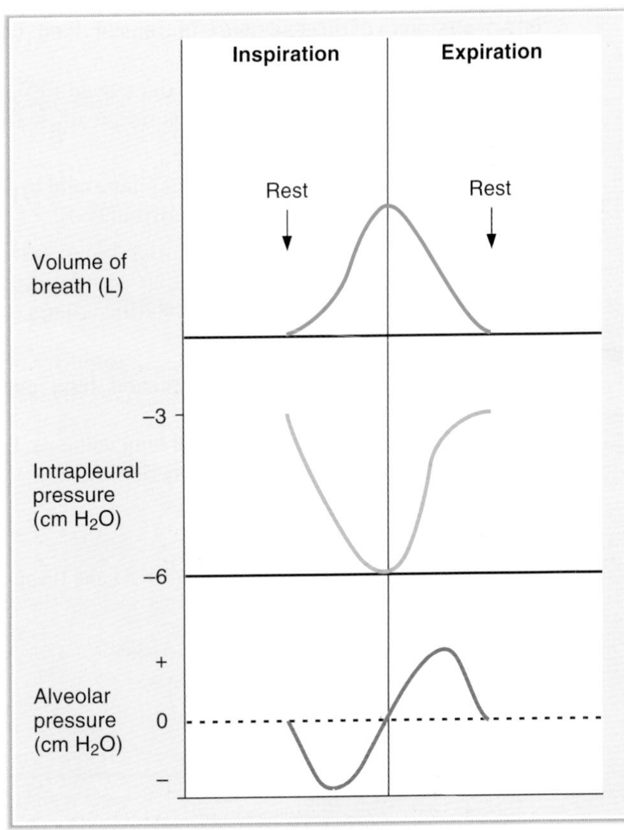

FIGURE 4.6. Volumes and pressures during the breathing cycle.

c. Lung volume increases by one V_T.

- At the peak of inspiration, lung volume is the FRC plus one V_T.

3. During expiration

a. Alveolar pressure becomes greater than atmospheric pressure.

- The alveolar pressure becomes greater (i.e., becomes positive) because alveolar gas is compressed by the elastic forces of the lung.
- Thus, alveolar pressure is now higher than atmospheric pressure, the pressure gradient is reversed, and air flows out of the lungs.

b. Intrapleural pressure returns to its resting value during a normal (passive) expiration.

- However, during a **forced expiration,** intrapleural pressure actually becomes positive. This positive intrapleural pressure compresses the airways and makes expiration more difficult.
- In **COPD,** in which airway resistance is increased, patients learn to expire slowly with **"pursed lips."** By creating a resistance at the mouth, airway pressure is increased and prevents the airway collapse that may occur with a forced expiration.

c. Lung volume returns to FRC.

G. Lung disease (Table 4.1)

1. Asthma

- is an **obstructive** disease in which expiration is impaired.
- is characterized by decreased FVC, decreased FEV_1, and **decreased FEV_1/FVC.**
- Air that should have been expired is not, leading to air trapping and **increased FRC.**

2. COPD

- is a combination of chronic bronchitis and emphysema.
- is an **obstructive** disease with **increased lung compliance** in which expiration is impaired.
- is characterized by decreased FVC, decreased FEV_1, and **decreased FEV_1/FVC.**
- Air that should have been expired is not, leading to air trapping, **increased FRC,** and a barrel-shaped chest.

a. "Pink puffers" (primarily emphysema) have **mild hypoxemia** and, because they maintain alveolar ventilation, **normocapnia** (normal P_{CO_2}).

b. "Blue bloaters" (primarily bronchitis) have **severe hypoxemia** with cyanosis and, because they do *not* maintain alveolar ventilation, **hypercapnia** (increased P_{CO_2}). They have right ventricular failure and systemic edema.

3. Fibrosis

- is a **restrictive** disease with **decreased lung compliance** in which inspiration is impaired.
- is characterized by a **decrease in all lung volumes.** Because FEV_1 is decreased less than is FVC, **FEV_1/FVC is increased** (or may be normal).

table 4.1 Characteristics of Lung Diseases

Disease	FEV_1	FVC	FEV_1/FVC	FRC
Asthma	↓↓	↓	↓	↑
COPD	↓↓	↓	↓	↑
Fibrosis	↓	↓↓	↑ (or normal)	↓

COPD = chronic obstructive pulmonary disease; FEV_1 = volume expired in first second of forced expiration; FRC = functional residual capacity; FVC = forced vital capacity.

III. GAS EXCHANGE

A. Dalton law of partial pressures

- can be expressed by the following equation:

$$\text{Partial pressure} = \text{Total pressure} \times \text{Fractional gas concentration}$$

1. **In dry inspired air,** the partial pressure of O_2 can be calculated as follows. Assume that total pressure is atmospheric and the fractional concentration of O_2 is 0.21.

$$\begin{aligned} P_{O_2} &= 760 \text{ mm Hg} \times 0.21 \\ &= 160 \text{ mm Hg} \end{aligned}$$

2. **In humidified tracheal air** at 37°C, the calculation is modified to correct for the partial pressure of H_2O, which is 47 mm Hg.

$$\begin{aligned} P_{Total} &= 760 \text{ mm Hg} - 47 \text{ mm Hg} \\ &= 713 \text{ mm Hg} \\ P_{O_2} &= 713 \text{ mm Hg} \times 0.21 \\ &= 150 \text{ mm Hg} \end{aligned}$$

B. Partial pressure of O_2 and CO_2 in inspired air, alveolar air, and blood (Table 4.2)

- Approximately 2% of the systemic cardiac output bypasses the pulmonary circulation (**"physiologic shunt"**). The resulting admixture of mixed venous blood with oxygenated arterial blood makes the Po_2 of arterial blood slightly lower than that of alveolar air.

C. Dissolved gases

- The amount of gas dissolved in a solution (such as blood) is proportional to its partial pressure. The units of concentration for a dissolved gas are mL gas/100 mL blood.
- The following calculation uses O_2 in arterial blood as an **example:**

$$\begin{aligned} \textbf{Dissolved } [\mathbf{O_2}] &= \mathbf{Po_2 \times} \textbf{ Solubility of } \mathbf{O_2} \textbf{ in blood} \\ &= \textbf{100 mm Hg} \times \textbf{0.003 mL } \mathbf{O_2}\textbf{/100 mL/mm Hg} \\ &= \textbf{0.3 mL } \mathbf{O_2}\textbf{/ 100 mL blood} \end{aligned}$$

where:
$[O_2]$ = O_2 concentration in blood
Po_2 = partial pressure of O_2 in blood
0.003 mL O_2/100 mL/mm Hg = solubility of O_2 in blood

table 4.2 Partial Pressures of O_2 and CO_2 (mm Hg)

Gas	Dry Inspired Air	Humidified Tracheal Air	Alveolar Air	Systemic Arterial Blood	Mixed Venous Blood
Po_2	160	150 Addition of H_2O decreases Po_2.	100 O_2 has diffused from alveolar air into pulmonary capillary blood, decreasing the Po_2 of alveolar air.	100* Blood has equilibrated with alveolar air (is "arterialized").	40 O_2 has diffused from arterial blood into tissues, decreasing the Po_2 of mixed venous blood.
Pco_2	0	0	40 CO_2 has diffused from pulmonary capillary blood into alveolar air, increasing the Pco_2 of alveolar air.	40 Blood has equilibrated with alveolar air.	46 CO_2 has diffused from the tissues into mixed venous blood, increasing the Pco_2 of mixed venous blood.

*Actually, slightly <100 mm Hg because of the "physiologic shunt."

D. Diffusion of gases such as O_2 and CO_2

- The diffusion rates of O_2 and CO_2 depend on the **partial pressure differences** across the membrane and the area available for diffusion.
- **For example,** the diffusion of O_2 from alveolar air into the pulmonary capillary depends on the partial pressure difference for O_2 between alveolar air and pulmonary capillary blood. Normally, capillary blood equilibrates with alveolar gas; when the partial pressures of O_2 become equal (see Table 4.2), there is no more net diffusion of O_2.
- Gas diffusion across the alveolar–pulmonary capillary barrier occurs according to **Fick law**:

$$\dot{V}_x = D_L \cdot \Delta P$$

where:

$\dot{V}_x$ = volume of gas transferred per minute (mL/min)
D_L = lung diffusing capacity (mL/min/mm Hg)
ΔP = partial pressure difference of gas (mm Hg)

- **D_L**, or lung diffusing capacity, is the equivalent of permeability of the alveolar–pulmonary capillary barrier and is proportional to diffusion coefficient of the gas and surface area and inversely proportional to thickness of the barrier. D_L is measured with carbon monoxide (i.e., $D_{L_{CO}}$).

1. **D_L increases** during exercise because there are more open capillaries and thus more surface area for diffusion.
2. **D_L decreases** in emphysema (because of decreased surface area) and in fibrosis and pulmonary edema (because of increased diffusion distance).

E. Perfusion-limited and diffusion-limited gas exchange (Table 4.3)

1. Perfusion-limited exchange

- is illustrated by **N_2O** and by **O_2 under normal conditions.**
- In perfusion-limited exchange, the gas **equilibrates** early along the length of the pulmonary capillary. The partial pressure of the gas in arterial blood becomes equal to the partial pressure in alveolar air.
- Thus, for a perfusion-limited process, diffusion of the gas can be increased only if blood flow increases.

2. Diffusion-limited exchange

- is illustrated by **CO** and by **O_2 during strenuous exercise.**
- is also illustrated in disease states. In **fibrosis,** the diffusion of O_2 is restricted because thickening of the alveolar membrane increases diffusion distance. In **emphysema,** the diffusion of O_2 is decreased because the surface area for diffusion of gases is decreased.
- In diffusion-limited exchange, the gas **does *not* equilibrate** by the time blood reaches the end of the pulmonary capillary. The partial pressure difference of the gas between alveolar air and pulmonary capillary blood is maintained. Diffusion continues as long as the partial pressure gradient is maintained.

table 4.3 Perfusion-Limited and Diffusion-Limited Gas Exchange

Perfusion-limited	Diffusion-limited
O_2 (normal conditions)	O_2 (emphysema, fibrosis, strenuous exercise)
CO_2	CO
N_2O	

IV. OXYGEN TRANSPORT

- O_2 is carried in blood in two forms: dissolved or bound to hemoglobin (most important).
- Hemoglobin, at its normal concentration, increases the O_2-carrying capacity of blood 70-fold.

A. Hemoglobin

1. Characteristics—globular protein of four subunits

- Each subunit contains a **heme moiety,** which is iron-containing porphyrin.
- The iron is in the ferrous state **(Fe^{2+})**, which binds O_2.
- Each subunit has a polypeptide chain. Two of the subunits have α chains and two of the subunits have β chains; thus, normal adult hemoglobin is called $\alpha_2\beta_2$.

2. Fetal hemoglobin [hemoglobin F (HbF)]

- In **fetal hemoglobin,** the **β chains are replaced by γ chains;** thus, fetal hemoglobin is called $\alpha_2\gamma_2$.
- The O_2 affinity of fetal hemoglobin is higher than the O_2 affinity of adult hemoglobin **(left shift)** because 2,3-diphosphoglycerate (DPG) binds less avidly to the γ chains of fetal hemoglobin than to the β chains of adult hemoglobin.
- Because the O_2 affinity of fetal hemoglobin is higher than the O_2 affinity of adult hemoglobin, O_2 movement from mother to fetus is facilitated (see IV C 2 b).

3. Methemoglobin

- Iron is in the Fe^{3+} state.
- does not bind O_2.

4. Hemoglobin S

- causes sickle cell disease.
- The α subunits are normal and the β subunits are abnormal, giving hemoglobin S the designation $\alpha_2^A\beta_2^S$.
- In the deoxygenated form, deoxyhemoglobin forms sickle-shaped rods that deform red blood cells (RBCs) and can cause occlusion of small blood vessels and pain.

5. O_2-binding capacity of hemoglobin

- is the maximum amount of O_2 that can be bound to hemoglobin.
- limits the amount of O_2 that can be carried in blood.
- is measured at 100% saturation.
- is expressed in units of mL O_2/g hemoglobin

6. O_2 content of blood

- is the total amount of O_2 carried in blood, including bound and dissolved O_2.
- depends on the hemoglobin concentration, the O_2-binding capacity of hemoglobin, the PO_2, and the P_{50} of hemoglobin.
- is given by the following equation:

$$\textbf{O}_2\ \textbf{content} = (\textbf{hemoglobin concentration} \times \textbf{O}_2\textbf{–binding capacity} \times \%\ \textbf{saturation}) + \textbf{Dissolved O}_2$$

where:

O_2 content = amount of O_2 in blood (mL O_2/100 mL blood)
Hemoglobin concentration = hemoglobin concentration (g/100 mL)
O_2–binding capacity = maximal amount of O_2 bound to hemoglobin at 100% saturation (mL O_2/g hemoglobin)
% saturation = % of heme groups bound to O_2 (%)
Dissolved O_2 = unbound O_2 in blood (mL O_2/100 mL blood)

B. Hemoglobin–O_2 dissociation curve (Figure 4.7)

1. Hemoglobin combines rapidly and reversibly with O_2 to form **oxyhemoglobin.**

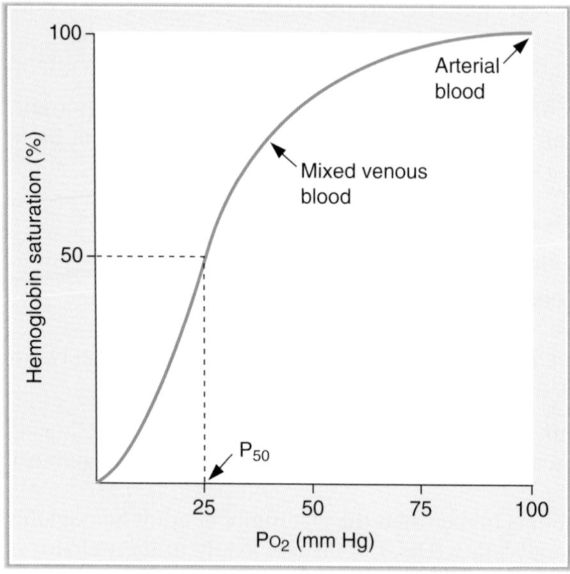

FIGURE 4.7. Hemoglobin–O_2 dissociation curve.

2. The hemoglobin–O_2 dissociation curve is a plot of percent saturation of hemoglobin as a function of Po_2.

   a. **At a Po_2 of 100 mm Hg** (e.g., arterial blood)

   - hemoglobin is 100% saturated; O_2 is bound to all four heme groups on all hemoglobin molecules.

   b. **At a Po_2 of 40 mm Hg** (e.g., mixed venous blood)

   - hemoglobin is 75% saturated, which means that, on average, three of the four heme groups on each hemoglobin molecule have O_2 bound.

   c. **At a Po_2 of 25 mm Hg**

   - hemoglobin is 50% saturated.
   - The Po_2 at 50% saturation is the **P_{50}**. Fifty percent saturation means that, on average, two of the four heme groups of each hemoglobin molecule have O_2 bound.

3. The **sigmoid shape** of the curve is the result of a change in the affinity of hemoglobin as each successive O_2 molecule binds to a heme site (called **positive cooperativity**).

   - Binding of the first O_2 molecule increases the affinity for the second O_2 molecule, and so forth.
   - The **affinity for the fourth O_2 molecule is the highest.**
   - This change in affinity facilitates the loading of O_2 in the lungs (flat portion of the curve) and the unloading of O_2 at the tissues (steep portion of the curve).

   a. **In the lungs**

   - Alveolar gas has a Po_2 of 100 mm Hg.
   - Pulmonary capillary blood is "arterialized" by the diffusion of O_2 from alveolar gas into blood so that the Po_2 of pulmonary capillary blood also becomes 100 mm Hg.
   - The very high affinity of hemoglobin for O_2 at a Po_2 of 100 mm Hg facilitates the diffusion process. By tightly binding O_2, the free O_2 concentration and O_2 partial pressure are kept low, thus maintaining the partial pressure gradient (that drives the diffusion of O_2).
   - The curve is almost **flat when the Po_2 is between 60 and 100 mm Hg.** Thus, humans can tolerate changes in atmospheric pressure (and Po_2) without compromising the O_2-carrying capacity of hemoglobin.

   b. **In the peripheral tissues**

   - O_2 diffuses from arterial blood to the cells.

- The gradient for O_2 diffusion is maintained because the cells consume O_2 for aerobic metabolism, keeping the tissue P_{O_2} low.
- The lower affinity of hemoglobin for O_2 in this steep portion of the curve facilitates the unloading of O_2 to the tissues.

4. Pulse oximetry

- measures % saturation of hemoglobin in arterial blood by dual-wavelength spectrophotometry.
- Pa_{O_2} can be estimated from the O_2–hemoglobin dissociation curve, using the measured % saturation.

C. Changes in the hemoglobin–O_2 dissociation curve (Figure 4.8)

1. Shifts to the right

- occur when the **affinity of hemoglobin for O_2 is decreased.**
- The **P_{50} is increased,** and unloading of O_2 from arterial blood to the tissues is facilitated.
- For any level of P_{O_2}, the percent saturation of hemoglobin, and thus the O_2 content of blood, is decreased.

a. Increases in P_{CO_2} or decreases in pH

- shift the curve to the right, decreasing the affinity of hemoglobin for O_2 and facilitating the unloading of O_2 in the tissues **(Bohr effect).**
- **For example,** during exercise, the tissues produce more CO_2, which decreases tissue pH and, through the Bohr effect, stimulates O_2 delivery to the exercising muscle.

b. Increases in temperature (e.g., during exercise)

- shift the curve to the right.
- The shift to the right decreases the affinity of hemoglobin for O_2 and facilitates the delivery of O_2 to the tissues during this period of high demand.

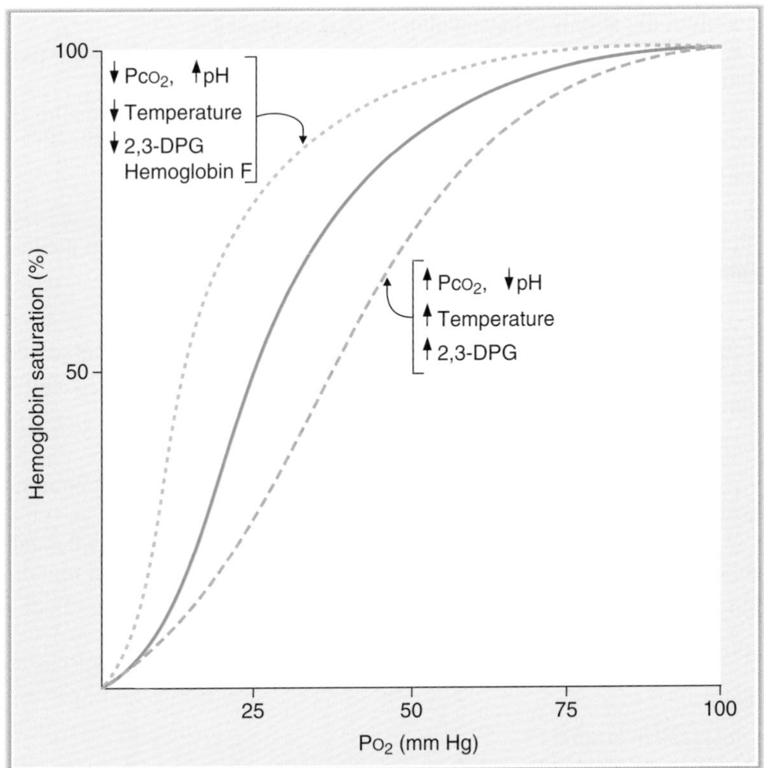

FIGURE 4.8. Changes in the hemoglobin–O_2 dissociation curve. Effects of P_{CO_2}, pH, temperature, 2,3-diphosphoglycerate (DPG), and fetal hemoglobin (hemoglobin F) on the hemoglobin–O_2 dissociation curve.

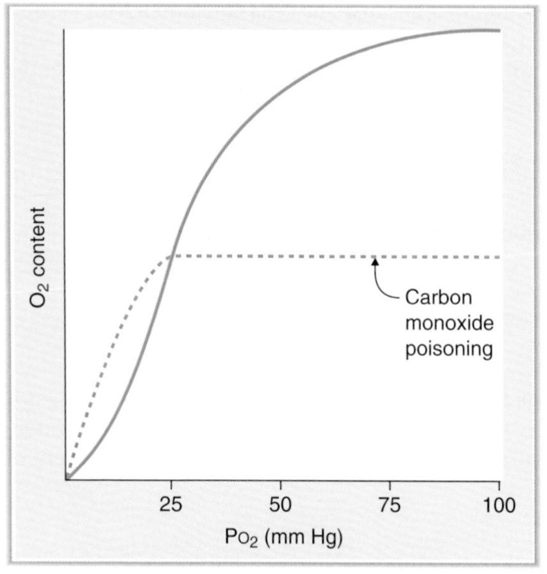

FIGURE 4.9. Effect of carbon monoxide on the hemoglobin–O_2 dissociation curve.

c. **Increases in 2,3-DPG concentration**

- shift the curve to the right by binding to the β chains of deoxyhemoglobin and decreasing the affinity of hemoglobin for O_2.
- The **adaptation to chronic hypoxemia** (e.g., living at high altitude) includes increased synthesis of 2,3-DPG, which binds to hemoglobin and facilitates unloading of O_2 in the tissues.

2. **Shifts to the left**

- occur when the **affinity of hemoglobin for O_2 is increased.**
- The **P_{50} is decreased,** and unloading of O_2 from arterial blood into the tissues is more difficult.
- For any level of PO_2, the percent saturation of hemoglobin, and thus the O_2 content of blood, is increased.

a. **Causes of a shift to the left**

- are the mirror image of those that cause a shift to the right.
- include **decreased PCO_2, increased pH, decreased temperature, and decreased 2,3-DPG concentration.**

b. **HbF**

- does not bind 2,3-DPG as strongly as does adult hemoglobin. Decreased binding of 2,3-DPG results in increased affinity of HbF for O_2, decreased P_{50}, and a **shift of the curve to the left.**

c. **Carbon monoxide (CO) poisoning** (Figure 4.9)

- CO competes for O_2-binding sites on hemoglobin. The affinity of hemoglobin for CO is 200 times its affinity for O_2.
- CO occupies O_2-binding sites on hemoglobin, thus **decreasing the O_2 content of blood.**
- In addition, binding of CO to hemoglobin increases the affinity of remaining sites for O_2, causing a **shift of the curve to the left.**

D. Causes of hypoxemia and hypoxia (Tables 4.4 and 4.5)

1. **Hypoxemia**

- is a **decrease in arterial PO_2.**
- is caused by decreased PA_{O_2}, diffusion defect, V/Q defects, and right-to-left shunts.

table 4.4 Causes of Hypoxemia

Cause	Pa_{O_2}	A–a Gradient
High altitude (↓ PB → PA_{O_2})	Decreased	Normal
Hypoventilation (↓ PA_{O_2})	Decreased	Normal
Diffusion defect (e.g., fibrosis)	Decreased	Increased
V/Q defect	Decreased	Increased
Right-to-left shunt	Decreased	Increased

A–a gradient = difference in PO_2 between alveolar gas and arterial blood; PB = barometric pressure; PA_{O_2} alveolar PO_2; Pa_{O_2} = arterial PO_2; V/Q = ventilation/perfusion ratio.

- **A–a gradient** can be used to compare causes of hypoxemia and is described by the following equation:

$$\textbf{A – a gradient} = PA_{O_2} - Pa_{O_2}$$

where:
A–a gradient = difference between alveolar PO_2 and arterial PO_2
PA_{O_2} = alveolar PO_2 (calculated from the alveolar gas equation)
Pa_{O_2} = arterial PO_2 (measured in arterial blood)

- Alveolar PO_2 is calculated from the **alveolar gas equation** as follows:

$$PA_{O_2} = PI_{O_2} - PA_{CO_2}/R$$

where:
PA_{O_2} = alveolar PO_2
PI_{O_2} = inspired PO_2
PA_{CO_2} = alveolar PCO_2 = arterial PCO_2 (measured in arterial blood)
R = respiratory exchange ratio (CO_2 production/O_2 consumption = 0.8)

- The A–a gradient is used to determine whether O_2 has equilibrated between alveolar gas and arterial blood.
  - **a.** The **normal A–a gradient is between 0 and 10 mm Hg.** Since O_2 normally equilibrates between alveolar gas and arterial blood, PA_{O_2} is approximately equal to Pa_{O_2}.
  - **b.** The **A–a gradient is increased** (>10 mm Hg) if O_2 does not equilibrate between alveolar gas and arterial blood (e.g., diffusion defect, V/Q defect, and right-to-left shunt) and PA_{O_2} is greater than Pa_{O_2}.

2. Hypoxia

- is **decreased O_2 delivery to the tissues.**

table 4.5 Causes of Hypoxia

Cause	Mechanisms
↓ cardiac output	↓ blood flow
Hypoxemia	↓ Pa_{O_2} causes ↓ % saturation of hemoglobin.
Anemia	↓ hemoglobin concentration causes ↓ O_2 content of blood.
Carbon monoxide poisoning	↓ O_2 content of blood and left shift of hemoglobin–O_2 dissociation curve
Cyanide poisoning	↓ O_2 utilization by tissues

Pa_{O_2} = arterial PO_2.

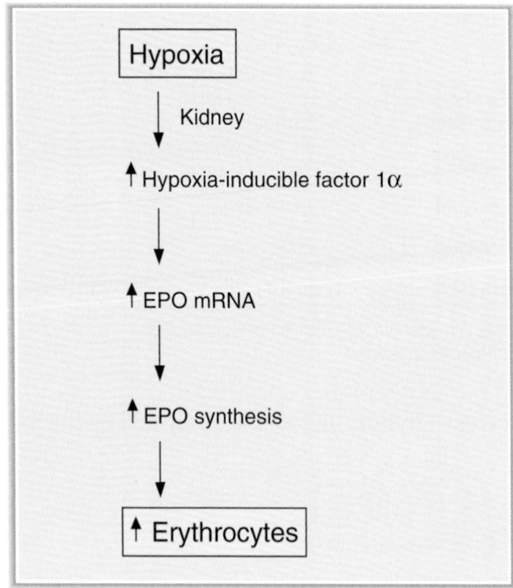

FIGURE 4.10. Hypoxia induces synthesis of erythropoietin. EPO = erythropoietin; mRNA = messenger RNA.

- is caused by decreased blood flow, hypoxemia, decreased hemoglobin concentration, CO poisoning, and cyanide poisoning.
- **O_2 delivery** is described by the following equation:

$$O_2 \text{ delivery} = \text{Cardiac output} \times O_2 \text{ content of blood}$$

- O_2 content of blood depends on hemoglobin concentration, O_2-binding capacity of hemoglobin, and Po_2 (which determines % saturation of hemoglobin by O_2).

E. Erythropoietin (EPO)

- is a growth factor that is synthesized in the kidneys in **response to hypoxia** (Figure 4.10).
- Decreased O_2 delivery to the kidneys causes increased production of **hypoxia-inducible factor 1α.**
- Hypoxia-inducible factor 1α directs synthesis of mRNA for EPO, which ultimately promotes development of mature red blood cells.

V. CO_2 TRANSPORT

A. Forms of CO_2 in blood

- CO_2 is produced in the tissues and carried to the lungs in the venous blood in three forms:

1. Dissolved CO_2 (small amount), which is free in solution
2. Carbaminohemoglobin (small amount), which is CO_2 bound to hemoglobin
3. HCO_3^- (from hydration of CO_2 in the RBCs), which is the **major form (90%)**

B. Transport of CO_2 as HCO_3^- (Figure 4.11)

1. **CO_2 is generated in the tissues** and diffuses freely into the venous plasma and then into the RBCs.
2. In the RBCs, CO_2 combines with H_2O to form H_2CO_3, a reaction that is catalyzed by **carbonic anhydrase.** H_2CO_3 dissociates into H^+ and HCO_3^-.
3. **HCO_3^-** leaves the RBCs in exchange for Cl^- **(chloride shift)** and is transported to the lungs in the plasma. HCO_3^- is the major form in which CO_2 is transported to the lungs.

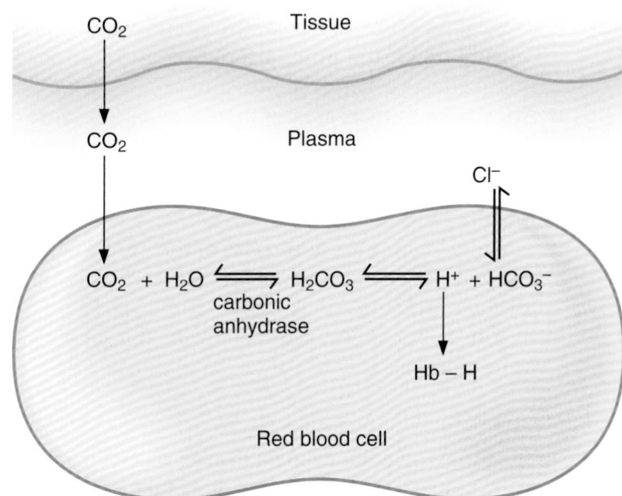

FIGURE 4.11. Transport of CO_2 from the tissues to the lungs in venous blood. H^+ is buffered by hemoglobin (Hb–H).

4. H^+ is buffered inside the RBCs by **deoxyhemoglobin.** Because deoxyhemoglobin is a better buffer for H^+ than is oxyhemoglobin, it is advantageous that hemoglobin has been deoxygenated by the time blood reaches the venous end of the capillaries (i.e., the site where CO_2 is being added).
5. **In the lungs**, all of the above reactions occur in reverse. HCO_3^- enters the RBCs in exchange for Cl^-. HCO_3^- recombines with H^+ to form H_2CO_3, which decomposes into CO_2 and H_2O. Thus, CO_2, originally generated in the tissues, is expired.

VI. PULMONARY CIRCULATION

A. Pressures and cardiac output in the pulmonary circulation

1. Pressures

- are **much lower** in the pulmonary circulation than in the systemic circulation.
- For example, pulmonary arterial pressure is 15 mm Hg (compared with aortic pressure of 100 mm Hg).

2. Resistance

- is **also much lower** in the pulmonary circulation than in the systemic circulation.

3. Cardiac output of the right ventricle

- is **pulmonary blood flow.**
- is equal to cardiac output of the left ventricle.
- Although pressures in the pulmonary circulation are low, they are sufficient to pump the cardiac output because resistance of the pulmonary circulation is proportionately low.

B. Distribution of pulmonary blood flow

- When a person is **supine,** blood flow is nearly uniform throughout the lung.
- When a person is **standing,** blood flow is unevenly distributed because of the **effect of gravity.** Blood flow is **lowest at the apex** of the lung (zone 1) and **highest at the base** of the lung (zone 3).

1. Zone 1—blood flow is lowest.

- Alveolar pressure > arterial pressure > venous pressure.
- The high alveolar pressure may compress the capillaries and reduce blood flow in zone 1. This situation can occur if arterial blood pressure is decreased as a result of **hemorrhage** or if alveolar pressure is increased because of **positive pressure ventilation.**

2. Zone 2—blood flow is medium.

- Arterial pressure > alveolar pressure > venous pressure.
- Moving down the lung, arterial pressure progressively increases because of gravitational effects on arterial pressure.
- Arterial pressure is greater than alveolar pressure in zone 2, and blood flow is driven by the difference between arterial pressure and alveolar pressure.

3. Zone 3—blood flow is highest.

- Arterial pressure > venous pressure > alveolar pressure.
- Moving down toward the base of the lung, arterial pressure is highest because of gravitational effects, and venous pressure finally increases to the point where it exceeds alveolar pressure.
- In zone 3, blood flow is driven by the difference between arterial and venous pressures, as in most vascular beds.

C. Regulation of pulmonary blood flow: hypoxic vasoconstriction

- In the lungs, alveolar **hypoxia causes vasoconstriction.**
- This response is the **opposite of that in other organs,** where hypoxia causes vasodilation.
- Physiologically, this effect is important because local vasoconstriction redirects blood away from poorly ventilated, hypoxic regions of the lung and toward well-ventilated regions.
- **Fetal pulmonary vascular resistance** is very high because of generalized hypoxic vasoconstriction; as a result, blood flow through the fetal lungs is low. With the first breath, the alveoli of the neonate are oxygenated, pulmonary vascular resistance decreases, and pulmonary blood flow increases and becomes equal to cardiac output (as occurs in the adult).

D. Shunts

1. Right-to-left shunts

- normally occur to a small extent because 2% of the cardiac output bypasses the lungs. May be as great as 50% of cardiac output in certain congenital abnormalities.
- are seen in **tetralogy of Fallot.**
- always result in a **decrease in arterial Po_2** because of the admixture of venous blood with arterial blood.
- The magnitude of a right-to-left shunt can be estimated by having the patient breathe 100% O_2 and measuring the degree of dilution of oxygenated arterial blood by nonoxygenated shunted (venous) blood.

2. Left-to-right shunts

- are **more common** than are right-to-left shunts because pressures are higher on the left side of the heart.
- are usually caused by congenital abnormalities (e.g., **patent ductus arteriosus**) or traumatic injury.
- ***do not* result in a decrease in arterial Po_2.** Instead, Po_2 will be elevated on the right side of the heart because there has been admixture of arterial blood with venous blood.

VII. V/Q DEFECTS

A. V/Q ratio

- is the **ratio of alveolar ventilation (V) to pulmonary blood flow (Q).** Ventilation and perfusion (blood flow) matching is important to achieve the ideal exchange of O_2 and CO_2.
- If the breathing rate, tidal volume, and cardiac output are normal, the V/Q ratio is approximately 0.8. This V/Q ratio results in an arterial Po_2 of 100 mm Hg and an arterial Pco_2 of 40 mm Hg.

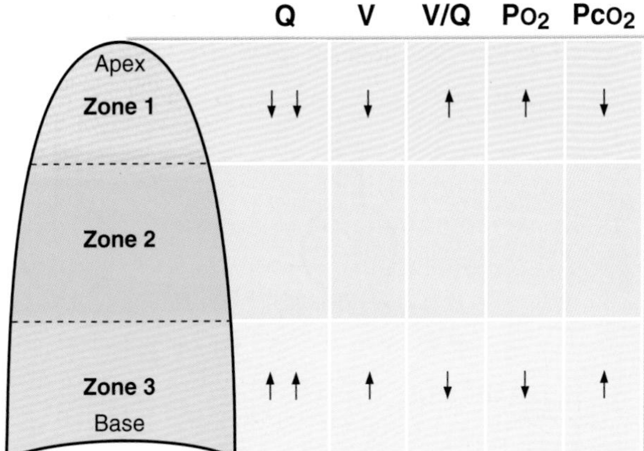

FIGURE 4.12. Regional variations in the lung of perfusion (blood flow [Q]), ventilation (V), V/Q, Po_2, and Pco_2.

B. V/Q ratios in different parts of the lung (Figure 4.12 and Table 4.6)

- Both ventilation and blood flow (perfusion) are nonuniformly distributed in the normal upright lung.

1. **Blood flow, or perfusion,** is lowest at the apex and highest at the base because of gravitational effects on arterial pressure.
2. **Ventilation** is lower at the apex and higher at the base because of gravitational effects in the upright lung. Importantly, however, the regional differences for ventilation are not as great as for perfusion.
3. **Therefore, the V/Q ratio is higher at the apex of the lung and lower at the base of the lung.**
4. **As a result of the regional differences in V/Q ratio,** there are corresponding differences in the efficiency of gas exchange and in the resulting pulmonary capillary Po_2 and Pco_2. Regional differences for Po_2 are greater than those for Pco_2.
   a. **At the apex** (higher V/Q), Po_2 is highest and Pco_2 is lowest because gas exchange is more efficient.
   b. **At the base** (lower V/Q), Po_2 is lowest and Pco_2 is highest because gas exchange is less efficient.

C. Changes in V/Q ratio (Figure 4.13)

1. V/Q ratio in airway obstruction

- If the airways are completely blocked (e.g., by a piece of steak caught in the trachea), then ventilation is zero. If blood flow is normal, then **V/Q is zero,** which is called a **right-to-left shunt.**
- There is ***no* gas exchange** in a lung that is perfused but not ventilated. The **Po_2 and Pco_2 of pulmonary capillary blood** (and, therefore, of systemic arterial blood) **will approach their values in mixed venous blood.**
- There is an **increased A–a gradient.**

table 4.6 V/Q Characteristics of Different Areas of the Lung

Area of Lung	Blood Flow	Ventilation	V/Q	Regional Arterial Po_2	Regional Arterial Pco_2
Apex	Lowest	Lower	Higher	Highest	Lower
Base	Highest	Higher	Lower	Lowest	Higher

V/Q = ventilation/perfusion ratio.

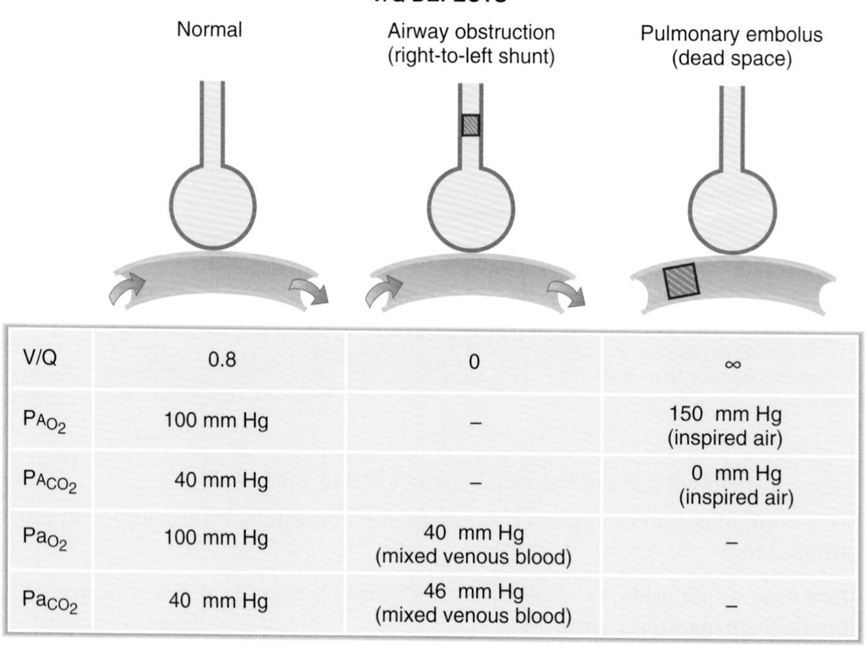

V/Q	0.8	0	∞
$P_{A_{O_2}}$	100 mm Hg	–	150 mm Hg (inspired air)
$P_{A_{CO_2}}$	40 mm Hg	–	0 mm Hg (inspired air)
$P_{a_{O_2}}$	100 mm Hg	40 mm Hg (mixed venous blood)	–
$P_{a_{CO_2}}$	40 mm Hg	46 mm Hg (mixed venous blood)	–

FIGURE 4.13. Effect of ventilation/perfusion (V/Q) defects on gas exchange. With airway obstruction, the composition of systemic arterial blood approaches that of mixed venous blood. With pulmonary embolus, the composition of alveolar gas approaches that of inspired air. $P_{A_{O2}}$ = alveolar P_{O_2}; $P_{A_{CO_2}}$ = alveolar P_{CO_2}; $P_{a_{O2}}$ = arterial P_{O_2}; $P_{a_{CO_2}}$ = arterial P_{CO_2}.

2. V/Q ratio in pulmonary embolism

- If blood flow to a lung is completely blocked (e.g., by an embolism occluding a pulmonary artery), then blood flow to that lung is zero. If ventilation is normal, then **V/Q is infinite,** which is called **dead space.**
- There is ***no* gas exchange** in a lung that is ventilated but not perfused. The **P_{O_2} and P_{CO_2} of alveolar gas** will approach their values in **inspired air.**

VIII. CONTROL OF BREATHING

- Sensory information (P_{CO_2}, lung stretch, irritants, muscle spindles, tendons, and joints) is coordinated in the **brainstem.**
- The output of the brainstem controls the respiratory muscles and the breathing cycle.

A. Central control of breathing (brainstem and cerebral cortex)

1. Medullary respiratory center

- is located in the **reticular formation.**

a. Dorsal respiratory group

- is primarily responsible for **inspiration** and generates the basic rhythm for breathing.
- **Input** to the dorsal respiratory group comes from the vagus and glossopharyngeal nerves. The vagus nerve relays information from peripheral chemoreceptors and mechanoreceptors in the lung. The glossopharyngeal nerve relays information from peripheral chemoreceptors.
- **Output** from the dorsal respiratory group travels, via the phrenic nerve, to the diaphragm.

b. Ventral respiratory group

- is primarily responsible for expiration.
- is not active during normal, quiet breathing, when expiration is passive.
- is activated, for example, during exercise, when expiration becomes an active process.

2. Apneustic center

- is located in the **lower pons.**
- **stimulates inspiration,** producing a deep and prolonged inspiratory gasp (apneusis).

3. Pneumotaxic center

- is located in the **upper pons.**
- **inhibits inspiration** and, therefore, regulates inspiratory volume and respiratory rate.

4. Cerebral cortex

- Breathing can be under voluntary control; therefore, a person can voluntarily hyperventilate or hypoventilate.
- Hypoventilation (breath-holding) is limited by the resulting increase in P_{CO_2} and decrease in P_{O_2}. A previous period of hyperventilation extends the period of breath-holding.

B. Chemoreceptors for CO_2, H^+, and O_2 (Table 4.7)

1. Central chemoreceptors in the medulla

- are sensitive to the **pH** of the cerebrospinal fluid (CSF). Decreases in the pH of the CSF produce increases in breathing rate (hyperventilation).
- H^+ does not cross the blood–brain barrier as well as CO_2 does.

a. CO_2 diffuses from arterial blood into the CSF because CO_2 is lipid soluble and readily crosses the blood–brain barrier.

b. In the CSF, CO_2 combines with H_2O to produce H^+ and HCO_3^-. The resulting **H^+ acts directly on the central chemoreceptors.**

c. Thus, increases in P_{CO_2} and $[H^+]$ stimulate breathing, and decreases in P_{CO_2} and $[H^+]$ inhibit breathing.

d. The resulting hyperventilation or hypoventilation then returns the arterial P_{CO_2} toward normal.

2. Peripheral chemoreceptors in the carotid and aortic bodies

- The carotid bodies are located at the bifurcation of the common carotid arteries.
- The aortic bodies are located above and below the aortic arch.

a. Decreases in arterial P_{O_2}

- stimulate the peripheral chemoreceptors and **increase breathing rate.**
- **P_{O_2} must decrease to low levels (<60 mm Hg)** before breathing is stimulated. When P_{O_2} is less than 60 mm Hg, breathing rate is exquisitely sensitive to P_{O_2}.

b. Increases in arterial P_{CO_2}

- stimulate peripheral chemoreceptors and **increase breathing rate.**
- potentiate the stimulation of breathing caused by hypoxemia.

table 4.7 Comparison of Central and Peripheral Chemoreceptors

Type of Chemoreceptor	Location	Stimuli That Increase Breathing Rate
Central	Medulla	↓ pH ↑ P_{CO_2}
Peripheral	Carotid and aortic bodies	↓ P_{O_2} (if <60 mm Hg) ↑ P_{CO_2} ↓ pH

- The response of the peripheral chemoreceptors to CO_2 is less important than is the response of the central chemoreceptors to CO_2 (or H^+).

c. Increases in arterial [H^+]

- stimulate the carotid body peripheral chemoreceptors directly, independent of changes in P_{CO_2}.
- In metabolic acidosis, breathing rate is increased (hyperventilation) because arterial [H^+] is increased and pH is decreased.

C. Other types of receptors for control of breathing

1. Lung stretch receptors

- are located in the smooth muscle of the airways.
- When these receptors are stimulated by distention of the lungs, they produce a reflex decrease in breathing frequency **(Hering–Breuer reflex).**

2. Irritant receptors

- are located between the airway epithelial cells.
- are stimulated by noxious substances (e.g., dust and pollen).

3. J (juxtacapillary) receptors

- are located in the alveolar walls, close to the capillaries.
- Engorgement of the pulmonary capillaries, such as that may occur with **left heart failure,** stimulates the J receptors, which then causes rapid, shallow breathing.

4. Joint and muscle receptors

- are activated during movement of the limbs.
- are involved in the early stimulation of breathing during exercise.

IX. INTEGRATED RESPONSES OF THE RESPIRATORY SYSTEM

A. Exercise (Table 4.8)

1. During exercise, there is an **increase in ventilatory rate** that matches the increase in O_2 consumption and CO_2 production by the body. The stimulus for the increased ventilation rate is not completely understood. However, joint and muscle receptors are activated and cause an increase in breathing rate at the beginning of exercise.

table 4.8 Summary of Respiratory Responses to Exercise

Parameter	Response
O_2 consumption	↑
CO_2 production	↑
Ventilation rate	↑ (matches O_2 consumption/CO_2 production)
Arterial P_{O_2} and P_{CO_2}	No change
Arterial pH	No change in moderate exercise
	↓ in strenuous exercise (lactic acidosis)
Venous P_{CO_2}	↑
Pulmonary blood flow (cardiac output)	↑
V/Q ratios	More evenly distributed in lung

V/Q = ventilation/perfusion ratio.

2. The ***mean* values for arterial Po_2 and Pco_2 do not change** during exercise.
   - **Arterial pH** does not change during moderate exercise, although it may decrease during strenuous exercise because of **lactic acidosis.**
3. On the other hand, **venous Pco_2 increases** during exercise because the excess CO_2 produced by the exercising muscle is carried to the lungs in venous blood.
4. **Pulmonary blood flow increases** because cardiac output increases during exercise. As a result, more pulmonary capillaries are perfused, and more gas exchange occurs. The **distribution of V/Q ratios** throughout the lung is **more even** during exercise than when at rest, and there is a resulting **decrease in the physiologic dead space.**

B. Adaptation to high altitude (Table 4.9)

1. **Alveolar Po_2 is decreased** at high altitude because the barometric pressure is decreased. As a result, arterial Po_2 is also decreased **(hypoxemia).**
2. **Hypoxemia stimulates the peripheral chemoreceptors** and increases the ventilation rate **(hyperventilation).** This hyperventilation produces **respiratory alkalosis,** which can be treated by administering **acetazolamide.**
3. **Hypoxemia also stimulates renal production of EPO,** which increases the production of RBCs. As a result, there is **increased hemoglobin concentration** and increased O_2 content of blood.
4. **2,3-DPG concentrations are increased,** shifting the hemoglobin–O_2 dissociation curve to the right. There is a resulting decrease in affinity of hemoglobin for O_2 that facilitates unloading of O_2 in the tissues.
5. **Pulmonary vasoconstriction** is a result of hypoxic vasoconstriction. Consequently, there is an increase in pulmonary arterial pressure, increased work of the right side of the heart against the higher resistance, and hypertrophy of the right ventricle.

table 4.9 Summary of Adaptation to High Altitude

Parameter	Response
Alveolar Po_2	↓ (resulting from ↓ barometric pressure)
Arterial Po_2	↓ (hypoxemia)
Ventilation rate	↑ (hyperventilation due to hypoxemia)
Arterial pH	↑ (respiratory alkalosis)
Hemoglobin concentration	↑ (↑ EPO)
2,3-DPG concentration	↑
Hemoglobin–O_2 curve	Shift to right; ↓ affinity; ↑ P_{50}
Pulmonary vascular resistance	↑ (hypoxic vasoconstriction)

DPG = diphosphoglycerate; EPO = erythropoietin.

Review Test

1. Which of the following lung volumes or capacities can be measured by spirometry?

(A) Functional residual capacity (FRC)
(B) Physiologic dead space
(C) Residual volume (RV)
(D) Total lung capacity (TLC)
(E) Vital capacity (VC)

2. An infant born prematurely in gestational week 25 has neonatal respiratory distress syndrome. Which of the following would be expected in this infant?

(A) Arterial P_{O_2} of 100 mm Hg
(B) Collapse of the small alveoli
(C) Increased lung compliance
(D) Normal breathing rate
(E) Lecithin:sphingomyelin ratio of greater than 2:1 in amniotic fluid

3. In which vascular bed does hypoxia cause vasoconstriction?

(A) Coronary
(B) Pulmonary
(C) Cerebral
(D) Muscle
(E) Skin

QUESTIONS 4 AND 5

A 12-year-old boy has a severe asthmatic attack with wheezing. He experiences rapid breathing and becomes cyanotic. His arterial P_{O_2} is 60 mm Hg and his P_{CO_2} is 30 mm Hg.

4. Which of the following statements about this patient is most likely to be true?

(A) Forced expiratory volume$_1$/forced vital capacity (FEV_1/FVC) is increased
(B) Ventilation/perfusion (V/Q) ratio is increased in the affected areas of his lungs
(C) His arterial P_{CO_2} is higher than normal because of inadequate gas exchange
(D) His arterial P_{CO_2} is lower than normal because hypoxemia is causing him to hyperventilate
(E) His residual volume (RV) is decreased

5. To treat this patient, the physician should administer

(A) an α_1-adrenergic antagonist
(B) a β_1-adrenergic antagonist
(C) a β_2-adrenergic agonist
(D) a muscarinic agonist
(E) a nicotinic agonist

6. Which of the following is true during inspiration?

(A) Intrapleural pressure is positive
(B) The volume in the lungs is less than the functional residual capacity (FRC)
(C) Alveolar pressure equals atmospheric pressure
(D) Alveolar pressure is higher than atmospheric pressure
(E) Intrapleural pressure is more negative than it is during expiration

7. Which volume remains in the lungs after a tidal volume (V_T) is expired?

(A) Tidal volume (V_T)
(B) Vital capacity (VC)
(C) Expiratory reserve volume (ERV)
(D) Residual volume (RV)
(E) Functional residual capacity (FRC)
(F) Inspiratory capacity
(G) Total lung capacity

8. A 35-year-old man has a vital capacity (VC) of 5 L, a tidal volume (V_T) of 0.5 L, an inspiratory capacity of 3.5 L, and a functional residual capacity (FRC) of 2.5 L. What is his expiratory reserve volume (ERV)?

(A) 4.5 L
(B) 3.9 L
(C) 3.6 L
(D) 3.0 L
(E) 2.5 L
(F) 2.0 L
(G) 1.5 L

9. When a person is standing, blood flow in the lungs is

(A) equal at the apex and the base
(B) highest at the apex owing to the effects of gravity on arterial pressure

(C) highest at the base because that is where the difference between arterial and venous pressure is greatest
(D) lowest at the base because that is where alveolar pressure is greater than arterial pressure

10. Which of the following is illustrated in the graph showing volume versus pressure in the lung–chest wall system?

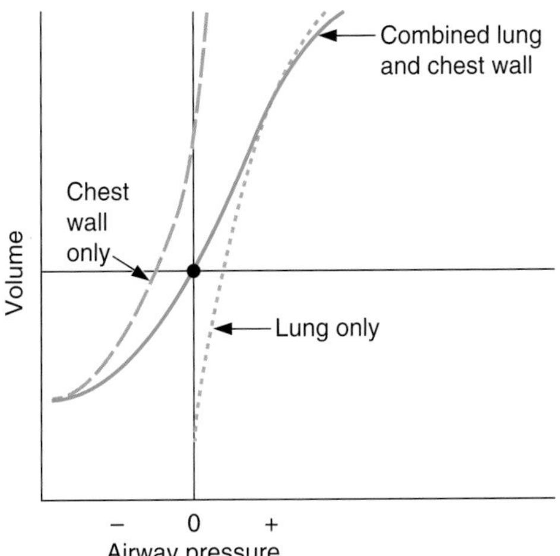

(A) The slope of each of the curves is resistance
(B) The compliance of the lungs alone is less than the compliance of the lungs plus chest wall
(C) The compliance of the chest wall alone is less than the compliance of the lungs plus chest wall
(D) When airway pressure is zero (atmospheric), the volume of the combined system is the functional residual capacity (FRC)
(E) When airway pressure is zero (atmospheric), intrapleural pressure is zero

11. Which of the following is the site of highest airway resistance?

(A) Trachea
(B) Largest bronchi
(C) Medium-sized bronchi
(D) Smallest bronchi
(E) Alveoli

12. A 49-year-old man has a pulmonary embolism that completely blocks blood flow to his left lung. As a result, which of the following will occur?

(A) Ventilation/perfusion (V/Q) ratio in the left lung will be zero
(B) Systemic arterial Po_2 will be elevated
(C) V/Q ratio in the left lung will be lower than in the right lung
(D) Alveolar Po_2 in the left lung will be approximately equal to the Po_2 in inspired air
(E) Alveolar Po_2 in the right lung will be approximately equal to the Po_2 in venous blood

QUESTIONS 13 AND 14

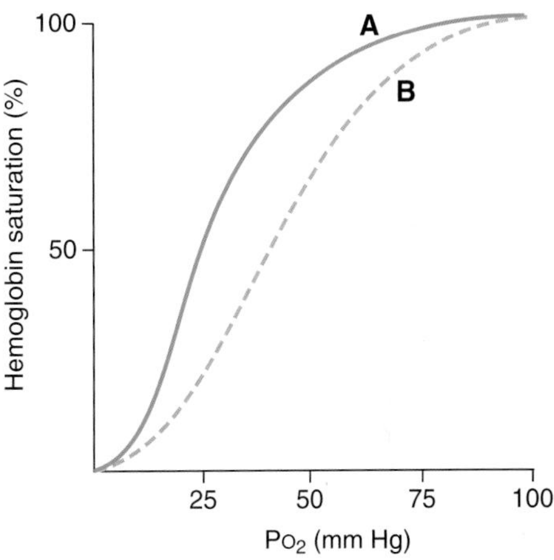

13. In the hemoglobin–O_2 dissociation curves shown above, the shift from curve A to curve B could be caused by

(A) increased pH
(B) decreased 2,3-diphosphoglycerate (DPG) concentration
(C) strenuous exercise
(D) fetal hemoglobin (HbF)
(E) carbon monoxide (CO) poisoning

14. The shift from curve A to curve B is associated with

(A) increased P_{50}
(B) increased affinity of hemoglobin for O_2
(C) impaired ability to unload O_2 in the tissues
(D) increased O_2-carrying capacity of hemoglobin
(E) decreased O_2-carrying capacity of hemoglobin

15. Which volume remains in the lungs after a maximal expiration?

(A) Tidal volume (V_T)
(B) Vital capacity (VC)
(C) Expiratory reserve volume (ERV)
(D) Residual volume (RV)
(E) Functional residual capacity (FRC)
(F) Inspiratory capacity
(G) Total lung capacity

16. Compared with the systemic circulation, the pulmonary circulation has a

(A) higher blood flow
(B) lower resistance
(C) higher arterial pressure
(D) higher capillary pressure
(E) higher cardiac output

17. A healthy 65-year-old man with a tidal volume (V_T) of 0.45 L has a breathing frequency of 16 breaths/min. His arterial P_{CO_2} is 41 mm Hg, and the P_{CO_2} of his expired air is 35 mm Hg. What is his alveolar ventilation?

(A) 0.066 L/min
(B) 0.38 L/min
(C) 5.0 L/min
(D) 6.14 L/min
(E) 8.25 L/min

18. Compared with the apex of the lung, the base of the lung has

(A) a higher pulmonary capillary P_{O_2}
(B) a higher pulmonary capillary P_{CO_2}
(C) a higher ventilation/perfusion (V/Q) ratio
(D) the same V/Q ratio

19. Hypoxemia produces hyperventilation by a direct effect on the

(A) phrenic nerve
(B) J receptors
(C) lung stretch receptors
(D) medullary chemoreceptors
(E) carotid and aortic body chemoreceptors

20. Which of the following changes occurs during strenuous exercise?

(A) Ventilation rate and O_2 consumption increase to the same extent
(B) Systemic arterial P_{O_2} decreases to about 70 mm Hg
(C) Systemic arterial P_{CO_2} increases to about 60 mm Hg
(D) Systemic venous P_{CO_2} decreases to about 20 mm Hg
(E) Pulmonary blood flow decreases at the expense of systemic blood flow

21. If an area of the lung is not ventilated because of bronchial obstruction, the pulmonary capillary blood serving that area will have a P_{O_2} that is

(A) equal to atmospheric P_{O_2}
(B) equal to mixed venous P_{O_2}
(C) equal to normal systemic arterial P_{O_2}
(D) higher than inspired P_{O_2}
(E) lower than mixed venous P_{O_2}

22. In the transport of CO_2 from the tissues to the lungs, which of the following occurs in venous blood?

(A) Conversion of CO_2 and H_2O to H^+ and HCO_3^- in the red blood cells (RBCs)
(B) Buffering of H^+ by oxyhemoglobin
(C) Shifting of HCO_3^- into the RBCs from plasma in exchange for Cl^-
(D) Binding of HCO_3^- to hemoglobin
(E) Alkalinization of the RBCs

23. Which of the following causes of hypoxia is characterized by a decreased arterial P_{O_2} and an increased A–a gradient?

(A) Hypoventilation
(B) Right-to-left cardiac shunt
(C) Anemia
(D) Carbon monoxide poisoning
(E) Ascent to high altitude

24. A 42-year-old woman with severe pulmonary fibrosis is evaluated by her physician and has the following arterial blood gases: pH = 7.48, Pa_{O_2} = 55 mm Hg, and Pa_{CO_2} = 32 mm Hg. Which statement best explains the observed value of Pa_{CO_2}?

(A) The increased pH stimulates breathing via peripheral chemoreceptors
(B) The increased pH stimulates breathing via central chemoreceptors
(C) The decreased Pa_{O_2} inhibits breathing via peripheral chemoreceptors
(D) The decreased Pa_{O_2} stimulates breathing via peripheral chemoreceptors
(E) The decreased Pa_{O_2} stimulates breathing via central chemoreceptors

25. A 38-year-old woman moves with her family from New York City (sea level) to Leadville, Colorado (10,200 feet above sea level). Which of the following will occur as a result of residing at high altitude?

(A) Hypoventilation
(B) Arterial P_{O_2} greater than 100 mm Hg
(C) Decreased 2,3-diphosphoglycerate (DPG) concentration
(D) Shift to the right of the hemoglobin–O_2 dissociation curve
(E) Pulmonary vasodilation
(F) Hypertrophy of the left ventricle
(G) Respiratory acidosis

26. The pH of venous blood is only slightly more acidic than the pH of arterial blood because

(A) CO_2 is a weak base
(B) there is no carbonic anhydrase in venous blood
(C) the H^+ generated from CO_2 and H_2O is buffered by HCO_3^- in venous blood
(D) the H^+ generated from CO_2 and H_2O is buffered by deoxyhemoglobin in venous blood
(E) oxyhemoglobin is a better buffer for H^+ than is deoxyhemoglobin

27. In a maximal expiration, the total volume expired is

(A) tidal volume (VT)
(B) vital capacity (VC)
(C) expiratory reserve volume (ERV)
(D) residual volume (RV)
(E) functional residual capacity (FRC)
(F) inspiratory capacity
(G) total lung capacity

28. A person with a ventilation/perfusion (V/Q) defect has hypoxemia and is treated with supplemental O_2. The supplemental O_2 will be *most* helpful if the person's predominant V/Q defect is

(A) dead space
(B) shunt
(C) high V/Q
(D) low V/Q
(E) V/Q = 0
(F) V/Q = ∞

29. Which person would be expected to have the largest A–a gradient?

(A) Person with pulmonary fibrosis
(B) Person who is hypoventilating due to morphine overdose
(C) Person at 12,000 feet above sea level
(D) Person with normal lungs breathing 50% O_2
(E) Person with normal lungs breathing 100% O_2

30. Which of the following sets of data would have the highest rate of O_2 transfer in the lungs?

	P_{IO_2} (mm Hg)	P_{VO_2} (mm Hg)	Surface Area (Relative)	Thickness (Relative)
(A)	150	40	1	1
(B)	150	40	2	2
(C)	300	40	1	2
(D)	150	80	1	1
(E)	190	80	2	2

31. A 48-year-old woman at sea level breaths a gas mixture containing 21% O_2. She has the following arterial blood gas values:

$$Pa_{O_2} = 60 \text{ mm Hg}$$
$$Pa_{CO_2} = 45 \text{ mm Hg}$$

Her measured DL_{CO} is normal. Which of the following is the cause of her hypoxemia?

(A) The values demonstrate normal lung function
(B) Hypoventilation
(C) Fibrosis
(D) Carbon monoxide poisoning
(E) Right-to-left shunt

32. A 62-year-old man at sea level breaths a gas mixture containing 21% O_2. He has the following arterial blood gas values:

$$Pa_{O_2} = 60 \text{ mm Hg}$$
$$Pa_{CO_2} = 70 \text{ mm Hg}$$

Which of the following is the cause of his hypoxemia?

(A) Hypoventilation
(B) Fibrosis
(C) V/Q defect
(D) Right-to-left shunt
(E) Anemia

Answers and Explanations

1. **The answer is E** [I A 4, 5, B 2, 3, 5]. Residual volume (RV) cannot be measured by spirometry. Therefore, any lung volume or capacity that includes the RV cannot be measured by spirometry. Measurements that include RV are functional residual capacity (FRC) and total lung capacity (TLC). Vital capacity (VC) does not include RV and is, therefore, measurable by spirometry. Physiologic dead space is not measurable by spirometry and requires sampling of arterial P_{CO_2} and expired CO_2.
2. **The answer is B** [II D 2]. Neonatal respiratory distress syndrome is caused by lack of adequate surfactant in the immature lung. Surfactant appears between the 24th and the 35th gestational week. In the absence of surfactant, the surface tension of the small alveoli is too high. When the pressure on the small alveoli is too high ($P = 2T/r$), the small alveoli collapse into larger alveoli. There is decreased gas exchange with the larger, collapsed alveoli; and ventilation/perfusion (V/Q) mismatch, hypoxemia, and cyanosis occur. The lack of surfactant also decreases lung compliance, making it harder to inflate the lungs, increasing the work of breathing, and producing dyspnea (shortness of breath). Generally, lecithin:sphingomyelin ratios greater than 2:1 signify mature levels of surfactant.
3. **The answer is B** [VI C]. Pulmonary blood flow is controlled locally by the P_{O_2} of alveolar air. Hypoxia causes pulmonary vasoconstriction and thereby shunts blood away from unventilated areas of the lung, where it would be wasted. In the coronary circulation, hypoxemia causes vasodilation. The cerebral, muscle, and skin circulations are not controlled directly by P_{O_2}.
4. **The answer is D** [VIII B 2 a]. The patient's arterial P_{CO_2} is lower than the normal value of 40 mm Hg because hypoxemia has stimulated peripheral chemoreceptors to increase his breathing rate; hyperventilation causes the patient to blow off extra CO_2 and results in respiratory alkalosis. In an obstructive disease, such as asthma, both forced expiratory volume (FEV_1) and forced vital capacity (FVC) are decreased, with the larger decrease occurring in FEV_1. Therefore, the FEV_1/FVC ratio is decreased. Poor ventilation of the affected areas decreases the ventilation/perfusion (V/Q) ratio and causes hypoxemia. The patient's residual volume (RV) is increased because he is breathing at a higher lung volume to offset the increased resistance of his airways.
5. **The answer is C** [II E 3 a (2)]. A cause of airway obstruction in asthma is bronchiolar constriction. β_2-adrenergic stimulation (β_2-adrenergic agonists) produces relaxation of the bronchioles.
6. **The answer is E** [II F 2]. During inspiration, intrapleural pressure becomes *more negative* than it is at rest or during expiration (when it returns to its less negative resting value). During inspiration, air flows into the lungs when alveolar pressure becomes lower (due to contraction of the diaphragm) than atmospheric pressure; if alveolar pressure were not lower than atmospheric pressure, air would not flow inward. The volume in the lungs during inspiration is the functional residual capacity (FRC) *plus* one tidal volume (V_T).
7. **The answer is E** [I B 2]. During normal breathing, the volume inspired and then expired is a tidal volume (V_T). The volume remaining in the lungs after expiration of a V_T is the functional residual capacity (FRC).
8. **The answer is G** [I A 3; Figure 4.1]. Expiratory reserve volume (ERV) equals vital capacity (VC) minus inspiratory capacity [inspiratory capacity includes tidal volume (V_T) and inspiratory reserve volume (IRV)].
9. **The answer is C** [VI B]. The distribution of blood flow in the lungs is affected by gravitational effects on arterial hydrostatic pressure. Thus, blood flow is highest at the base,

where arterial hydrostatic pressure is greatest and the difference between arterial and venous pressure is also greatest. This pressure difference drives the blood flow.

10. **The answer is D** [II C 2; Figure 4.3]. By convention, when airway pressure is equal to atmospheric pressure, it is designated as zero pressure. Under these equilibrium conditions, there is no airflow because there is no pressure gradient between the atmosphere and the alveoli, and the volume in the lungs is the functional residual capacity (FRC). The slope of each curve is compliance, not resistance; the steeper the slope is, the greater the volume change is for a given pressure change, or the greater compliance is. The compliance of the lungs alone or the chest wall alone is greater than that of the combined lung–chest wall system (the slopes of the individual curves are steeper than the slope of the combined curve, which means higher compliance). When airway pressure is zero (equilibrium conditions), intrapleural pressure is negative because of the opposing tendencies of the chest wall to spring out and the lungs to collapse.

11. **The answer is C** [II E 4]. The medium-sized bronchi actually constitute the site of highest resistance along the bronchial tree. Although the small radii of the alveoli might predict that they would have the highest resistance, they do not because of their parallel arrangement. In fact, early changes in resistance in the small airways may be "silent" and go undetected because of their small overall contribution to resistance.

12. **The answer is D** [VII B 2]. Alveolar Po_2 in the left lung will equal the Po_2 in inspired air. Because there is no blood flow to the left lung, there can be no gas exchange between the alveolar air and the pulmonary capillary blood. Consequently, O_2 is not added to the capillary blood. The ventilation/perfusion (V/Q) ratio in the left lung will be infinite (not zero or lower than that in the normal right lung) because Q (the denominator) is zero. Systemic arterial Po_2 will, of course, be decreased because the left lung has no gas exchange. Alveolar Po_2 in the right lung is unaffected.

13. **The answer is C** [IV C 1; Figure 4.8]. Strenuous exercise increases the temperature and decreases the pH of skeletal muscle; both effects would cause the hemoglobin–O_2 dissociation curve to shift to the right, making it easier to unload O_2 in the tissues to meet the high demand of the exercising muscle. 2,3-Diphosphoglycerate (DPG) binds to the β chains of adult hemoglobin and reduces its affinity for O_2, shifting the curve to the right. In fetal hemoglobin, the β chains are replaced by γ chains, which do not bind 2,3-DPG, so the curve is shifted to the left. Because carbon monoxide (CO) increases the affinity of the remaining binding sites for O_2, the curve is shifted to the left.

14. **The answer is A** [IV C 1; Figure 4.8]. A shift to the right of the hemoglobin–O_2 dissociation curve represents decreased affinity of hemoglobin for O_2. At any given Po_2, the percent saturation is decreased, the P_{50} is increased (read the Po_2 from the graph at 50% hemoglobin saturation), and unloading of O_2 in the tissues is facilitated. The O_2-carrying capacity of hemoglobin is the mL of O_2 that can be bound to a gram of hemoglobin at 100% saturation and is unaffected by the shift from curve A to curve B.

15. **The answer is D** [I A 3]. During a forced maximal expiration, the volume expired is a tidal volume (V_T) plus the expiratory reserve volume (ERV). The volume remaining in the lungs is the residual volume (RV).

16. **The answer is B** [VI A]. Blood flow (or cardiac output) in the systemic and pulmonary circulations is nearly equal; pulmonary flow is slightly less than systemic flow because about 2% of the systemic cardiac output bypasses the lungs. The pulmonary circulation is characterized by both lower pressure and lower resistance than the systemic circulation, so flows through the two circulations are approximately equal (flow = pressure/resistance).

17. **The answer is D** [I A 5 b, 6 b]. Alveolar ventilation is the difference between tidal volume (V_T) and dead space multiplied by breathing frequency. V_T and breathing frequency are given, but dead space must be calculated. Dead space is V_T multiplied by the difference between arterial Pco_2 and expired Pco_2 divided by arterial Pco_2. Thus, dead space = $0.45 \times (41 - 35/41) = 0.066$ L. Alveolar ventilation is then calculated as $(0.45 \text{ L} - 0.066 \text{ L}) \times 16$ breaths/min = 6.14 L/min.

18. The answer is B [VII C; Figure 4.10; Table 4.5]. Ventilation and perfusion of the lung are not distributed uniformly. Both are lowest at the apex and highest at the base. However, the differences for ventilation are not as great as for perfusion, making the ventilation/perfusion (V/Q) ratios higher at the apex and lower at the base. As a result, gas exchange is more efficient at the apex and less efficient at the base. Therefore, blood leaving the apex will have a higher Po_2 and a lower Pco_2.

19. The answer is E [VIII B 2]. Hypoxemia stimulates breathing by a direct effect on the peripheral chemoreceptors in the carotid and aortic bodies. Central (medullary) chemoreceptors are stimulated by CO_2 (or H^+). The J receptors and lung stretch receptors are not chemoreceptors. The phrenic nerve innervates the diaphragm, and its activity is determined by the output of the brainstem breathing center.

20. The answer is A [IX A]. During exercise, the ventilation rate increases to match the increased O_2 consumption and CO_2 production. This matching is accomplished without a change in mean arterial Po_2 or Pco_2. Venous Pco_2 increases because extra CO_2 is being produced by the exercising muscle. Because this CO_2 will be blown off by the hyperventilating lungs, it does not increase the arterial Pco_2. Pulmonary blood flow (cardiac output) increases manifold during strenuous exercise.

21. The answer is B [VII B 1]. If an area of lung is not ventilated, there can be no gas exchange in that region. The pulmonary capillary blood serving that region will not equilibrate with alveolar Po_2 but will have a Po_2 equal to that of mixed venous blood.

22. The answer is A [V B; Figure 4.9]. CO_2 generated in the tissues is hydrated to form H^+ and HCO_3^- in red blood cells (RBCs). H^+ is buffered inside the RBCs by deoxyhemoglobin, which *acidifies* the RBCs. HCO_3^- leaves the RBCs in exchange for Cl^- and is carried to the lungs in the plasma. A small amount of CO_2 (not HCO_3^-) binds directly to hemoglobin (carbaminohemoglobin).

23. The answer is B [IV A 4; IV D; Table 4.4; Table 4.5]. Hypoxia is defined as decreased O_2 delivery to the tissues. It occurs as a result of decreased blood flow or decreased O_2 content of the blood. Decreased O_2 content of the blood is caused by decreased hemoglobin concentration (anemia), decreased O_2-binding capacity of hemoglobin (carbon monoxide poisoning), or decreased arterial Po_2 (hypoxemia). Hypoventilation, right-to-left cardiac shunt, and ascent to high altitude all cause hypoxia by decreasing arterial Po_2. Of these, only right-to-left cardiac shunt is associated with an increased A–a gradient, reflecting a lack of O_2 equilibration between alveolar gas and systemic arterial blood. In right-to-left shunt, a portion of the right heart output, or pulmonary blood flow, is not oxygenated in the lungs and thereby "dilutes" the Po_2 of the normally oxygenated blood. With hypoventilation and ascent to high altitude, both alveolar and arterial Po_2 are decreased, but the A–a gradient is normal.

24. The answer is D [VIII B; Table 4.7]. The patient's arterial blood gases show increased pH, decreased Pa_{O_2}, and decreased Pa_{CO_2}. The decreased Pa_{O_2} causes hyperventilation (stimulates breathing) via the peripheral chemoreceptors, but not via the central chemoreceptors. The decreased Pa_{CO_2} results from hyperventilation (increased breathing) and causes increased pH, which *inhibits* breathing via the peripheral and central chemoreceptors.

25. The answer is D [IX B; Table 4.9]. At high altitudes, the Po_2 of alveolar air is decreased because barometric pressure is decreased. As a result, arterial Po_2 is decreased (<100 mm Hg), and hypoxemia occurs and causes hyperventilation by an effect on peripheral chemoreceptors. Hyperventilation leads to respiratory alkalosis. 2,3-Diphosphoglycerate (DPG) levels increase adaptively; 2,3-DPG binds to hemoglobin and causes the hemoglobin–O_2 dissociation curve to shift to the right to improve unloading of O_2 in the tissues. The pulmonary vasculature vasoconstricts in response to alveolar hypoxia, resulting in increased pulmonary arterial pressure and hypertrophy of the right ventricle (not the left ventricle).

26. The answer is D [V B]. In venous blood, CO_2 combines with H_2O and produces the weak acid H_2CO_3, catalyzed by carbonic anhydrase. The resulting H^+ is buffered by deoxyhemoglobin, which is such an effective buffer for H^+ (meaning that the pK is within 1.0 unit of the pH of blood) that the pH of venous blood is only slightly more acid than the pH of arterial blood. Oxyhemoglobin is a less effective buffer than is deoxyhemoglobin.

27. The answer is B [I B 3]. The volume expired in a forced maximal expiration is forced vital capacity, or vital capacity (VC).

28. The answer is D [VII]. Supplemental O_2 (breathing inspired air with a high Po_2) is most helpful in treating hypoxemia associated with a ventilation/perfusion (V/Q) defect if the predominant defect is low V/Q. Regions of low V/Q have the highest blood flow. Thus, breathing high Po_2 air will raise the Po_2 of a large volume of blood and have the greatest influence on the total blood flow leaving the lungs (which becomes systemic arterial blood). Dead space (i.e., $V/Q = \infty$) has no blood flow, so supplemental O_2 has no effect on these regions. Shunt (i.e., $V/Q = 0$) has no ventilation, so supplemental O_2 has no effect. Regions of high V/Q have little blood flow, thus raising the Po_2 of a small volume of blood will have little overall effect on systemic arterial blood.

29. The answer is A [IV D]. Increased A–a gradient signifies lack of O_2 equilibration between alveolar gas (A) and systemic arterial blood (a). In pulmonary fibrosis, there is thickening of the alveolar/pulmonary capillary barrier and increased diffusion distance for O_2, which results in lack of equilibration of O_2, hypoxemia, and increased A–a gradient. Hypoventilation and ascent to 12,000 feet also cause hypoxemia, because systemic arterial blood is equilibrated with a lower alveolar Po_2 (normal A–a gradient). Persons breathing 50% or 100% O_2 will have elevated alveolar Po_2, and their arterial Po_2 will equilibrate with this higher value (normal A–a gradient).

30. The answer is C [III D]. The diffusion of O_2 from alveolar gas to pulmonary capillary blood is proportional to the partial pressure difference for O_2 between inspired air and mixed venous blood entering the pulmonary capillaries, proportional to the surface area for diffusion and inversely proportional to diffusion distance or thickness of the barrier.

31. The answer is D [IV D]. Since the woman is hypoxemic at sea level and breathing a mixture containing a normal % of O_2, she cannot have normal lung function. Also, because she is hypoxemic, she does not have carbon monoxide poisoning (which would decrease O_2 content of blood but would not decrease Pa_{O_2}). The remaining choices of hypoventilation, fibrosis, and right-to-left shunt cause hypoxemia. Fibrosis can be eliminated because it causes a diffusion defect and decreased DL_{CO}. Hypoventilation can be eliminated because it would cause greatly increased Pa_{CO_2}. Right-to-left shunt as the cause of the woman's hypoxemia is further supported by calculating the A–a gradient as follows. PI_{O_2} = (760 mm Hg − 47 mm Hg) × 0.21 = 150 mm Hg. PA_{O_2} = 150 mm Hg − 45 mm Hg/0.8 = 94 mm Hg. A–a gradient = 94 mm Hg − 60 mm Hg = 34 mm Hg, which is increased and consistent with right-to-left shunt.

32. The answer is A [IV D]. Anemia (or decreased hemoglobin concentration) is eliminated because it causes decreased O_2 content of blood but does not cause hypoxemia. The remaining choices all cause hypoxemia. Calculating the A–a gradient distinguishes between these causes as follows: PI_{O_2} = (760 mm Hg − 47 mm Hg) × 0.21 = 150 mm Hg. PA_{O_2} = 150 mm Hg − 70 mm Hg/0.8 = 63 mm Hg. A–a gradient = 63 mm Hg − 60 mm Hg = 3 mm Hg, which is normal. Among the choices, the only cause of hypoxemia with a normal A–a gradient is hypoventilation, whereby PA_{O_2} is lowered by hypoventilation and Pa_{O_2} equilibrates with that lowered value; since PA_{O_2} and Pa_{O_2} are equilibrated (but lower than normal), they are essentially equal and A–a is close to zero, or normal. Fibrosis, V/Q defect, and right-to-left shunt all cause decreased Pa_{O_2} that is not equilibrated with PA_{O_2} and thus cause increased A–a gradient.

Chapter 5

Renal and Acid–Base Physiology

I. BODY FLUIDS

- Total body water (TBW) is approximately **60% of body weight.**
- Body water is inversely proportional to body fat.
- The percentage of TBW is **highest in newborns and adult males** and **lowest in adult females** and in adults with a large amount of adipose tissue.

A. Distribution of water (Figure 5.1 and Table 5.1)

1. Intracellular fluid (ICF)

- is **two-thirds of TBW.**
- The major cations of ICF are $\mathbf{K^+}$ and $\mathbf{Mg^{2+}}$.
- The major anions of ICF are **protein and organic phosphates** (adenosine triphosphate [ATP], adenosine diphosphate [ADP], and adenosine monophosphate [AMP]).

2. Extracellular fluid (ECF)

- **is one-third of TBW.**
- is composed of interstitial fluid and plasma. The major cation of ECF is $\mathbf{Na^+}$.
- The major anions of ECF are $\mathbf{Cl^-}$ and $\mathbf{HCO_3^-}$.

a. Plasma is one-fourth of the ECF. Thus, it is one-twelfth of TBW ($1/4 \times 1/3$).

- The major **plasma proteins** are albumin and globulins.

b. Interstitial fluid is three-fourths of the ECF. Thus, it is one-fourth of TBW ($3/4 \times 1/3$).

- The composition of interstitial fluid is the same as that of plasma except that it has **little protein.** Thus, interstitial fluid is an **ultrafiltrate of plasma.**

3. 60-40-20 rule

- TBW is **60%** of body weight.
- ICF is **40%** of body weight.
- ECF is **20%** of body weight.

B. Measuring the volumes of the fluid compartments (see Table 5.1)

1. Dilution method

a. A **known amount** of a substance is given whose volume of distribution is the body fluid compartment of interest.

- For example:

(1) *Tritiated water* is a marker for TBW that distributes wherever water is found.
(2) *Mannitol* is a marker for ECF because it is a large molecule that cannot cross cell membranes and is therefore excluded from the ICF.
(3) *Evans blue* is a marker for plasma volume because it is a dye that binds to serum albumin and is therefore confined to the plasma compartment.

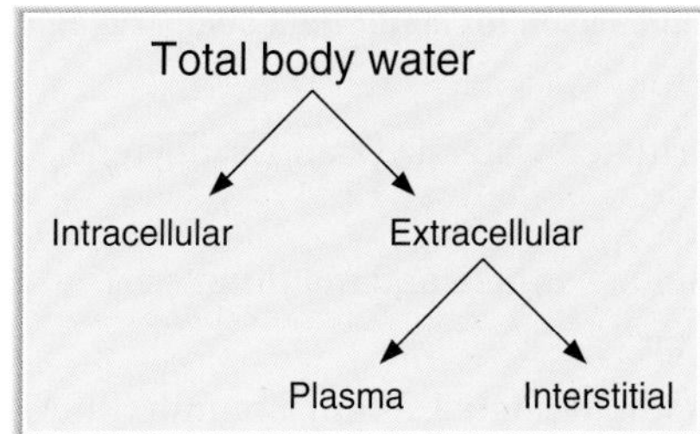

FIGURE 5.1. Body fluid compartments.

b. The substance is allowed to **equilibrate.**

c. The **concentration** of the substance is measured in plasma, and the **volume of distribution is calculated** as follows:

$$\textbf{Volume} = \frac{\textbf{Amount}}{\textbf{Concentration}}$$

where:

Volume = volume of distribution, or volume of the body fluid compartment (L)

Amount = amount of substance present (mg)

Concentration = concentration in plasma (mg/L)

d. Sample calculation

- A patient is injected with 500 mg of mannitol. After a 2-hour equilibration period, the concentration of mannitol in plasma is 3.2 mg/100 mL. During the equilibration period, 10% of the injected mannitol is excreted in urine. What is the patient's ECF volume?

$$\begin{aligned}\text{Volume} &= \frac{\text{Amount}}{\text{Concentration}} \\ &= \frac{\text{Amount injected} - \text{Amount excreted}}{\text{Concentration}} \\ &= \frac{500\text{ mg} - 50\text{ mg}}{3.2\text{ mg}/100\text{ mL}} \\ &= 14.1\text{ L}\end{aligned}$$

table 5.1 Body Water and Body Fluid Compartments

Body Fluid Compartment	Fraction of TBW*	Markers Used to Measure Volume	Major Cations	Major Anions
TBW	1.0	Tritiated H_2O D_2O Antipyrene		
ECF	1/3	Sulfate Inulin Mannitol	Na^+	Cl^- HCO_3^-
Plasma	1/12 (1/4 of ECF)	RISA Evans blue	Na^+	Cl^- HCO_3^- Plasma protein
Interstitial	1/4 (3/4 of ECF)	ECF–plasma volume (indirect)	Na^+	Cl^- HCO_3^-
ICF	2/3	TBW–ECF (indirect)	K^+	Organic phosphates Protein

*Total body water (TBW) is approximately 60% of total body weight, or 42 L in a 70-kg man. ECF = extracellular fluid; ICF = intracellular fluid; RISA = radioiodinated serum albumin.

2. Substances used for major fluid compartments (see Table 5.1)

a. TBW

- Tritiated water, D_2O, and antipyrene

b. ECF

- Sulfate, inulin, and mannitol

c. Plasma

- Radioiodinated serum albumin (RISA) and Evans blue

d. Interstitial

- Measured indirectly (ECF volume–plasma volume)

e. ICF

- Measured indirectly (TBW–ECF volume)

C. Shifts of water between compartments

1. Basic principles

a. Osmolarity is concentration of solute particles.

b. Plasma osmolarity (P_{osm}) is estimated as:

$$P_{osm} = 2 \times Na^+ + Glucose/18 + BUN/2.8$$

where:

P_{osm} = plasma osmolarity (mOsm/L)
Na^+ = plasma Na^+ concentration (mEq/L)
Glucose = plasma glucose concentration (mg/dL)
BUN = blood urea nitrogen concentration (mg/dL)

c. At steady state, **ECF osmolarity and ICF osmolarity are equal.**

d. To achieve this equality, **water shifts** between the ECF and ICF compartments.

e. It is assumed that solutes such as NaCl and mannitol do not cross cell membranes and are confined to ECF.

2. Examples of shifts of water between compartments (Figure 5.2 and Table 5.2)

a. Infusion of isotonic NaCl—addition of isotonic fluid

- is also called **isosmotic volume expansion.**

(1) *ECF volume increases*, but **no change occurs in the osmolarity** of ECF or ICF. Because osmolarity is unchanged, water does not shift between the ECF and ICF compartments.

(2) *Plasma protein concentration and hematocrit decrease* because the addition of fluid to the ECF dilutes the protein and red blood cells (RBCs). Because ECF osmolarity is unchanged, the RBCs will not shrink or swell.

(3) *Arterial blood pressure increases* because ECF volume increases.

b. Diarrhea—loss of isotonic fluid

- is also called **isosmotic volume contraction.**

(1) *ECF volume decreases*, but **no change occurs in the osmolarity** of ECF or ICF. Because osmolarity is unchanged, water does not shift between the ECF and ICF compartments.

(2) *Plasma protein concentration and hematocrit increase* because the loss of ECF concentrates the protein and RBCs. Because ECF osmolarity is unchanged, the RBCs will not shrink or swell.

(3) *Arterial blood pressure* decreases because ECF volume decreases.

c. Excessive NaCl intake—addition of NaCl

- is also called **hyperosmotic volume expansion.**

(1) *The osmolarity of ECF increases* because osmoles (NaCl) have been added to the ECF.

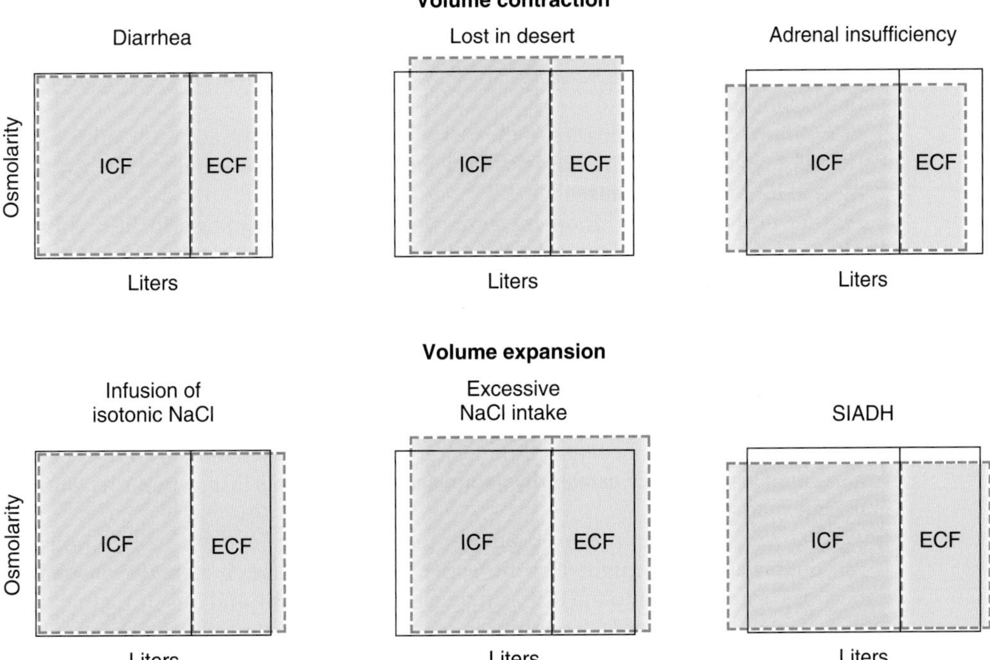

FIGURE 5.2. Shifts of water between body fluid compartments. Volume and osmolarity of normal extracellular fluid (ECF) and intracellular fluid (ICF) are indicated by the *solid lines*. Changes in volume and osmolarity in response to various situations are indicated by the *dashed lines*. SIADH = syndrome of inappropriate antidiuretic hormone.

(2) *Water shifts from ICF to ECF.* As a result of this shift, **ICF osmolarity increases** until it equals that of ECF.

(3) As a result of the shift of water out of the cells, **ECF volume increases** (volume expansion) and **ICF volume decreases.**

(4) Plasma protein concentration and hematocrit decrease because of the increase in ECF volume.

table 5.2 Changes in Volume and Osmolarity of Body Fluids

Type	Key Examples	ECF Volume	ICF Volume	ECF Osmolarity	Hct and Serum [Na⁺]
Isosmotic volume expansion	Isotonic NaCl infusion	↑	No change	No change	↓ Hct –[Na⁺]
Isosmotic volume contraction	Diarrhea	↓	No change	No change	↑ Hct –[Na⁺]
Hyperosmotic volume expansion	High NaCl intake	↑	↓	↑	↓ Hct ↑ [Na⁺]
Hyperosmotic volume contraction	Sweating Fever Diabetes insipidus	↓	↓	↑	–Hct ↑ [Na⁺]
Hyposmotic volume expansion	SIADH	↑	↑	↓	–Hct ↓ [Na⁺]
Hyposmotic volume contraction	Adrenal insufficiency	↓	↑	↓	↑ Hct ↓ [Na⁺]

– = no change; ECF = extracellular fluid; Hct = hematocrit; ICF = intracellular fluid; SIADH = syndrome of inappropriate antidiuretic hormone.

d. Sweating in a desert—loss of water

- is also called **hyperosmotic volume contraction.**

(1) *The osmolarity of ECF increases* because sweat is hyposmotic (relatively more water than salt is lost).

(2) *ECF volume decreases* because of the loss of volume in the sweat. Water shifts out of ICF; as a result of the shift, **ICF osmolarity increases** until it is equal to ECF osmolarity, and **ICF volume decreases.**

(3) *Plasma protein concentration increases* because of the decrease in ECF volume. Although **hematocrit** might also be expected to increase, it remains **unchanged** because water shifts out of the RBCs, decreasing their volume and offsetting the concentrating effect of the decreased ECF volume.

e. Syndrome of inappropriate antidiuretic hormone (SIADH)—gain of water

- is also called **hyposmotic volume expansion.**

(1) *The osmolarity of ECF decreases* because excess water is retained.

(2) *ECF volume increases* because of the water retention. Water shifts into the cells; as a result of this shift, **ICF osmolarity decreases** until it equals ECF osmolarity, and **ICF volume increases.**

(3) *Plasma protein concentration decreases* because of the increase in ECF volume. Although **hematocrit** might also be expected to decrease, it remains **unchanged** because water shifts into the RBCs, increasing their volume and offsetting the diluting effect of the gain of ECF volume.

f. Adrenocortical insufficiency—loss of NaCl

- is also called **hyposmotic volume contraction.**

(1) *The osmolarity of ECF decreases.* As a result of the lack of aldosterone in adrenocortical insufficiency, there is decreased NaCl reabsorption, and the kidneys excrete more NaCl than water.

(2) *ECF volume decreases.* Water shifts into the cells; as a result of this shift, **ICF osmolarity decreases** until it equals ECF osmolarity, and **ICF volume increases.**

(3) *Plasma protein concentration increases* because of the decrease in ECF volume. **Hematocrit increases** because of the decreased ECF volume and because the RBCs swell as a result of water entry.

(4) *Arterial blood pressure decreases* because of the decrease in ECF volume.

II. RENAL CLEARANCE, RENAL BLOOD FLOW (RBF), AND GLOMERULAR FILTRATION RATE (GFR)

A. Clearance equation

- gives the volume of plasma cleared of a substance per unit time.
- The units of clearance are **mL/min** or **mL/24 h.**

$$C = \frac{UV}{P}$$

where:

C = clearance (mL/min or mL/24 h)
U = urine concentration (mg/mL)
V = urine volume/time (mL/min)
P = plasma concentration (mg/mL)

- **Example:** If the plasma [Na^+] is 140 mEq/L, the urine [Na^+] is 700 mEq/L, and the urine flow rate is 1 mL/min, what is the clearance of Na^+?

$$C_{Na^+} = \frac{[U]_{Na^+} \times V}{[P]_{Na^+}} = \frac{700\ \text{mEq/L} \times 1\ \text{mL/min}}{140\ \text{mEq/L}} = 5\ \text{mL/min}$$

B. RBF

- is **25% of the cardiac output.**
- is directly proportional to the pressure difference between the renal artery and the renal vein and is inversely proportional to the resistance of the renal vasculature.
- **Vasoconstriction** of renal arterioles, which leads to a decrease in RBF, is produced by activation of the **sympathetic nervous system** and angiotensin II. At low concentrations, **angiotensin II** preferentially constricts efferent arterioles, thereby "protecting" (increasing) the GFR. **Angiotensin-converting enzyme (ACE) inhibitors** dilate efferent arterioles and produce a decrease in GFR; these drugs reduce hyperfiltration and the occurrence of diabetic nephropathy in diabetes mellitus.
- **Vasodilation** of renal arterioles, which leads to an increase in RBF, is produced by **prostaglandins E_2 and I_2,** bradykinin, nitric oxide, and **dopamine.**
- **Atrial natriuretic peptide (ANP)** causes vaso*dilation* of afferent arterioles and, to a lesser extent, vaso*constriction* of efferent arterioles; overall, ANP increases RBF and GFR.

1. Autoregulation of RBF

- is accomplished by **changing renal vascular resistance.** If arterial pressure changes, a proportional change occurs in renal vascular resistance to maintain a constant RBF.
- RBF remains constant over the range of arterial pressures from 80 to 200 mm Hg **(autoregulation).**
- The mechanisms for autoregulation include:

a. Myogenic mechanism, in which the renal afferent arterioles contract in response to stretch. Thus, increased renal arterial pressure stretches the arterioles, which contract and increase resistance to maintain constant blood flow.

b. Tubuloglomerular feedback, in which increased renal arterial pressure leads to increased delivery of fluid to the **macula densa.** The macula densa senses the increased load and causes constriction of the nearby afferent arteriole, increasing resistance to maintain constant blood flow.

- High-protein diet increases GFR by increasing Na^+ and Cl^- reabsorption, decreasing Na^+ and Cl^- delivery to the macula densa, and thereby increasing GFR via tubuloglomerular feedback.

2. Measurement of renal plasma flow (RPF)—clearance of para-aminohippuric acid (PAH)

- PAH is **filtered and secreted** by the renal tubules.
- Clearance of PAH is used to measure RPF.
- Clearance of PAH measures **effective RPF** and underestimates true RPF by 10%. (Clearance of PAH does not measure renal plasma flow to regions of the kidney that do not filter and secrete PAH, such as adipose tissue.)

$$\mathbf{RPF = C_{PAH} = \frac{[U]_{PAH}\, V}{[P]_{PAH}}}$$

where:

RPF = renal plasma flow (mL/min or mL/24 h)
C_{PAH} = clearance of PAH (mL/min or mL/24 h)
$[U]_{PAH}$ = urine concentration of PAH (mg/mL)
V = urine flow rate (mL/min or mL/24 h)
$[P]_{PAH}$ = plasma concentration of PAH (mg/mL)

3. Measurement of RBF

$$\mathbf{RBF} = \frac{\mathbf{RPF}}{\mathbf{1\text{-}Hematocrit}}$$

- Note that the denominator in this equation, 1 – hematocrit, is the fraction of blood volume occupied by plasma.

C. GFR

1. Measurement of GFR—clearance of inulin

- Inulin is **filtered, but not reabsorbed or secreted** by the renal tubules.
- The clearance of inulin is used to measure GFR, as shown in the following equation:

$$\mathbf{GFR} = \frac{[\mathbf{U}]_{\mathbf{inulin}}\,\mathbf{V}}{[\mathbf{P}]_{\mathbf{inulin}}}$$

where:

GFR = glomerular filtration rate (mL/min or mL/24 h)
$[U]_{inulin}$ = urine concentration of inulin (mg/mL)
V = urine flow rate (mL/min or mL/24 h)
$[P]_{inulin}$ = plasma concentration of inulin (mg/mL)

- **Example of calculation of GFR:** Inulin is infused in a patient to achieve a steady-state plasma concentration of 1 mg/mL. A urine sample collected during 1 hour has a volume of 60 mL and an inulin concentration of 120 mg/mL. What is the patient's GFR?

$$\begin{aligned} \text{GFR} &= \frac{[U]_{inulin}\,V}{[P]_{inulin}} \\ &= \frac{120\ \text{mg/mL} \times 60\ \text{mL/h}}{1\ \text{mg/mL}} \\ &= \frac{120\ \text{mg/mL} \times 1\ \text{mL/min}}{1\ \text{mg/mL}} \\ &= 120\ \text{mL/min} \end{aligned}$$

2. Estimates of GFR with blood urea nitrogen (BUN) and serum [creatinine]

- Both BUN and serum [creatinine] increase when GFR decreases.
- In **prerenal azotemia** (hypovolemia), BUN increases more than serum creatinine (because hypovolemia increases urea reabsorption in proximal tubule) and there is an **increased BUN/creatinine ratio** (>20:1).
- **GFR decreases with age**, although serum [creatinine] remains constant because of decreased muscle mass.

3. Filtration fraction

- is the fraction of RPF filtered across the glomerular capillaries, as shown in the following equation:

$$\textbf{Filtration fraction} = \frac{\mathbf{GFR}}{\mathbf{RPF}}$$

- is **normally about 0.20.** Thus, 20% of the RPF is filtered. The remaining 80% leaves the glomerular capillaries by the efferent arterioles and becomes the peritubular capillary circulation.
- **Increases in the filtration fraction** produce increases in the protein concentration of peritubular capillary blood, which leads to increased reabsorption in the proximal tubule.
- **Decreases in the filtration fraction** produce decreases in the protein concentration of peritubular capillary blood and decreased reabsorption in the proximal tubule.

4. Determining GFR–Starling forces (Figure 5.3)

- The driving force for glomerular filtration is the **net ultrafiltration pressure** across the glomerular capillaries.

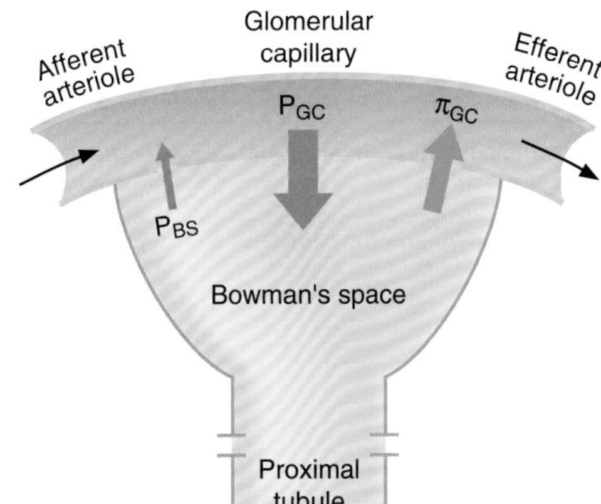

FIGURE 5.3. Starling forces across the glomerular capillaries. *Heavy arrows* indicate the driving forces across the glomerular capillary wall. P_{BS} = hydrostatic pressure in Bowman space; P_{GC} = hydrostatic pressure in the glomerular capillary; π_{GC} = colloid osmotic pressure in the glomerular capillary.

- **Filtration is always favored** in glomerular capillaries because the net ultrafiltration pressure always favors the movement of fluid out of the capillary.
- GFR can be expressed by the **Starling equation:**

$$GFR = K_f[(P_{GC} - P_{BS}) - (\pi_{GC} - \pi_{BS})]$$

a. GFR is filtration across the glomerular capillaries.

b. K_f is the filtration coefficient of the glomerular capillaries.

- The glomerular barrier consists of the capillary endothelium, basement membrane, and filtration slits of the podocytes.
- Normally, **anionic glycoproteins line the filtration barrier** and restrict the filtration of plasma proteins, which are also negatively charged.
- In **glomerular disease**, the anionic charges on the barrier may be removed, resulting in proteinuria.

c. P_{GC} is glomerular capillary hydrostatic pressure, which is constant along the length of the capillary.

- It is **increased by dilation of the afferent arteriole or constriction of the efferent arteriole.** Increases in π_{GC} cause increases in net ultrafiltration pressure and GFR.

d. P_{BS} is Bowman space hydrostatic pressure and is analogous to P_i in systemic capillaries.

- It is **increased by constriction of the ureters.** Increases in P_{BS} cause decreases in net ultrafiltration pressure and GFR.

e. π_{GC} is glomerular capillary oncotic pressure. It normally **increases along the length of the glomerular capillary** because filtration of water increases the protein concentration of glomerular capillary blood.

- It is **increased by increases in protein concentration.** Increases in π_{GC} cause decreases in net ultrafiltration pressure and GFR.

f. π_{BS} is Bowman space oncotic pressure. It is usually **zero**, and therefore ignored, because only a small amount of protein is normally filtered.

5. Sample calculation of ultrafiltration pressure with the Starling equation

- At the afferent arteriolar end of a glomerular capillary, P_{GC} is 45 mm Hg, P_{BS} is 10 mm Hg, and π_{GC} is 27 mm Hg. What are the value and direction of the net ultrafiltration pressure?

$$\text{Net pressure} = (P_{GC} - P_{BS}) - \pi_{GC}$$
$$\text{Net pressure} = (45\text{ mm Hg} - 10\text{ mm Hg}) - 27\text{ mm Hg}$$
$$= +8\text{ mm Hg (favoring filtration)}$$

6. Changes in Starling forces—effect on GFR and filtration fraction (Table 5.3)

table 5.3 Effect of Changes in Starling Forces on GFR, RPF, and Fraction Filtration

	Effect on GFR	Effect on RPF	Effect on Filtration Fraction
Constriction of afferent arteriole (e.g., sympathetic)	↓ (caused by ↓ P_{GC})	↓	No change
Constriction of efferent arteriole (e.g., angiotensin II)	↑ (caused by ↑ P_{GC})	↓	↑ (↑ GFR/↓ RPF)
Increased plasma (protein)	↓ (caused by ↑ π_{GC})	No change	↓ (↓ GFR/unchanged RPF)
Ureteral stone	↓ (caused by ↑ P_{BS})	No change	↓ (↓ GFR/unchanged RPF)

GFR = glomerular filtration rate; RPF = renal plasma flow.

III. REABSORPTION AND SECRETION (FIGURE 5.4)

A. Calculation of reabsorption and secretion rates

- The reabsorption or secretion rate is the difference between the amount filtered across the glomerular capillaries and the amount excreted in urine. It is calculated with the following equations:

$$\text{Filtered load} = \text{GFR} \times [\text{plasma}]$$
$$\text{Excretion rate} = \text{V} \times [\text{urine}]$$
$$\text{Reabsorption rate} = \text{Filtered load} - \text{Excretion rate}$$
$$\text{Secretion rate} = \text{Excretion rate} - \text{Filtered load}$$

- If the filtered load is greater than the excretion rate, then **net reabsorption** of the substance has occurred. If the filtered load is less than the excretion rate, then **net secretion** of the substance has occurred.
- **Example:** A woman with untreated diabetes mellitus has a GFR of 120 mL/min, a plasma glucose concentration of 400 mg/dL, a urine glucose concentration of 2500 mg/dL, and a urine flow rate of 4 mL/min. What is the reabsorption rate of glucose?

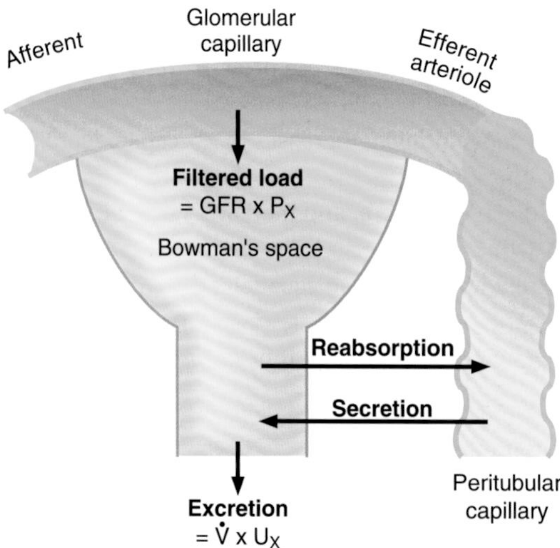

FIGURE 5.4. Processes of filtration, reabsorption, and secretion. The sum of the three processes is excretion.

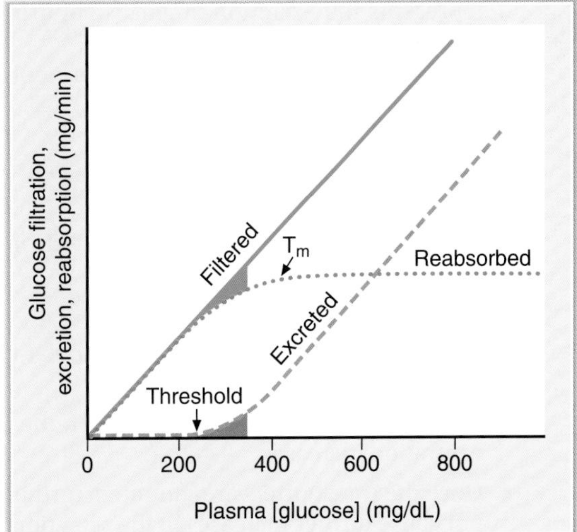

FIGURE 5.5. Glucose titration curve. Glucose filtration, excretion, and reabsorption are shown as a function of plasma [glucose]. Shaded area indicates the "splay." T_m = transport maximum.

$$
\begin{aligned}
\text{Filtered load} &= \text{GFR} \times \text{Plasma}\left[\text{glucose}\right] \\
&= 120 \text{ mL/min} \times 400 \text{ mg/dL} \\
&= 480 \text{ mg/min} \\
\text{Excretion} &= \text{V} \times \text{Urine}\left[\text{glucose}\right] \\
&= 4 \text{ mL/min} \times 2500 \text{ mg/dL} \\
&= 100 \text{ mg/min} \\
\text{Reabsorption} &= 480 \text{ mg/min} - 100 \text{ mg/min} \\
&= 380 \text{ mg/min}
\end{aligned}
$$

B. Transport maximum (T_m) curve for glucose—a reabsorbed substance (Figure 5.5)

1. Filtered load of glucose

- increases in direct proportion to the plasma glucose concentration (filtered load of glucose = GFR × $[P]_{glucose}$).

2. Reabsorption of glucose

a. **Na^+–glucose cotransport** in the early **proximal tubule** reabsorbs glucose from tubular fluid into the blood. There are a limited number of Na^+–glucose transporters.

b. At plasma glucose concentrations less than 250 mg/dL, all of the filtered glucose can be reabsorbed because plenty of carriers are available; in this range, the line for reabsorption is the same as that for filtration.

c. At plasma glucose concentrations greater than 350 mg/dL, the carriers are saturated. Therefore, increases in plasma concentration above 350 mg/dL do not result in increased rates of reabsorption. The reabsorptive rate at which the carriers are saturated is the **T_m**.

3. Excretion of glucose

a. At plasma concentrations less than 250 mg/dL, all of the filtered glucose is reabsorbed and excretion is zero. **Threshold** (defined as the plasma concentration at which glucose first appears in the urine) is approximately 250 mg/dL.

b. At plasma concentrations greater than 350 mg/dL, reabsorption is saturated (T_m). Therefore, as the plasma concentration increases, the additional filtered glucose cannot be reabsorbed and is excreted in the urine.

4. Splay

- is the region of the glucose curves **between threshold and T_m**.
- occurs between plasma glucose concentrations of approximately 250 and 350 mg/dL.

- represents the excretion of glucose in urine before saturation of reabsorption (T_m) is fully achieved.
- is explained by the heterogeneity of nephrons and the relatively low affinity of the Na^+–glucose transporters.

C. T_m curve for PAH—a secreted substance (Figure 5.6)

1. Filtered load of PAH

- As with glucose, the filtered load of PAH increases in direct proportion to the plasma PAH concentration.

2. Secretion of PAH

a. Secretion of PAH occurs from peritubular capillary blood into tubular fluid (urine) via transporters in the **proximal tubule.**

b. At low plasma concentrations of PAH, the secretion rate increases as the plasma concentration increases.

c. Once the transporters are saturated, further increases in plasma PAH concentration do not cause further increases in the secretion rate **(T_m).**

3. Excretion of PAH

a. Excretion of PAH is the **sum of filtration** across the glomerular capillaries **plus secretion** from peritubular capillary blood.

b. The curve for excretion is steepest at low plasma PAH concentrations (lower than at T_m). Once the T_m for secretion is exceeded and all of the carriers for secretion are saturated, the excretion curve flattens and becomes parallel to the curve for filtration.

c. **RPF** is measured by the **clearance of PAH** at plasma concentrations of PAH that are lower than at T_m.

D. Relative clearances of substances

1. Substances with the highest clearances

- are those that are both filtered across the glomerular capillaries and secreted from the peritubular capillaries into urine (e.g., PAH).

2. Substances with the lowest clearances

- are those that either are not filtered (e.g., protein) or are filtered and subsequently reabsorbed into peritubular capillary blood (e.g., Na^+, glucose, amino acids, HCO_3^-, Cl^-).

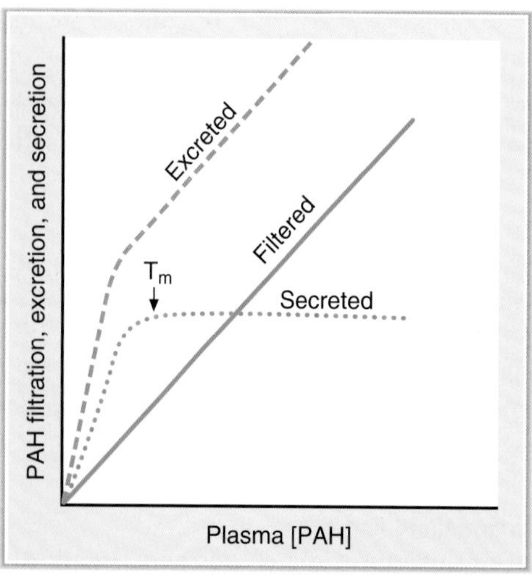

FIGURE 5.6. Para-aminohippuric acid (PAH) titration curve. PAH filtration, excretion, and secretion are shown as a function of plasma [PAH]. T_m = transport maximum.

3. Substances with clearances equal to GFR

- are **glomerular markers.**
- are those that are freely filtered, but not reabsorbed or secreted (e.g., inulin).

4. Relative clearances

- PAH > K^+ (high-K^+ diet) > inulin > urea > Na^+ > glucose, amino acids, and HCO_3^-.

E. Nonionic diffusion

1. Weak acids

- have an HA form and an A^- form.
- The HA form, which is uncharged and lipid soluble, can "back-diffuse" from urine to blood.
- The A^- form, which is charged and not lipid soluble, cannot back-diffuse.
- At **acidic urine pH**, the HA form predominates, there is more back-diffusion, and there is decreased excretion of the weak acid.
- At **alkaline urine pH**, the A^- form predominates, there is less back-diffusion, and there is increased excretion of the weak acid. For example, the excretion of **salicylic acid** (a weak acid) can be increased by alkalinizing the urine.

2. Weak bases

- have a BH^+ form and a B form.
- The B form, which is uncharged and lipid soluble, can "back-diffuse" from urine to blood.
- The BH^+ form, which is charged and not lipid soluble, cannot back-diffuse.
- At **acidic urine pH**, the BH^+ form predominates, there is less back-diffusion, and there is increased excretion of the weak base. For example, the excretion of **morphine** (a weak base) can be increased by acidifying the urine.
- At **alkaline urine pH**, the B form predominates, there is more back-diffusion, and there is decreased excretion of the weak base.

IV. NaCl REGULATION

A. Single nephron terminology

- **Tubular fluid (TF) is urine** at any point along the nephron.
- **Plasma (P) is systemic plasma.** It is considered to be constant.

1. TF/P_x ratio

- compares the concentration of a substance in tubular fluid at any point along the nephron with the concentration in plasma.

a. If TF/P = 1.0, then *either* there has been no reabsorption of the substance *or* reabsorption of the substance has been exactly proportional to the reabsorption of water.

- **For example,** if TF/P_{Na^+} = 1.0, the $[Na^+]$ in tubular fluid is identical to the $[Na^+]$ in plasma.
- For any freely filtered substance, TF/P = 1.0 in Bowman space (before any reabsorption or secretion has taken place to modify the tubular fluid).

b. If TF/P < 1.0, then reabsorption of the substance has been greater than the reabsorption of water and the concentration in tubular fluid is less than that in plasma.

- **For example,** if TF/P_{Na^+} = 8.0, then the $[Na^+]$ in tubular fluid is 80% of the $[Na^+]$ in plasma.

c. If TF/P > 1.0, then *either* reabsorption of the substance has been less than the reabsorption of water *or* there has been secretion of the substance.

2. TF/P_{inulin}

- is used as a marker for water reabsorption along the nephron.
- increases as water is reabsorbed.
- Because inulin is freely filtered, but not reabsorbed or secreted, its concentration in tubular fluid is determined solely by how much water remains in the tubular fluid.
- The following equation shows how to calculate the **fraction of the filtered water that has been reabsorbed:**

$$\text{Fraction of filtered } H_2O \text{ reabsorbed} = 1 - \frac{1}{[TF/P]_{inulin}}$$

- **For example,** if 50% of the filtered water has been reabsorbed, the $TF/P_{inulin} = 2.0$. For another example, if $TF/P_{inulin} = 3.0$, then 67% of the filtered water has been reabsorbed (i.e., 1 – 1/3).

3. $[TF/P]_x/[TF/P]_{inulin}$ ratio

- corrects the TF/P_x ratio for water reabsorption. This double ratio gives the **fraction of the filtered load remaining at any point along the nephron.**
- **For example,** if $[TF/P]_{K^+}/[TF/P]_{inulin} = 0.3$ at the end of the proximal tubule, then 30% of the filtered K^+ remains in the tubular fluid and 70% has been reabsorbed into the blood.

B. General information about Na^+ reabsorption

- Na^+ is freely filtered across the glomerular capillaries; therefore, the $[Na^+]$ in the tubular fluid of Bowman space equals that in plasma (i.e., $TF/P_{Na^+} = 1.0$).
- Na^+ is reabsorbed along the entire nephron, and very little is excreted in urine (<1% of the filtered load).

C. Na^+ reabsorption along the nephron (Figure 5.7)

1. Proximal tubule

- **reabsorbs two-thirds, or 67%, of the filtered Na^+ and H_2O,** more than any other part of the nephron.
- is the site of **glomerulotubular balance.**

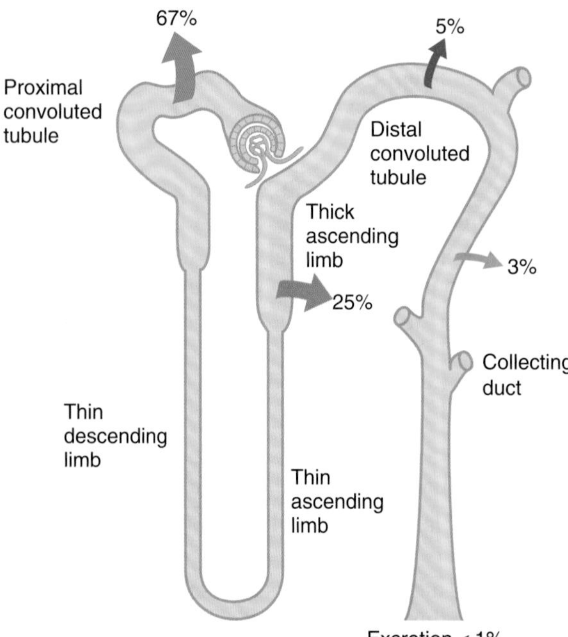

FIGURE 5.7. Na^+ handling along the nephron. *Arrows* indicate reabsorption of Na^+. *Numbers* indicate the percentage of the filtered load of Na^+ that is reabsorbed or excreted.

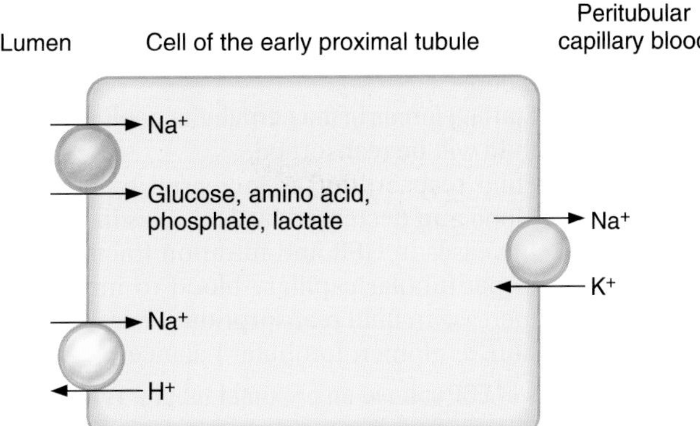

FIGURE 5.8. Mechanisms of Na^+ reabsorption in the cells of the early proximal tubule.

- The process is **isosmotic.** The reabsorption of Na^+ and H_2O in the proximal tubule is exactly proportional. Therefore, both TF/P_{Na^+} and $TF/P_{osm} = 1.0$.

a. **Early proximal tubule—special features** (Figure 5.8)
   - reabsorbs Na^+ and H_2O with HCO_3^-, glucose, amino acids, phosphate, and lactate.
   - Na^+ is reabsorbed by **cotransport** with glucose, amino acids, phosphate, and lactate. These cotransport processes account for the reabsorption of all of the filtered glucose and amino acids.
   - Na^+ is also reabsorbed by **countertransport** via **Na^+–H^+ exchange**, which is linked directly to the reabsorption of filtered HCO_3^-.
   - **Carbonic anhydrase inhibitors** (e.g., acetazolamide) are diuretics that act in the early proximal tubule by inhibiting the reabsorption of filtered HCO_3^-.

b. **Late proximal tubule—special features**
   - Filtered glucose, amino acids, and HCO_3^- have already been completely reabsorbed from the tubular fluid in the early proximal tubule.
   - In the late proximal tubule, **Na^+ is reabsorbed with Cl^-.**

c. **Glomerulotubular balance in the proximal tubule**
   - maintains **constant fractional reabsorption** (two-thirds, or 67%) of the filtered Na^+ and H_2O.

   (1) **For example,** if GFR spontaneously increases, the filtered load of Na^+ also increases. Without a change in reabsorption, this increase in GFR would lead to increased Na^+ excretion. However, glomerulotubular balance functions such that Na^+ reabsorption also will increase, ensuring that a constant fraction is reabsorbed.

   (2) *The mechanism of glomerulotubular balance* is based on Starling forces in the peritubular capillaries, which alter the reabsorption of Na^+ and H_2O in the proximal tubule (Figure 5.9).

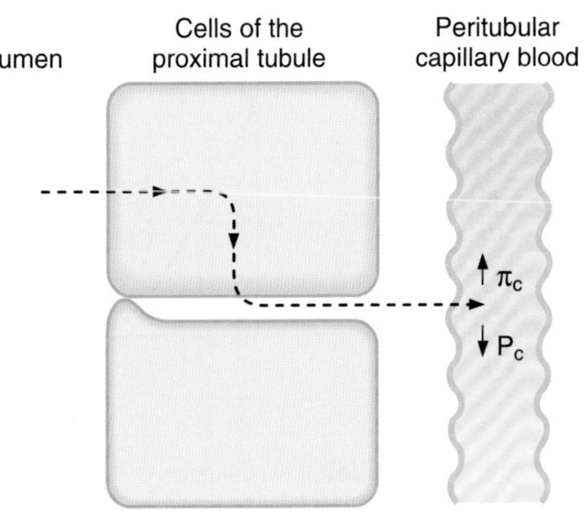

FIGURE 5.9. Mechanism of isosmotic reabsorption in the proximal tubule. The *dashed arrow* shows the pathway. Increases in π_c and decreases in P_c cause increased rates of isosmotic reabsorption.

- The route of isosmotic fluid reabsorption is from the lumen, to the proximal tubule cell, to the lateral intercellular space, and then to the peritubular capillary blood.
- **Starling forces in the peritubular capillary blood** govern how much of this isosmotic fluid will be reabsorbed.
- Fluid reabsorption is increased by increases in π_c of the peritubular capillary blood and decreased by decreases in π_c.
- Increases in GFR and filtration fraction cause the protein concentration and π_c of peritubular capillary blood to increase. This increase, in turn, produces an increase in fluid reabsorption. Thus, there is matching of filtration and reabsorption, or glomerulotubular balance.

d. Effects of ECF volume on proximal tubular reabsorption

(1) *ECF volume contraction increases reabsorption.* Volume contraction increases peritubular capillary protein concentration and π_c and decreases peritubular capillary P_c. Together, these changes in Starling forces in peritubular capillary blood **cause an increase in proximal tubular reabsorption.**

(2) *ECF volume expansion decreases reabsorption.* Volume expansion decreases peritubular capillary protein concentration and π_c and increases P_c. Together, these changes in Starling forces in peritubular capillary blood **cause a decrease in proximal tubular reabsorption.**

e. TF/P ratios along the proximal tubule (Figure 5.10)

- At the beginning of the proximal tubule (i.e., Bowman space), TF/P for freely filtered substances is 1.0, since no reabsorption or secretion has taken place yet.
- Moving along the proximal tubule, TF/P for **Na^+ and osmolarity** remain at 1.0 because Na^+ and total solute are reabsorbed proportionately with water, that is, isosmotically. **Glucose, amino acids, and HCO_3^-** are reabsorbed proportionately more than water, so their TF/P values fall below 1.0. In the early proximal tubule, **Cl^-** is reabsorbed proportionately less than water, so its TF/P value is greater than 1.0. **Inulin** is not reabsorbed, so its TF/P value increases steadily above 1.0, as water is reabsorbed and inulin is "left behind" and becomes concentrated.

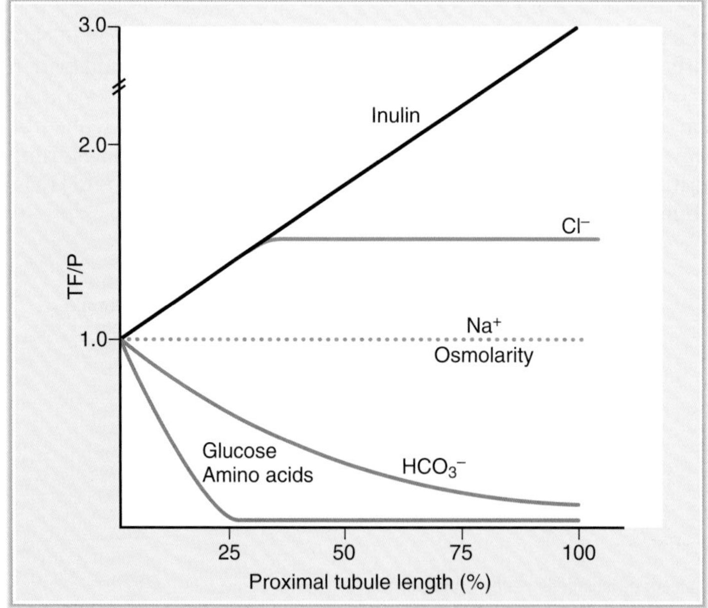

FIGURE 5.10. Changes in TF/P concentration ratios for various solutes along the proximal tubule.

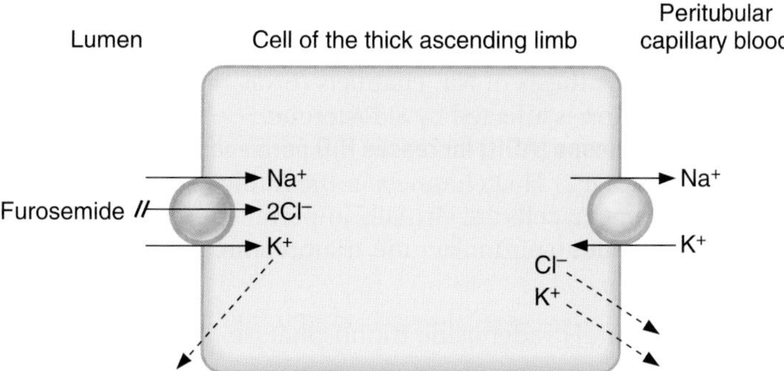

FIGURE 5.11. Mechanism of ion transport in the thick ascending limb of the loop of Henle.

2. **Thick ascending limb of the loop of Henle** (Figure 5.11)
   - **reabsorbs 25% of the filtered Na^+.**
   - contains a **Na^+–K^+–$2Cl^-$ cotransporter** in the luminal membrane.
   - is the site of action of the **loop diuretics** (furosemide, ethacrynic acid, bumetanide), which inhibit the Na^+–K^+–$2Cl^-$ cotransporter.
   - is **impermeable to water.** Thus, NaCl is reabsorbed without water. As a result, tubular fluid [Na^+] and tubular fluid osmolarity decrease to less than their concentrations in plasma (i.e., TF/P_{Na+} and $TF/P_{osm} < 1.0$). This segment, therefore, is called the **diluting segment.**
   - has a **lumen-positive potential difference.** Although the Na^+–K^+–$2Cl^-$ cotransporter appears to be electroneutral, some K^+ diffuses back into the lumen, making the lumen electrically positive.
3. **Distal tubule and collecting duct**
   - together **reabsorb 8% of the filtered Na^+.**

   a. **Early distal tubule—special features** (Figure 5.12)
      - reabsorbs NaCl by a **Na^+–Cl^- cotransporter.**
      - is the site of action of **thiazide diuretics.**
      - is **impermeable to water**, as is the thick ascending limb. Thus, reabsorption of NaCl occurs without water, which further dilutes the tubular fluid.
      - is called the **cortical diluting segment.**

   b. **Late distal tubule and collecting duct—special features**
      - have two cell types.

      (1) *Principal cells*
         - **reabsorb Na^+ and H_2O.**
         - **secrete K^+.**

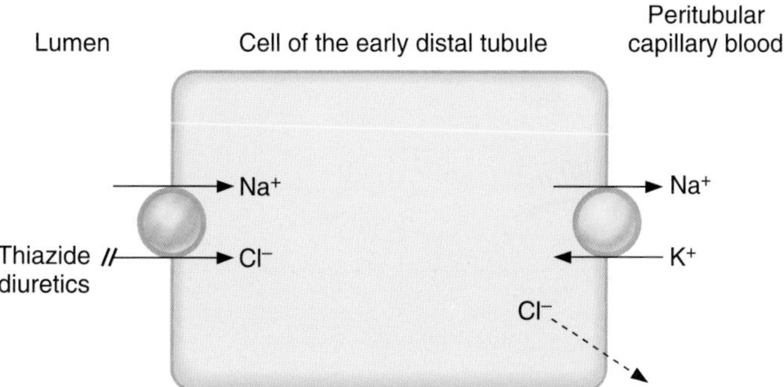

FIGURE 5.12. Mechanisms of ion transport in the early distal tubule.

- **Aldosterone increases Na^+ reabsorption and increases K^+ secretion.** Like other steroid hormones, the action of aldosterone takes several hours to develop because new protein synthesis of Na^+ channels (ENaC) is required. About 2% of overall Na^+ reabsorption is affected by aldosterone.
- **Antidiuretic hormone (ADH) increases H_2O permeability** by directing the insertion of aquaporin 2 (AQP2) H_2O channels in the luminal membrane. In the absence of ADH, the principal cells are virtually impermeable to water.
- **K^+-sparing diuretics** (spironolactone, triamterene, amiloride) **decrease K^+ secretion.**

(2) *α-Intercalated cells*

- **secrete H^+** by an H^+-adenosine triphosphatase (ATPase), which is stimulated by **aldosterone.**
- **reabsorb K^+** by an H^+, K^+-ATPase.

V. K^+ REGULATION

A. Shifts of K^+ between the ICF and ECF (Figure 5.13 and Table 5.4)

- Most of the body's K^+ is located in the ICF.
- A **shift of K^+ out of cells** causes **hyperkalemia.**
- A **shift of K^+ into cells** causes **hypokalemia.**

B. Renal regulation of K^+ balance (Figure 5.14)

- **K^+** is **filtered, reabsorbed**, and **secreted** by the nephron.
- **K^+ balance** is achieved when urinary excretion of K^+ exactly equals intake of K^+ in the diet.
- K^+ excretion can vary widely from 1% to 110% of the filtered load, depending on dietary K^+ intake, aldosterone levels, and acid–base status.

1. Glomerular capillaries

- **Filtration** occurs freely across the glomerular capillaries. Therefore, TF/P_{K^+} in Bowman space is 1.0.

2. Proximal tubule

- **reabsorbs 67%** of the filtered K^+ along with Na^+ and H_2O.

3. Thick ascending limb of the loop of Henle

- **reabsorbs 20%** of the filtered K^+.
- Reabsorption involves the **Na^+–K^+–$2Cl^-$ cotransporter** in the luminal membrane of cells in the thick ascending limb (see Figure 5.11).

4. Distal tubule and collecting duct

- either reabsorb or secrete K^+, depending on dietary K^+ intake.

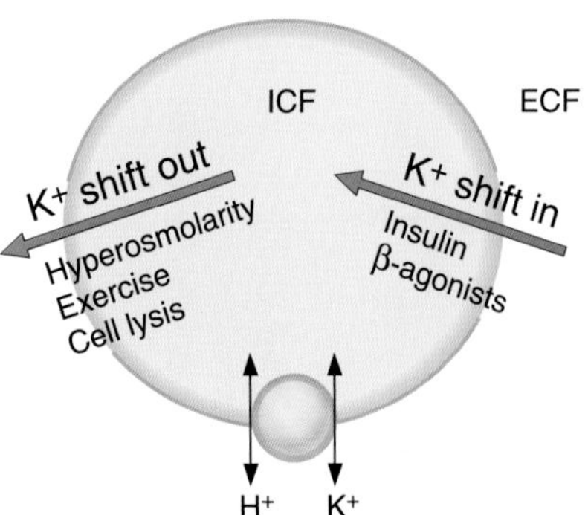

FIGURE 5.13. Internal K^+ balance. ECF = extracellular fluid; ICF = intracellular fluid.

table 5.4 Shifts of K^+ between ECF and ICF

Causes of Shift of K^+ Out of Cells→Hyperkalemia	Causes of Shift of K^+ into Cells→Hypokalemia
Insulin deficiency	Insulin
β-Adrenergic antagonists	β-Adrenergic agonists
Acidosis (exchange of extracellular H^+ for intracellular K^+)	Alkalosis (exchange of intracellular H^+ for extracellular K^+)
Hyperosmolarity (H_2O flows out of the cell; K^+ diffuses out with H_2O)	Hyposmolarity (H_2O flows into the cell; K^+ diffuses in with H_2O)
Inhibitors of Na^+–K^+ pump (e.g., digitalis) (when pump is blocked, K^+ is not taken up into cells)	
Exercise	
Cell lysis	

ECF = extracellular fluid; ICF = intracellular fluid.

a. Reabsorption of K^+

- involves an **H^+, K^+-ATPase** in the luminal membrane of the α-intercalated cells.
- occurs only on a **low-K^+** diet (K^+ depletion). Under these conditions, K^+ excretion can be as low as 1% of the filtered load because the kidney conserves as much K^+ as possible.

b. Secretion of K^+

- occurs in the **principal cells.**
- is **variable** and accounts for the wide range of urinary K^+ excretion.
- depends on factors such as dietary K^+, aldosterone levels, acid–base status, and urine flow rate.

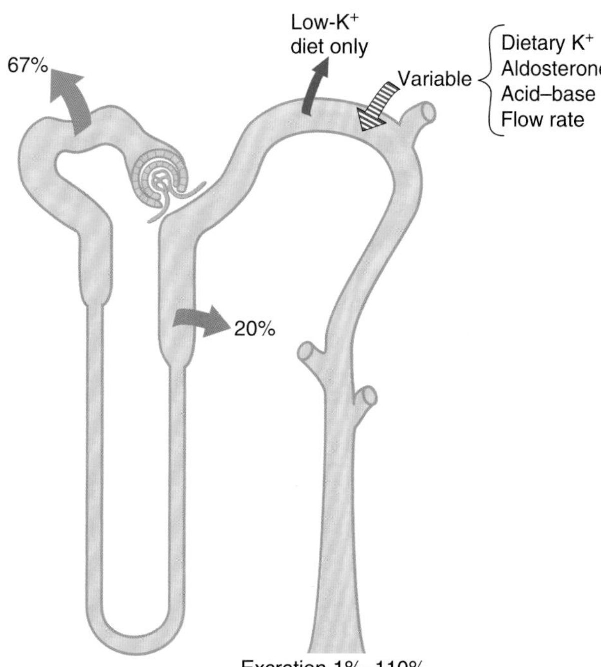

FIGURE 5.14. K^+ handling along the nephron. *Arrows* indicate reabsorption of secretion of K^+. *Numbers* indicate the percentage of the filtered load of K^+ that is reabsorbed, secreted, or excreted.

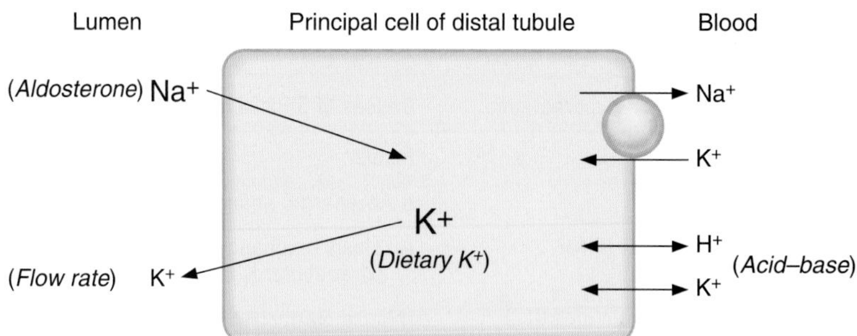

FIGURE 5.15. Mechanism of K^+ secretion in the principal cell of the distal tubule.

(1) *Mechanism of distal K^+ secretion* (Figure 5.15)

(a) At the basolateral membrane, K^+ is actively transported into the cell by the Na^+–K^+ pump. As in all cells, this mechanism maintains a high intracellular K^+ concentration.

(b) At the luminal membrane, K^+ is passively secreted into the lumen through K^+ channels. The magnitude of this passive secretion is **determined by the chemical and electrical driving forces on K^+ across the luminal membrane.**

- Maneuvers that increase the intracellular K^+ concentration or decrease the luminal K^+ concentration will increase K^+ secretion by increasing the driving force.
- Maneuvers that decrease the intracellular K^+ concentration will decrease K^+ secretion by decreasing the driving force.

(2) *Factors that change distal K^+ secretion* (see Figure 5.15 and Table 5.5)

- Distal K^+ secretion by the principal cells is increased when the electrochemical driving force for K^+ across the luminal membrane is increased. Secretion is decreased when the electrochemical driving force is decreased.

(a) Dietary K^+

- A diet high in K^+ increases K^+ secretion, and a diet low in K^+ decreases K^+ secretion.
- On a **high-K^+ diet**, intracellular K^+ increases so that the driving force for K^+ secretion also increases.
- On a **low-K^+ diet**, intracellular K^+ decreases so that the driving force for K^+ secretion decreases. Also, the α-intercalated cells are stimulated to reabsorb K^+ by the H^+, K^+-ATPase.

(b) Aldosterone

- **increases K^+ secretion.**
- The **mechanism** involves increased Na^+ entry into the cells across the luminal membrane and increased pumping of Na^+ out of the cells by the Na^+–K^+

table 5.5 Changes in Distal K^+ Secretion

Causes of Increased Distal K^+ Secretion	Causes of Decreased Distal K^+ Secretion
High-K^+ diet	Low-K^+ diet
Hyperaldosteronism	Hypoaldosteronism
Alkalosis	Acidosis
Thiazide diuretics (↑ urine flow rate)	K^+-sparing diuretics
Loop diuretics (↑ urine flow rate)	
Luminal anions (↑ lumen negativity)	

pump. Stimulation of the Na^+–K^+ pump simultaneously increases K^+ uptake into the principal cells, increasing the intracellular K^+ concentration and the driving force for K^+ secretion. Aldosterone also increases the number of luminal membrane K^+ channels.

- **Hyperaldosteronism** increases K^+ secretion and causes **hypokalemia.**
- **Hypoaldosteronism** decreases K^+ secretion and causes **hyperkalemia.**

(c) Acid–base

- Effectively, H^+ and K^+ exchange for each other across the basolateral cell membrane.
- **Acidosis decreases K^+ secretion.** The blood contains excess H^+; therefore, H^+ enters the cell across the basolateral membrane and K^+ leaves the cell. As a result, the intracellular K^+ concentration and the driving force for K^+ secretion decrease.
- **Alkalosis increases K^+ secretion.** The blood contains too little H^+; therefore, H^+ leaves the cell across the basolateral membrane and K^+ enters the cell. As a result, the intracellular K^+ concentration and the driving force for K^+ secretion increase.

(d) Loop and thiazide diuretics

- **increase K^+ secretion.**
- Loop and thiazide diuretics that increase **flow rate** through the late distal tubule and collecting ducts cause dilution of the luminal K^+ concentration, increasing the driving force for K^+ secretion.
- Loop and thiazide diuretics also increase Na^+ delivery to the late distal tubule and collecting ducts, which leads to increased Na^+ entry across the luminal membrane of principle cells, increased Na^+ pumping out of the cells by the Na^+–K^+ pump, increased intracellular K^+ concentration, and increased driving force for K^+ secretion.
- Also, as a result of increased K^+ secretion, these diuretics cause **hypokalemia.**

(e) K^+-sparing diuretics

- **decrease K^+ secretion.** If used alone, they cause **hyperkalemia.**
- Spironolactone is an antagonist of aldosterone; triamterene and amiloride act directly on the principal cells.
- The most important use of the K^+-sparing diuretics is in combination with thiazide or loop diuretics to offset (reduce) urinary K^+ losses.

(f) Luminal anions

- Excess anions (e.g., HCO_3^-) in the lumen cause an increase in K^+ secretion by increasing the negativity of the lumen and increasing the driving force for K^+ secretion.

VI. RENAL REGULATION OF UREA, PHOSPHATE, CALCIUM, AND MAGNESIUM

A. Urea

- Urea is reabsorbed and secreted in the nephron by **diffusion**, either simple or facilitated, depending on the segment of the nephron.
- Fifty percent of the filtered urea is reabsorbed in the proximal tubule by simple diffusion.
- Urea is secreted into the thin descending limb of the loop of Henle by simple diffusion (from the high concentration of urea in the medullary interstitial fluid).
- The distal tubule, cortical collecting ducts, and outer medullary collecting ducts are impermeable to urea; thus, no urea is reabsorbed by these segments.
- **ADH** stimulates a facilitated diffusion transporter for urea (**UT1**) in the **inner medullary collecting ducts.** In the presence of ADH, urea reabsorption from inner medullary collecting ducts contributes to **urea recycling in the inner medulla** and to the addition of urea to the corticopapillary osmotic gradient.

- **Urea excretion varies with urine flow rate.** At high levels of water reabsorption (low urine flow rate), there is greater urea reabsorption and decreased urea excretion. At low levels of water reabsorption (high urine flow rate), there is less urea reabsorption and increased urea excretion.

B. Phosphate

- **Eighty-five percent of the filtered phosphate is reabsorbed** in the proximal tubule by **Na^+–phosphate cotransport.** Because distal segments of the nephron do not reabsorb phosphate, 15% of the filtered load is excreted in urine.
- **Parathyroid hormone (PTH) inhibits phosphate reabsorption** in the early proximal tubule by activating adenylate cyclase, generating cyclic AMP (cAMP), and inhibiting Na^+–phosphate cotransport. Therefore, PTH causes **phosphaturia** and increased **urinary cAMP.**
- Phosphate is a urinary buffer for H^+; excretion of $H_2PO_4^-$ is called **titratable acid.**
- **Fibroblast growth factor (FGF23),** which is secreted by bone, inhibits Na^+–phosphate cotransport in the early proximal tubule.

C. Calcium (Ca^{2+})

- **Sixty percent of the plasma Ca^{2+} is filtered** across the glomerular capillaries.
- Together, the **proximal tubule and thick ascending limb** reabsorb more than 90% of the filtered Ca^{2+} by passive processes that are coupled to Na^+ reabsorption.
- **Loop diuretics (e.g., furosemide)** cause increased urinary Ca^{2+} excretion. Because Ca^{2+} reabsorption is driven by the lumen-positive potential difference in the loop of Henle, inhibiting the Na^+–$2Cl^-$–K^+ cotransporter reabsorption with a loop diuretic inhibits the lumen-positive potential difference and thereby inhibits Ca^{2+} reabsorption. If volume is replaced, loop diuretics can be used in the **treatment of hypercalcemia.**
- Together, the **distal tubule and collecting duct** reabsorb 8% of the filtered Ca^{2+} by an active process.

1. **PTH increases Ca^{2+} reabsorption** by activating adenylate cyclase in the distal tubule.
2. **Thiazide diuretics increase Ca^{2+} reabsorption** in the early distal tubule and therefore decrease Ca^{2+} excretion. For this reason, thiazides are used in the **treatment of idiopathic hypercalciuria.**

D. Magnesium (Mg^{2+})

- is **reabsorbed** in the proximal tubule, thick ascending limb of the loop of Henle, and distal tubule.
- In the **thick ascending limb**, Mg^{2+} and Ca^{2+} compete for reabsorption; therefore, hypercalcemia causes an increase in Mg^{2+} excretion (by inhibiting Mg^{2+} reabsorption). Likewise, hypermagnesemia causes an increase in Ca^{2+} excretion (by inhibiting Ca^{2+} reabsorption).

VII. CONCENTRATION AND DILUTION OF URINE

A. Regulation of plasma osmolarity

- is accomplished by varying the amount of water excreted relative to the amount of solute excreted (i.e., by varying urine osmolarity).

1. **Response to water deprivation** (Figure 5.16)
2. **Response to water intake** (Figure 5.17)

B. Production of concentrated urine (Figure 5.18)

- is also called **hyperosmotic urine,** in which urine osmolarity > blood osmolarity.
- is produced when circulating ADH levels are high (e.g., **water deprivation, volume depletion, SIADH**).

1. **Corticopapillary osmotic gradient—high ADH**
   - is the gradient of osmolarity from the cortex (300 mOsm/L) to the papilla (1200 mOsm/L) and is composed primarily of NaCl and urea.
   - is established by countercurrent multiplication in the loops of Henle and urea recycling in the inner medullary collecting ducts.
   - is maintained by countercurrent exchange in the vasa recta.

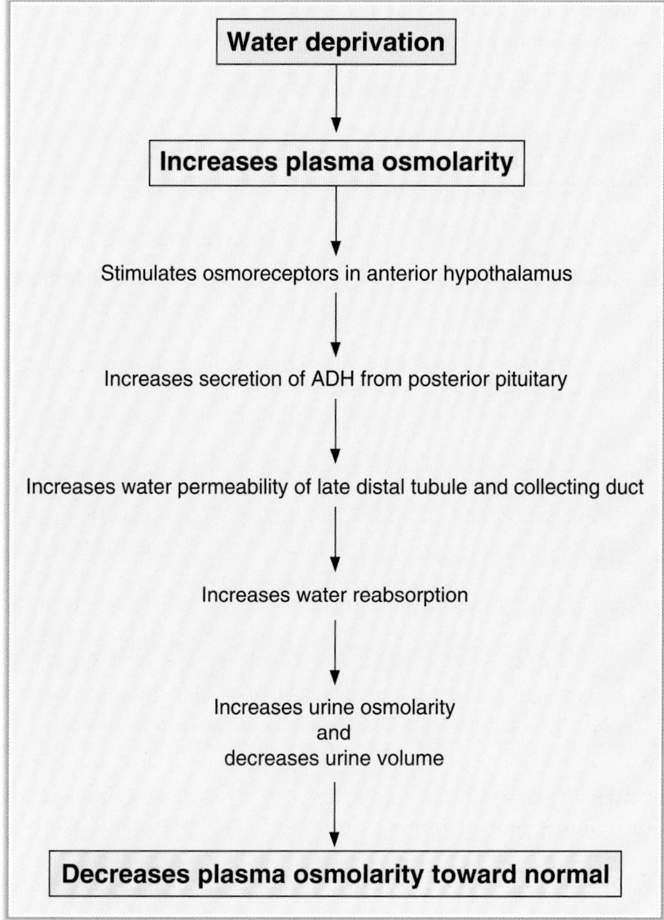

FIGURE 5.16. Responses to water deprivation. ADH = antidiuretic hormone.

a. **Countercurrent multiplication in the loop of Henle**
   - depends on **NaCl reabsorption in the thick ascending limb** and **countercurrent flow** in the descending and ascending limbs of the loop of Henle.
   - is **augmented by ADH**, which stimulates NaCl reabsorption in the thick ascending limb. Therefore, the presence of ADH increases the size of the corticopapillary osmotic gradient.

b. **Urea recycling** from the inner medullary collecting ducts into the medullary interstitial fluid also is **augmented by ADH** (by stimulating the UT1 transporter).

c. **Vasa recta** are the capillaries that supply the loop of Henle. They maintain the corticopapillary gradient by serving as **osmotic exchangers.** Vasa recta blood equilibrates osmotically with the interstitial fluid of the medulla and papilla.

2. **Proximal tubule—high ADH**
   - The osmolarity of the glomerular filtrate is identical to that of plasma (300 mOsm/L).
   - Two-thirds of the filtered H_2O is reabsorbed **isosmotically** (with Na^+, Cl^-, HCO_3^-, glucose, amino acids, and so forth) in the proximal tubule.
   - **TF/P_{osm} = 1.0** throughout the proximal tubule because H_2O is reabsorbed isosmotically with solute.

3. **Thick ascending limb of the loop of Henle—high ADH**
   - is called the **diluting segment.**
   - reabsorbs NaCl by the **Na^+–K^+–$2Cl^-$ cotransporter.**

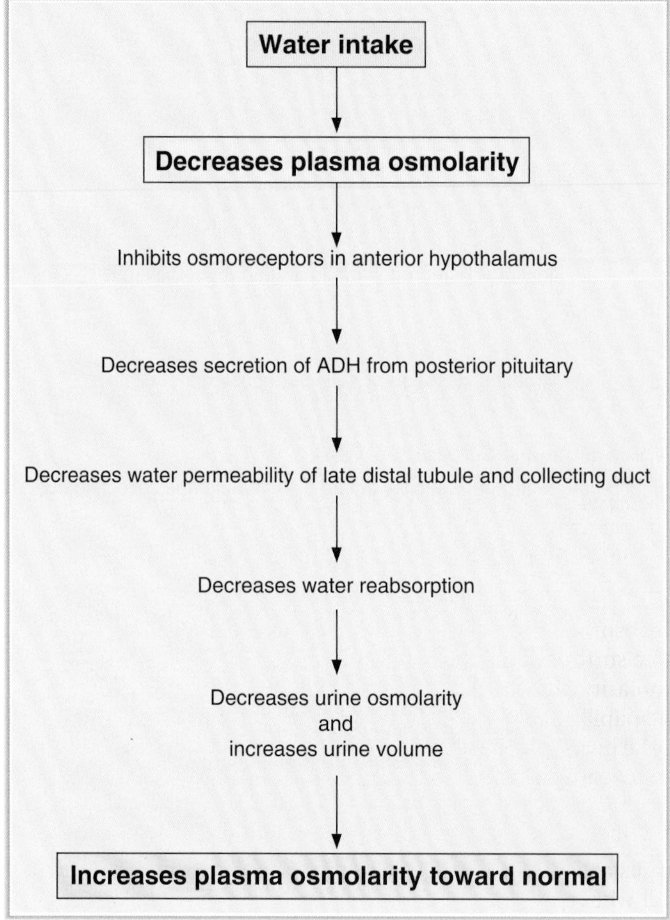

FIGURE 5.17. Responses to water intake. ADH = antidiuretic hormone.

- is **impermeable to H_2O.** Therefore, H_2O is not reabsorbed with NaCl, and the tubular fluid becomes dilute.
- The fluid that leaves the thick ascending limb has an osmolarity of 100 mOsm/L and **TF/P_{osm} < 1.0** as a result of the dilution process.

4. Early distal tubule—high ADH

- is called the **cortical diluting segment.**
- Like the thick ascending limb, the early distal tubule reabsorbs NaCl but is **impermeable to water.** Consequently, tubular fluid is further diluted.

5. Late distal tubule—high ADH

- **ADH increases the H_2O permeability of the principal cells** of the late distal tubule.
- H_2O is reabsorbed from the tubule until the osmolarity of distal tubular fluid equals that of the surrounding interstitial fluid in the renal cortex (300 mOsm/L).
- **TF/P_{osm} = 1.0** at the end of the distal tubule because osmotic equilibration occurs in the presence of ADH.

6. Collecting ducts—high ADH

- As in the late distal tubule, **ADH increases the H_2O permeability of the principal cells** of the collecting ducts.
- As tubular fluid flows through the collecting ducts, it passes through the corticopapillary gradient (regions of increasingly higher osmolarity), which was previously established by countercurrent multiplication and urea recycling.

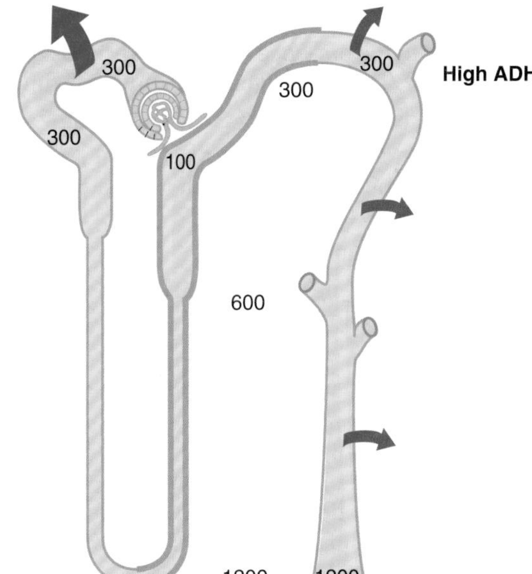

FIGURE 5.18. Mechanisms for producing hyperosmotic (concentrated) urine in the presence of antidiuretic hormone (ADH). *Numbers* indicate osmolarity. *Heavy arrows* indicate water reabsorption. The *thick outline* shows the water-impermeable segments of the nephron.

- H_2O is reabsorbed from the collecting ducts until the osmolarity of tubular fluid equals that of the surrounding interstitial fluid.
- The osmolarity of the final urine equals that at the bend of the loop of Henle and the tip of the papilla (1200 mOsm/L).
- **TF/P_{osm} > 1.0** because osmotic equilibration occurs with the corticopapillary gradient in the presence of ADH.

C. Production of dilute urine (Figure 5.19)

- is called **hyposmotic urine**, in which urine osmolarity < blood osmolarity.
- is produced when circulating levels of ADH are low (e.g., **water intake, central diabetes insipidus**) or when ADH is ineffective **(nephrogenic diabetes insipidus).**

1. Corticopapillary osmotic gradient—no ADH

- is **smaller** than in the presence of ADH because ADH stimulates both countercurrent multiplication and urea recycling.

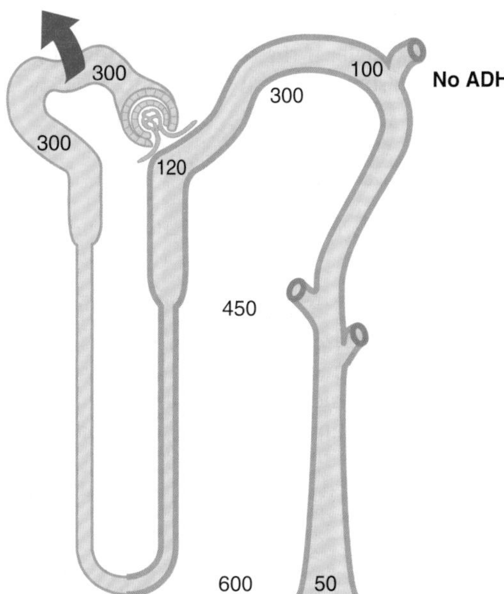

FIGURE 5.19. Mechanisms for producing hyposmotic (dilute) urine in the absence of antidiuretic hormone (ADH). *Numbers* indicate osmolarity. *Heavy arrow* indicates water reabsorption. The *thick outline* shows the water-impermeable segments of the nephron.

2. **Proximal tubule—no ADH**
   - As in the presence of ADH, two-thirds of the filtered water is reabsorbed **isosmotically.**
   - TF/P_{osm} **= 1.0** throughout the proximal tubule.
3. **Thick ascending limb of the loop of Henle—no ADH**
   - As in the presence of ADH, NaCl is reabsorbed without water, and the tubular fluid becomes dilute (although not quite as dilute as in the presence of ADH).
   - $TF/P_{osm} < 1.0$.
4. **Early distal tubule—no ADH**
   - As in the presence of ADH, NaCl is reabsorbed without H_2O and the tubular fluid is further diluted.
   - $TF/P_{osm} < 1.0$.
5. **Late distal tubule and collecting ducts—no ADH**
   - In the absence of ADH, the cells of the late distal tubule and collecting ducts are **impermeable to H_2O.**
   - Thus, even though the tubular fluid flows through the corticopapillary osmotic gradient, osmotic equilibration does not occur.
   - The osmolarity of the final urine will be dilute with an osmolarity as low as 50 mOsm/L.
   - $TF/P_{osm} < 1.0$.

D. Free-water clearance (C_{H_2O})

- is used to **estimate the ability to concentrate or dilute the urine.**
- Free water, or solute-free water, is produced in the diluting segments of the kidney (i.e., thick ascending limb and early distal tubule), where NaCl is reabsorbed and free water is left behind in the tubular fluid.
- In the **absence of ADH**, this solute-free water is excreted and **C_{H_2O} is positive.**
- In the **presence of ADH**, this solute-free water is not excreted but is reabsorbed by the late distal tubule and collecting ducts and **C_{H_2O} is negative.**

1. **Calculation of C_{H_2O}**

$$C_{H_2O} = V - C_{osm}$$

where:

C_{H_2O} = free-water clearance (mL/min)
V = urine flow rate (mL/min)
C_{osm} = osmolar clearance ($U_{osm}V/P_{osm}$) (mL/min)

   - **Example:** If the urine flow rate is 10 mL/min, urine osmolarity is 100 mOsm/L, and plasma osmolarity is 300 mOsm/L, what is the free-water clearance?

$$\begin{aligned} C_{H_2O} &= V - C_{osm} \\ &= 10\ \text{mL/min} - \frac{100\ \text{mOsm/L} \times 10\ \text{mL/min}}{300\ \text{mOsm/L}} \\ &= 10\ \text{mL/min} - 3.33\ \text{mL/min} \\ &= +6.7\ \text{mL/min} \end{aligned}$$

2. **Urine that is isosmotic to plasma (isosthenuric)**
   - **C_{H_2O} is zero.**
   - is produced during treatment with a **loop diuretic**, which inhibits NaCl reabsorption in the thick ascending limb, inhibiting both dilution in the thick ascending limb and production of the corticopapillary osmotic gradient. Therefore, the urine cannot be diluted during high water intake (because a diluting segment is inhibited) or concentrated during water deprivation (because the corticopapillary gradient has been abolished).

table 5.6 Summary of ADH Pathophysiology

	Serum ADH	Serum Osmolarity/ Serum [Na^+]	Urine Osmolarity	Urine Flow Rate	C_{H_2O}
Primary polydipsia	↓	Decreased	Hyposmotic	High	Positive
Central diabetes insipidus	↓	Increased (because of excretion of too much H_2O)	Hyposmotic	High	Positive
Nephrogenic diabetes insipidus	↑ (because of increased plasma osmolarity)	Increased (because of excretion of too much H_2O)	Hyposmotic	High	Positive
Water deprivation	↑	High–normal	Hyperosmotic	Low	Negative
SIADH	↑↑	Decreased (because of reabsorption of too much H_2O)	Hyperosmotic	Low	Negative

ADH = antidiuretic hormone; C_{H_2O} = free-water clearance; SIADH = syndrome of inappropriate antidiuretic hormone.

3. Urine that is hyposmotic to plasma (low ADH)

- **C_{H_2O} is positive.**
- is produced with **high water intake** (in which ADH release from the posterior pituitary is suppressed), **central diabetes insipidus** (in which pituitary ADH is insufficient), or **nephrogenic diabetes insipidus** (in which the collecting ducts are unresponsive to ADH).

4. Urine that is hyperosmotic to plasma (high ADH)

- **C_{H_2O} is negative.**
- is produced in **water deprivation** (ADH release from the pituitary is stimulated) or **SIADH.**

E. Clinical disorders related to the concentration or dilution of urine (Table 5.6)

VIII. RENAL HORMONES

- See Table 5.7 for a summary of renal hormones (see Chapter 7 for a discussion of hormones).

IX. ACID–BASE BALANCE

A. Acid production

- Two types of acid are produced in the body: volatile acid and nonvolatile acids.

1. Volatile acid

- is **CO_2**.
- is produced from the aerobic metabolism of cells.
- CO_2 combines with H_2O to form the weak acid H_2CO_3, which dissociates into H^+ and HCO_3^- by the following reactions:

$$CO_2 + H_2O \leftrightarrow H_2CO_3 \leftrightarrow H^+ + HCO_3^-$$

- **Carbonic anhydrase**, which is present in most cells, catalyzes the reversible reaction between CO_2 and H_2O.

2. Nonvolatile acids

- are also called **fixed acids.**

table 5.7 Summary of Hormones That Act on the Kidney

Hormone	Stimulus for Secretion	Time Course	Mechanism of Action	Actions on the Kidneys
PTH	↓ plasma $[Ca^{2+}]$	Fast	Basolateral receptor Adenylate cyclase cAMP→urine	↓ phosphate reabsorption (early proximal tubule) ↑ Ca^{2+} reabsorption (distal tubule) Stimulates 1α-hydroxylase (proximal tubule)
ADH	↑ plasma osmolarity ↓ blood volume	Fast	Basolateral V_2 receptor Adenylate cyclase cAMP (Note: V_1 receptors are on blood vessels; mechanism is Ca^{2+}–IP_3)	↑ H_2O permeability (late distal tubule and collecting duct principal cells)
Aldosterone	↓ blood volume (via renin–angiotensin II) ↑ plasma $[K^+]$	Slow	New protein synthesis	↑ Na^+ reabsorption (ENaC, distal tubule principal cells) ↑ K^+ secretion (distal tubule principal cells) ↑ H^+ secretion (distal tubule α-intercalated cells)
ANP	↑ atrial pressure	Fast	Guanylate cyclase cGMP	↑ GFR ↓ Na^+ reabsorption
Angiotensin II	↓ blood volume (via renin)	Fast		↑ Na^+–H^+ exchange and HCO_3^- reabsorption (proximal tubule)

ADH = antidiuretic hormone; ANP = atrial natriuretic peptide; cAMP = cyclic adenosine monophosphate; cGMP = cyclic guanosine monophosphate; GFR = glomerular filtration rate; PTH = parathyroid hormone; EnaC = epithelial Na^+ channel.

- include **sulfuric acid** (a product of protein catabolism) and phosphoric acid (a product of phospholipid catabolism).
- are normally produced at a rate of **40 to 60 mmoles/d.**
- Other fixed acids that may be overproduced in disease include **ketoacids and lactic acid.** Other fixed acids may be ingested including **salicylic acid, formic acid** (methanol poisoning), and **oxalic and glycolic acids** (ethylene glycol poisoning).

B. Buffers

- prevent a change in pH when H^+ ions are added to or removed from a solution.
- are **most effective within 1.0 pH unit of the pK** of the buffer (i.e., in the linear portion of the titration curve).

1. Extracellular buffers

a. The major extracellular buffer is **HCO_3^-**, which is produced from CO_2 and H_2O.

- The **pK** of the CO_2/HCO_3^- buffer pair is 6.1.

b. Phosphate is a minor extracellular buffer.

- The **pK** of the $H_2PO_4^-/HPO_4^{-2}$ buffer pair is 6.8.
- Phosphate is most important as a **urinary buffer;** excretion of H^+ as $H_2PO_4^-$ is called **titratable acid.**

2. Intracellular buffers

a. Organic phosphates (e.g., AMP, ADP, ATP, 2,3-diphosphoglycerate [DPG])

b. Proteins

- Imidazole and α-amino groups on proteins have pKs that are within the physiologic pH range.

- **Hemoglobin** is a major intracellular buffer.
- In the physiologic pH range, **deoxyhemoglobin is a better buffer than oxyhemoglobin.**

3. Using the Henderson-Hasselbalch equation to calculate pH

$$\mathbf{pH = pK + log\frac{[A^-]}{[HA]}}$$

where:

pH = $-\log_{10}$ [H$^+$] (pH units)
pK = $-\log_{10}$ equilibrium constant (pH units)
[A$^-$] = concentration of base form of buffer (mM)
[HA] = concentration of acid form of buffer (mM)

- A$^-$, the base form of the buffer, is the H$^+$ acceptor.
- HA, the acid form of the buffer, is the H$^+$ donor.
- When the concentrations of A$^-$ and HA are equal, the **pH of the solution equals the pK of the buffer**, as calculated by the Henderson-Hasselbalch equation.
- **Example:** The pK of the $H_2PO_4^-/HPO_4^{-2}$ buffer pair is 6.8. What are the relative concentrations of $H_2PO_4^-$ and HPO_4^{-2} in a urine sample that has a pH of 4.8?

$$pH = pK + \log\frac{HPO_4^{-2}}{H_2PO_4^-}$$

$$4.8 = 6.8 + \log\frac{HPO_4^{-2}}{H_2PO_4^-}$$

$$\log\frac{HPO_4^{-2}}{H_2PO_4^-} = -2.0$$

$$\frac{HPO_4^{-2}}{H_2PO_4^-} = 0.01$$

$$\frac{H_2PO_4^-}{HPO_4^{-2}} = 100$$

For this buffer pair, HPO_4^{-2} is A$^-$ and $H_2PO_4^-$ is HA. Thus, the Henderson-Hasselbalch equation can be used to calculate that the concentration of $H_2PO_4^-$ is 100 times that of HPO_4^{-2} in a urine sample of pH 4.8.

4. Titration curves (Figure 5.20)

- describe how the pH of a buffered solution changes as H$^+$ ions are added to it or removed from it.
- As H$^+$ ions are added to the solution, the HA form is produced; as H$^+$ ions are removed, the A$^-$ form is produced.

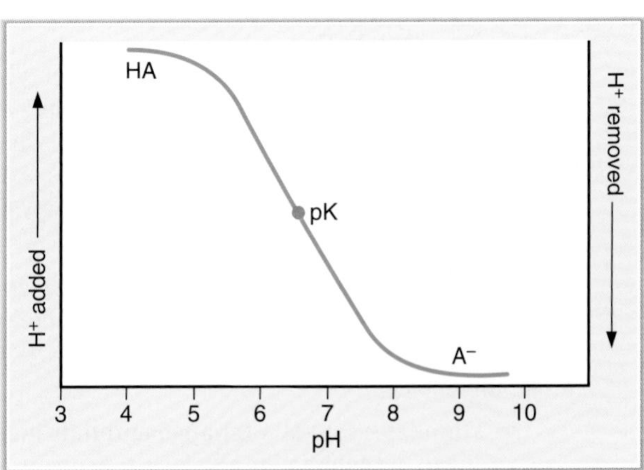

FIGURE 5.20. Titration curve for a weak acid (HA) and its conjugate base (A$^-$).

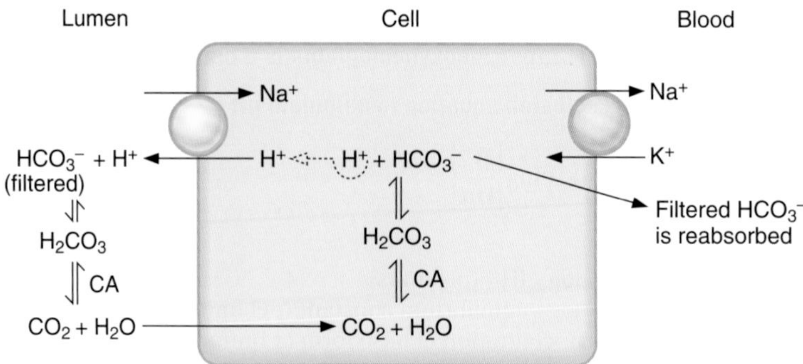

FIGURE 5.21. Mechanism for reabsorption of filtered HCO_3^- in the proximal tubule. CA = carbonic anhydrase.

- A buffer is **most effective in the linear portion** of the titration curve, where the addition or removal of H^+ causes little change in pH.
- According to the Henderson-Hasselbalch equation, when the **pH of the solution equals the pK, the concentrations of HA and A^- are equal.**

C. Renal acid–base

1. Reabsorption of filtered HCO_3^- (Figure 5.21)

- occurs primarily in the **early proximal tubule.**

a. Key features of reabsorption of filtered HCO_3^-

(1) H^+ and HCO_3^- are produced in the proximal tubule cells from CO_2 and H_2O. CO_2 and H_2O combine to form H_2CO_3, catalyzed by **intracellular carbonic anhydrase;** H_2CO_3 dissociates into H^+ and HCO_3^-. H^+ is secreted into the lumen via the Na^+–H^+ exchange mechanism in the luminal membrane. The HCO_3^- is reabsorbed.

(2) In the lumen, the secreted H^+ combines with filtered HCO_3^- to form H_2CO_3, which dissociates into CO_2 and H_2O, catalyzed by **brush border carbonic anhydrase.** CO_2 and H_2O diffuse into the cell to start the cycle again.

(3) The process results in **net reabsorption of filtered HCO_3^-**. However, **it *does not* result in net secretion of H^+.**

b. Regulation of reabsorption of filtered HCO_3^-

(1) *Filtered load*

- Increases in the filtered load of HCO_3^- result in increased rates of HCO_3^- reabsorption. However, if the plasma HCO_3^- concentration becomes very high (e.g., metabolic alkalosis), the filtered load will exceed the reabsorptive capacity, and HCO_3^- will be excreted in the urine.

(2) *P_{CO_2}*

- **Increases in P_{CO_2}** result in increased rates of HCO_3^- reabsorption because the supply of intracellular H^+ for secretion is increased. This mechanism is the basis for the **renal compensation for respiratory acidosis.**
- **Decreases in P_{CO_2}** result in decreased rates of HCO_3^- reabsorption because the supply of intracellular H^+ for secretion is decreased. This mechanism is the basis for the **renal compensation for respiratory alkalosis.**

(3) *ECF volume*

- **ECF volume expansion** results in decreased HCO_3^- reabsorption.
- **ECF volume contraction** results in increased HCO_3^- reabsorption (contraction alkalosis).

(4) *Angiotensin II*

- stimulates Na^+–H^+ exchange and thus increases HCO_3^- reabsorption, contributing to the **contraction alkalosis** that occurs secondary to ECF volume contraction.

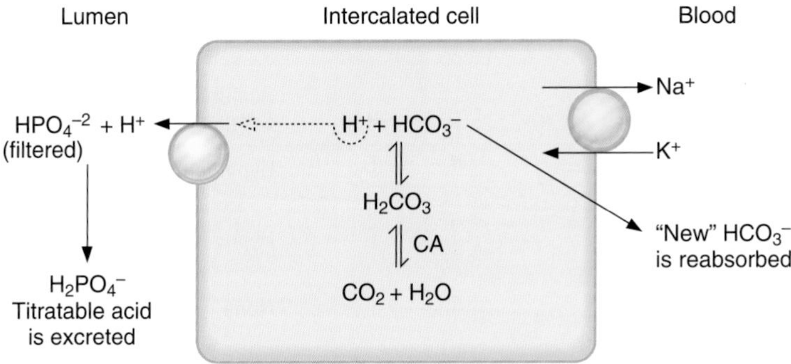

FIGURE 5.22. Mechanism for excretion of H^+ as titratable acid. CA = carbonic anhydrase.

2. Excretion of fixed H^+

- Fixed H^+ produced from the catabolism of protein and phospholipid is excreted by two mechanisms, titratable acid and NH_4^+.

a. Excretion of H^+ as titratable acid ($H_2PO_4^-$) (Figure 5.22)

- The amount of H^+ excreted as titratable acid depends on the **amount of urinary buffer** present (usually HPO_4^{-2}) and the **pK of the buffer.**

(1) H^+ and HCO_3^- are produced in the intercalated cells from CO_2 and H_2O. The H^+ is secreted into the lumen by an H^+-ATPase, and the HCO_3^- is reabsorbed into the blood ("new" HCO_3^-). In the urine, the secreted H^+ combines with filtered HPO_4^{-2} to form $H_2PO_4^-$, which is excreted as **titratable acid.** The H^+-ATPase is increased by **aldosterone.**

(2) This process results in **net secretion of H^+** and **net reabsorption of newly synthesized HCO_3^-.**

(3) As a result of H^+ secretion, the pH of urine becomes progressively lower. **The minimum urinary pH is 4.4.**

(4) The amount of H^+ excreted as titratable acid is determined by the **amount of urinary buffer** and the **pK of the buffer.**

b. Excretion of H^+ as NH_4^+ (Figure 5.23)

- The amount of H^+ excreted as NH_4^+ depends on both the **amount of NH_3 synthesized** by renal cells and the **urine pH.**

(1) NH_3 is produced in renal cells from **glutamine**. It diffuses down its concentration gradient from the cells into the lumen.

(2) H^+ and HCO_3^- are produced in the intercalated cells from CO_2 and H_2O. The H^+ is secreted into the lumen via an H^+-ATPase and combines with NH_3 to form NH_4^+, which is excreted **(diffusion trapping).** The HCO_3^- is reabsorbed into the blood ("new" HCO_3^-).

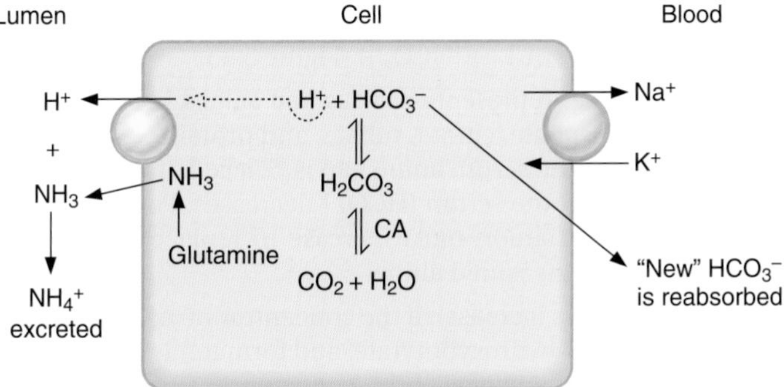

FIGURE 5.23. Mechanism for excretion of H^+ as NH_4^+. CA = carbonic anhydrase.

table 5.8 Summary of Acid–Base Disorders

Disorder	$CO_2 + H_2O$	↔	H^+	HCO_3^-	Respiratory Compensation	Renal Compensation
Metabolic acidosis	↓ (respiratory compensation)		↑	**↓**	Hyperventilation	
Metabolic alkalosis	↑ (respiratory compensation)		↓	**↑**	Hypoventilation	
Respiratory acidosis	**↑**		↑	↑	None	↑ H^+ excretion ↑ HCO_3^- reabsorption
Respiratory alkalosis	**↓**		↓	↓	None	↓ H^+ excretion ↓ HCO_3^- reabsorption

Heavy arrows indicate *primary* disturbance.

(3) The lower the pH of the tubular fluid, the greater the excretion of H^+ as NH_4^+; at low urine pH, there is more NH_4^+ relative to NH_3 in the urine, thus increasing the gradient for NH_3 diffusion.

(4) In acidosis, an **adaptive increase in NH_3 synthesis** occurs and aids in the excretion of excess H^+.

(5) **Hyperkalemia inhibits NH_3 synthesis**, which produces a decrease in H^+ excretion as NH_4^+ **(type 4 renal tubular acidosis [RTA])**. For example, **hypoaldosteronism** causes hyperkalemia and thus also causes type 4 RTA. Conversely, hypokalemia stimulates NH_3 synthesis, which produces an increase in H^+ excretion.

D. Acid–base disorders (Tables 5.8 and 5.9 and Figure 5.24)

- The expected compensatory responses to simple acid–base disorders can be calculated as shown in Table 5.10. If the actual response equals the calculated (predicted) response, then one acid–base disorder is present. If the actual response differs from the calculated response, then more than one acid–base disorder is present.

1. Metabolic acidosis

a. Overproduction or ingestion of fixed acid or loss of base produces a **decrease in arterial [HCO_3^-]**. This decrease is the primary disturbance in metabolic acidosis.

b. Decreased HCO_3^- concentration causes a **decrease in blood pH** (acidemia).

c. Acidemia causes **hyperventilation (Kussmaul breathing)**, which is the **respiratory compensation** for metabolic acidosis.

d. Correction of metabolic acidosis consists of increased excretion of the excess fixed H^+ as titratable acid and NH_4^+ and increased reabsorption of "new" HCO_3^-, which replenishes the blood HCO_3^- concentration.

- In chronic metabolic acidosis, an **adaptive increase in NH_3 synthesis** aids in the excretion of excess H^+.

e. **Serum anion gap** = $[Na^+] - ([Cl^-] + [HCO_3^-])$ (Figure 5.25)

- The serum anion gap represents **unmeasured anions** in serum. These unmeasured anions include phosphate, citrate, sulfate, and protein.
- The normal value of the serum anion gap is **12 mEq/L** (range 8 to 16 mEq/L)
- In metabolic acidosis, the serum [HCO_3^-] decreases. For electroneutrality, the concentration of another anion must increase to replace HCO_3^-. That anion can be Cl^- or it can be an unmeasured anion.

(1) **The serum anion gap is increased** if the concentration of an unmeasured anion (e.g., phosphate, lactate, β-hydroxybutyrate, and formate) is increased to replace HCO_3^-.

(2) **The serum anion gap is normal** if the concentration of Cl^- is increased to replace HCO_3^- **(hyperchloremic metabolic acidosis).**

table 5.9 Causes of Acid–Base Disorders

	Example	Comments
Metabolic acidosis	Ketoacidosis	Accumulation of β-OH-butyric acid and acetoacetic acid ↑ anion gap
	Lactic acidosis	Accumulation of lactic acid during hypoxia ↑ anion gap
	Chronic renal failure	Failure to excrete H^+ as titratable acid and NH_4^+ ↑ anion gap
	Salicylate intoxication	Also causes respiratory alkalosis ↑ anion gap
	Methanol/formaldehyde intoxication	Produces formic acid ↑ anion gap
	Ethylene glycol intoxication	Produces glycolic and oxalic acids ↑ anion gap
	Diarrhea	GI loss of HCO_3^- Normal anion gap
	Type 2 RTA	Renal loss of HCO_3^- Normal anion gap
	Type 1 RTA	Failure to excrete titratable acid and NH_4^+; failure to acidify urine Normal anion gap
	Type 4 RTA	Hypoaldosteronism; failure to excrete NH_4^+ Hyperkalemia caused by lack of aldosterone inhibits NH_3 synthesis Normal anion gap
Metabolic alkalosis	Vomiting	Loss of gastric H^+; leaves HCO_3^- behind in blood Worsened by volume contraction Hypokalemia May have ↑ anion gap because of production of ketoacids (starvation)
	Hyperaldosteronism	Increased H^+ secretion by distal tubule; increased new HCO_3^- reabsorption
	Loop or thiazide diuretics	Volume contraction alkalosis
Respiratory acidosis	Opiates; sedatives; anesthetics	Inhibition of medullary respiratory center
	Guillain-Barré syndrome, polio, ALS, multiple sclerosis	Weakening of respiratory muscles
	Airway obstruction	↓ CO_2 exchange in lungs
	Adult respiratory distress syndrome, COPD	↓ CO_2 exchange in lungs
Respiratory alkalosis	Pneumonia, pulmonary embolus	Hypoxemia causes ↑ ventilation rate
	High altitude	Hypoxemia causes ↑ ventilation rate
	Psychogenic	
	Salicylate intoxication	Direct stimulation of medullary respiratory center; also causes metabolic acidosis

ALS = amyotrophic lateral sclerosis; COPD = chronic obstructive pulmonary disease; GI = gastrointestinal; RTA = renal tubular acidosis.

2. Metabolic alkalosis

a. Loss of fixed H^+ or gain of base produces an **increase in arterial [HCO_3^-]**. This increase is the primary disturbance in metabolic alkalosis.

- For example, in **vomiting**, H^+ is lost from the stomach, HCO_3^- remains behind in the blood, and the [HCO_3^-] increases.

b. Increased HCO_3^- concentration causes an **increase in blood pH** (alkalemia).

c. Alkalemia causes **hypoventilation**, which is the **respiratory compensation** for metabolic alkalosis.

d. Correction of metabolic alkalosis consists of increased excretion of HCO_3^- because the filtered load of HCO_3^- exceeds the ability of the renal tubule to reabsorb it.

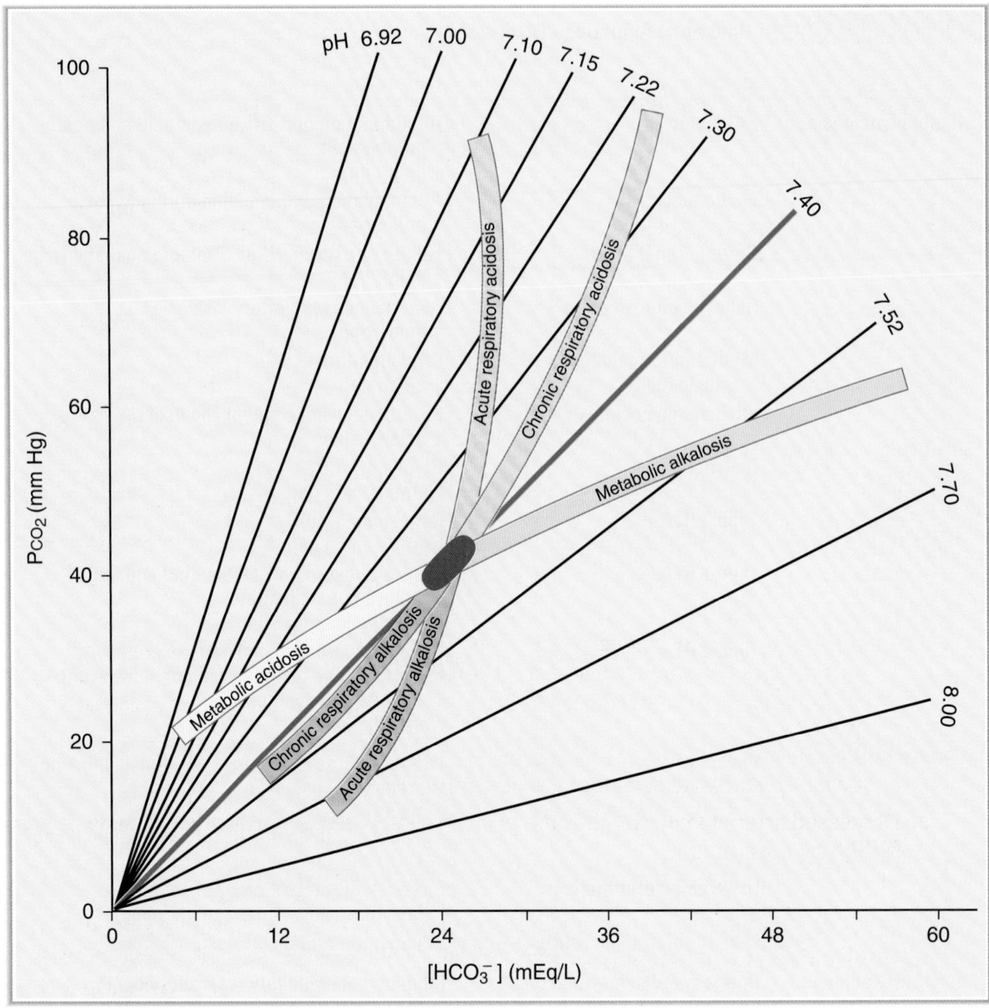

FIGURE 5.24. Acid–base map with values for simple acid–base disorders superimposed. The relationships are shown between arterial P_{CO_2}, $[HCO_3^-]$, and pH. The ellipse in the center shows the normal range of values. Shaded areas show the range of values associated with simple acid–base disorders. Two shaded areas are shown for each respiratory disorder: one for the acute phase and one for the chronic phase.

- If metabolic alkalosis is accompanied by **ECF volume contraction** (e.g., vomiting), the reabsorption of HCO_3^- increases (secondary to ECF volume contraction and activation of the renin–angiotensin II–aldosterone system), worsening the metabolic alkalosis (i.e., **contraction alkalosis**).

3. Respiratory acidosis

- is **caused by decreased alveolar ventilation and retention of CO_2.**

a. Increased arterial P_{CO_2}, which is the primary disturbance, causes an **increase in $[H^+]$ and $[HCO_3^-]$** by mass action.

b. There is **no respiratory compensation** for respiratory acidosis.

c. **Renal compensation** consists of increased excretion of H^+ as titratable acid and NH_4^+ and increased reabsorption of “new” HCO_3^-. This process is aided by the increased P_{CO_2}, which supplies more H^+ to the renal cells for secretion. The resulting further increase in serum $[HCO_3^-]$ helps to normalize the pH.

table 5.10 Calculating Compensatory Responses to Simple Acid–Base Disorders

Acid–Base Disturbance	Primary Disturbance	Compensation	Predicted Compensatory Response
Metabolic acidosis	↓ $[HCO_3^-]$	↓ P_{CO_2}	1 mEq/L decrease in HCO_3^- → 1.3 mm Hg decrease in P_{CO_2}
Metabolic alkalosis	↑ $[HCO_3^-]$	↑ P_{CO_2}	1 mEq/L increase in HCO_3^- → 0.7 mm Hg increase in P_{CO_2}
Respiratory acidosis			
Acute	↑ P_{CO_2}	↑ $[HCO_3^-]$	1 mm Hg increase in P_{CO_2} → 0.1 mEq/L increase in HCO_3^-
Chronic	↑ P_{CO_2}	↑ $[HCO_3^-]$	1 mm Hg increase in P_{CO_2} → 0.4 mEq/L increase in HCO_3^-
Respiratory alkalosis			
Acute	↓ P_{CO_2}	↓ $[HCO_3^-]$	1 mm Hg decrease in P_{CO_2} → 0.2 mEq/L decrease in HCO_3^-
Chronic	↓ P_{CO_2}	↓ $[HCO_3^-]$	1 mm Hg decrease in P_{CO_2} → 0.4 mEq/L decrease in HCO_3^-

- In **acute respiratory acidosis,** renal compensation has not yet occurred.
- In **chronic respiratory acidosis,** renal compensation (increased HCO_3^- reabsorption) has occurred. Thus, arterial pH is increased toward normal (i.e., a compensation).

4. Respiratory alkalosis

- is **caused by increased alveolar ventilation** and **loss of CO_2.**

a. Decreased arterial P_{CO_2}, which is the primary disturbance, causes a **decrease in $[H^+]$ and $[HCO_3^-]$** by mass action.

b. There is **no respiratory compensation** for respiratory alkalosis.

c. Renal compensation consists of decreased excretion of H^+ as titratable acid and NH_4^+ and decreased reabsorption of "new" HCO_3^-. This process is aided by the decreased P_{CO_2}, which causes a deficit of H^+ in the renal cells for secretion. The resulting further decrease in serum $[HCO_3^-]$ helps to normalize the pH.

- In **acute respiratory alkalosis,** renal compensation has not yet occurred.
- In **chronic respiratory alkalosis,** renal compensation (decreased HCO_3^- reabsorption) has occurred. Thus, arterial pH is decreased toward normal (i.e., a compensation).

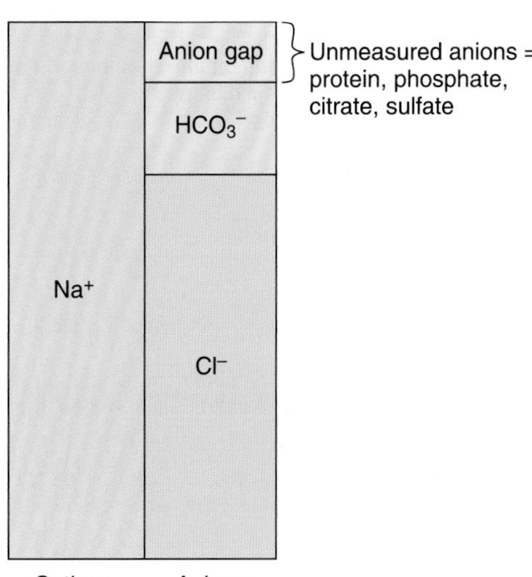

FIGURE 5.25. Serum anion gap.

d. Symptoms of **hypocalcemia** (e.g., tingling, numbness, muscle spasms) may occur because H^+ and Ca^{2+} compete for binding sites on plasma proteins. Decreased $[H^+]$ causes increased protein binding of Ca^{2+} and decreased free ionized Ca^{2+}.

X. DIURETICS (TABLE 5.11)

XI. INTEGRATIVE EXAMPLES

A. Hypoaldosteronism

1. Case study

- A woman has a history of weakness, weight loss, orthostatic hypotension, increased pulse rate, and increased skin pigmentation. She has decreased serum $[Na^+]$, decreased serum osmolarity, increased serum $[K^+]$, and arterial blood gases consistent with metabolic acidosis.

2. Explanation of hypoaldosteronism

a. The **lack of aldosterone** has three direct effects on the kidney: decreased Na^+ reabsorption, decreased K^+ secretion, and decreased H^+ secretion. As a result, there is **ECF volume contraction** (caused by decreased Na^+ reabsorption), **hyperkalemia** (caused by decreased K^+ secretion), and **metabolic acidosis** (caused by decreased H^+ secretion).

table 5.11 Effects of Diuretics on the Nephron

Class of Diuretic	Site of Action	Mechanism	Major Effect
Carbonic anhydrase inhibitors (acetazolamide)	Early proximal tubule	Inhibition of carbonic anhydrase	↑ HCO_3^- excretion
Loop diuretics (furosemide, ethacrynic acid, bumetanide)	Thick ascending limb of the loop of Henle	Inhibition of Na^+–K^+– $2Cl^-$ cotransport	↑ NaCl excretion ↑ K^+ excretion (↑ distal tubule flow rate) ↑ Ca^{2+} excretion (decreased lumen-positive potential, treat hypercalcemia) ↓ ability to concentrate urine (↓ corticopapillary gradient) ↓ ability to dilute urine (inhibition of diluting segment)
Thiazide diuretics (chlorothiazide, hydrochlorothiazide)	Early distal tubule (cortical diluting segment)	Inhibition of Na^+–Cl^- cotransport	↑ NaCl excretion ↑ K^+ excretion (↑ distal tubule flow rate) ↓ Ca^{2+} excretion (treatment of idiopathic hypercalciuria) ↓ ability to dilute urine (inhibition of cortical diluting segment) No effect on ability to concentrate urine
K^+-sparing diuretics (spironolactone, triamterene, amiloride)	Late distal tubule and collecting duct	Inhibition of Na^+ reabsorption Inhibition of K^+ secretion Inhibition of H^+ secretion	↑ Na^+ excretion (small effect) ↓ K^+ excretion (used in combination with loop or thiazide diuretics) ↓ H^+ excretion

b. The ECF volume contraction is responsible for this woman's **orthostatic hypotension.** The decreased arterial pressure produces an **increased pulse rate** via the baroreceptor mechanism.

c. The ECF volume contraction also stimulates **ADH secretion from the posterior pituitary** via volume receptors. ADH causes increased water reabsorption from the collecting ducts, which results in decreased serum [Na^+] **(hyponatremia)** and decreased serum osmolarity. Thus, ADH released by a volume mechanism is "inappropriate" for the serum osmolarity in this case.

d. **Hyperpigmentation** is caused by adrenal insufficiency. Decreased levels of cortisol produce increased secretion of adrenocorticotropic hormone (ACTH) by negative feedback. ACTH has pigmenting effects similar to those of melanocyte-stimulating hormone.

B. Vomiting

1. Case study

- A man is admitted to the hospital for evaluation of severe epigastric pain. He has had persistent nausea and vomiting for 4 days. Upper gastrointestinal (GI) endoscopy shows a pyloric ulcer with partial gastric outlet obstruction. He has orthostatic hypotension, decreased serum [K^+], decreased serum [Cl^-], arterial blood gases consistent with metabolic alkalosis, and decreased ventilation rate.

2. Responses to vomiting (Figure 5.26)

a. Loss of H^+ from the stomach by vomiting causes increased blood [HCO_3^-] and **metabolic alkalosis.** Because Cl^- is lost from the stomach along with H^+, **hypochloremia** and **ECF volume contraction** occur.

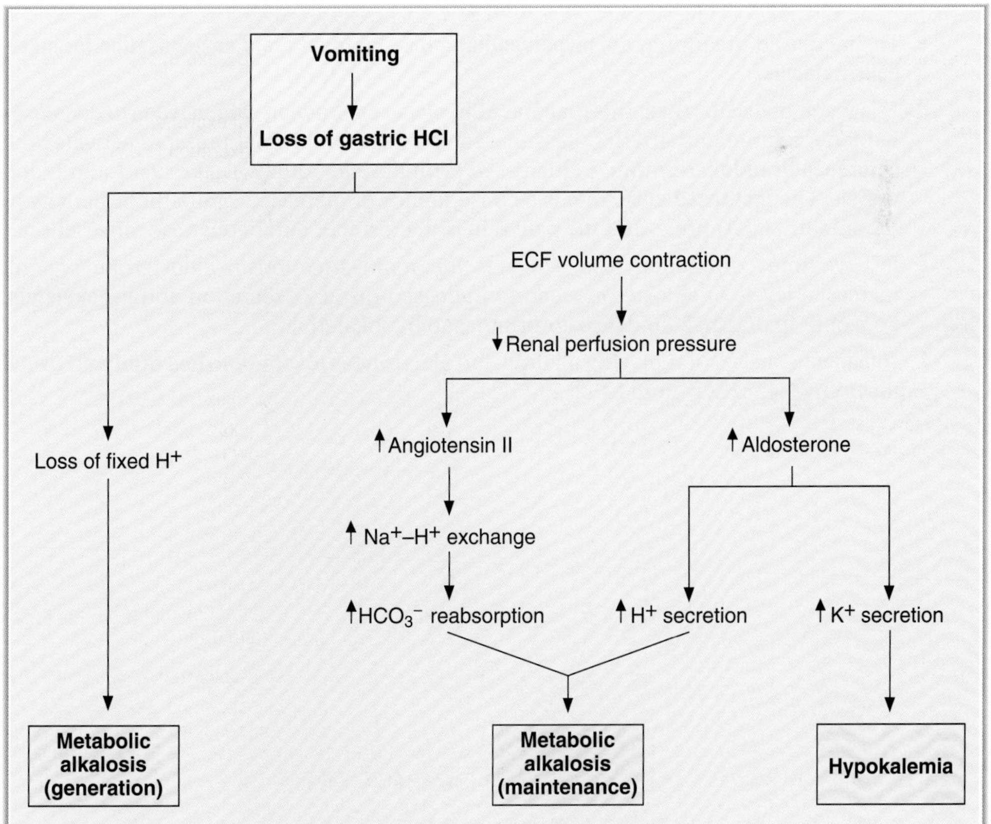

FIGURE 5.26. Metabolic alkalosis caused by vomiting. ECF = extracellular fluid.

b. The decreased ventilation rate is the **respiratory compensation for metabolic alkalosis.**

c. ECF volume contraction is associated with decreased blood volume and **decreased renal perfusion pressure.** As a result, renin secretion is increased, production of angiotensin II is increased, and **secretion of aldosterone** is increased. Thus, the ECF volume contraction worsens the metabolic alkalosis because angiotensin II increases HCO_3^- reabsorption in the proximal tubule **(contraction alkalosis).**

d. The increased levels of aldosterone (secondary to ECF volume contraction) cause increased distal K^+ secretion and **hypokalemia.** Increased aldosterone also causes increased distal H^+ secretion, further worsening the metabolic alkalosis.

e. Treatment consists of NaCl infusion to correct ECF volume contraction (which is maintaining the metabolic alkalosis and causing hypokalemia) and administration of K^+ to replace K^+ lost in the urine.

C. Diarrhea

1. Case study

- A man returns from a trip abroad with "traveler's diarrhea." He has weakness, weight loss, orthostatic hypotension, increased pulse rate, increased breathing rate, pale skin, a serum [Na^+] of 132 mEq/L, a serum [Cl^-] of 111 mEq/L, and a serum [K^+] of 2.3 mEq/L. His arterial blood gases are pH 7.25, P_{CO_2} 24 mm Hg, and HCO_3^- 10.2 mEq/L.

2. Explanation of responses to diarrhea

a. Loss of HCO_3^- from the GI tract causes a decrease in the blood [HCO_3^-] and, according to the Henderson-Hasselbalch equation, a decrease in blood pH. Thus, this man has **metabolic acidosis.**

b. To maintain electroneutrality, the HCO_3^- lost from the body is replaced by Cl^-, a measured anion; thus, there is a **normal anion gap.** The serum anion gap = $[Na^+] - ([Cl^-] + [HCO_3^-]) = 132 - (111 + 10.2) = 10.8$ mEq/L.

c. The increased breathing rate **(hyperventilation)** is the **respiratory compensation for metabolic acidosis.**

d. As a result of his diarrhea, this man has **ECF volume contraction**, which leads to decreases in blood volume and arterial pressure. The decrease in arterial pressure activates the **baroreceptor reflex**, resulting in increased sympathetic outflow to the heart and blood vessels. The **increased pulse rate** is a consequence of increased sympathetic activity in the sinoatrial (SA) node, and the pale skin is the result of cutaneous vasoconstriction.

e. ECF volume contraction also activates the renin–angiotensin–aldosterone system. Increased levels of aldosterone lead to increased distal K^+ secretion and **hypokalemia.** Loss of K^+ in diarrhea fluid also contributes to hypokalemia.

f. Treatment consists of replacing all fluid and electrolytes lost in diarrhea fluid and urine, including Na^+, HCO_3^-, and K^+.

Review Test

1. Secretion of K^+ by the distal tubule will be decreased by

(A) metabolic alkalosis
(B) a high-K^+ diet
(C) hyperaldosteronism
(D) spironolactone administration
(E) thiazide diuretic administration

2. Jared and Adam both weigh 70 kg. Jared drinks 2 L of distilled water, and Adam drinks 2 L of isotonic NaCl. As a result of these ingestions, Adam will have a

(A) greater change in intracellular fluid (ICF) volume
(B) higher positive free-water clearance (C_{H_2O})
(C) greater change in plasma osmolarity
(D) higher urine osmolarity
(E) higher urine flow rate

QUESTIONS 3 AND 4

A 45-year-old woman develops severe diarrhea while on vacation. She has the following arterial blood values:

pH = 7.25
P_{CO_2} = 24 mm Hg
$[HCO_3^-]$ = 10 mEq/L
Venous blood samples show decreased blood $[K^+]$ and a normal anion gap.

3. The correct diagnosis for this patient is

(A) metabolic acidosis
(B) metabolic alkalosis
(C) respiratory acidosis
(D) respiratory alkalosis
(E) normal acid–base status

4. Which of the following statements about this patient is correct?

(A) She is hypoventilating
(B) The decreased arterial $[HCO_3^-]$ is a result of buffering of excess H^+ by HCO_3^-
(C) The decreased blood $[K^+]$ is a result of exchange of intracellular H^+ for extracellular K^+
(D) The decreased blood $[K^+]$ is a result of increased circulating levels of aldosterone
(E) The decreased blood $[K^+]$ is a result of decreased circulating levels of antidiuretic hormone (ADH)

5. Use the values below to answer the following question.

Glomerular capillary hydrostatic pressure = 47 mm Hg
Bowman space hydrostatic pressure = 10 mm Hg
Bowman space oncotic pressure = 0 mm Hg
At what value of glomerular capillary oncotic pressure would glomerular filtration stop?

(A) 57 mm Hg
(B) 47 mm Hg
(C) 37 mm Hg
(D) 10 mm Hg
(E) 0 mm Hg

6. The reabsorption of filtered HCO_3^-

(A) results in reabsorption of less than 50% of the filtered load when the plasma concentration of HCO_3^- is 24 mEq/L
(B) acidifies tubular fluid to a pH of 4.4
(C) is directly linked to excretion of H^+ as NH_4^+
(D) is inhibited by decreases in arterial P_{CO_2}
(E) can proceed normally in the presence of a renal carbonic anhydrase inhibitor

7. The following information was obtained in a 20-year-old college student who was participating in a research study in the Clinical Research Unit:

Plasma	Urine
[Inulin] = 1 mg/mL	[Inulin] = 150 mg/mL
[X] = 2 mg/mL	[X] = 100 mg/mL
	Urine flow rate = 1 mL/min

Assuming that X is freely filtered, which of the following statements is most correct?

(A) There is net secretion of X
(B) There is net reabsorption of X
(C) There is both reabsorption and secretion of X
(D) The clearance of X could be used to measure the glomerular filtration rate (GFR)
(E) The clearance of X is greater than the clearance of inulin

8. To maintain normal H^+ balance, total daily excretion of H^+ should equal the daily

(A) fixed acid production plus fixed acid ingestion
(B) HCO_3^- excretion
(C) HCO_3^- filtered load
(D) titratable acid excretion
(E) filtered load of H^+

9. One gram of mannitol was injected into a woman. After equilibration, a plasma sample had a mannitol concentration of 0.08 g/L. During the equilibration period, 20% of the injected mannitol was excreted in the urine. The woman's

(A) extracellular fluid (ECF) volume is 1 L
(B) intracellular fluid (ICF) volume is 1 L
(C) ECF volume is 10 L
(D) ICF volume is 10 L
(E) interstitial volume is 12.5 L

10. A 58-year-old man is given a glucose tolerance test. In the test, the plasma glucose concentration is increased and glucose reabsorption and excretion are measured. When the plasma glucose concentration is higher than occurs at transport maximum (T_m), the

(A) clearance of glucose is zero
(B) excretion rate of glucose equals the filtration rate of glucose
(C) reabsorption rate of glucose equals the filtration rate of glucose
(D) excretion rate of glucose increases with increasing plasma glucose concentrations
(E) renal vein glucose concentration equals the renal artery glucose concentration

11. A negative free-water clearance $\left(-C_{H_2O}\right)$ will occur in a person who

(A) drinks 2 L of distilled water in 30 minutes
(B) begins excreting large volumes of urine with an osmolarity of 100 mOsm/L after a severe head injury
(C) is receiving lithium treatment for depression and has polyuria that is unresponsive to the administration of antidiuretic hormone (ADH)
(D) has an oat cell carcinoma of the lung, and excretes urine with an osmolarity of 1000 mOsm/L

12. A buffer pair (HA/A^-) has a pK of 5.4. At a blood pH of 7.4, the concentration of HA is

(A) 1/100 that of A^-
(B) 1/10 that of A^-
(C) equal to that of A^-
(D) 10 times that of A^-
(E) 100 times that of A^-

13. Which of the following would produce an increase in the reabsorption of isosmotic fluid in the proximal tubule?

(A) Increased filtration fraction
(B) Extracellular fluid (ECF) volume expansion
(C) Decreased peritubular capillary protein concentration
(D) Increased peritubular capillary hydrostatic pressure
(E) Oxygen deprivation

14. Which of the following substances or combinations of substances could be used to measure interstitial fluid volume?

(A) Mannitol
(B) D_2O alone
(C) Evans blue
(D) Inulin and D_2O
(E) Inulin and radioactive albumin

15. At plasma para-aminohippuric acid (PAH) concentrations below the transport maximum (T_m), PAH

(A) reabsorption is not saturated
(B) clearance equals inulin clearance
(C) secretion rate equals PAH excretion rate
(D) concentration in the renal vein is close to zero
(E) concentration in the renal vein equals PAH concentration in the renal artery

16. Compared with a person who ingests 2 L of distilled water, a person with water deprivation will have a

(A) higher free-water clearance $\left(C_{H_2O}\right)$
(B) lower plasma osmolarity
(C) lower circulating level of antidiuretic hormone (ADH)
(D) higher tubular fluid/plasma (TF/P) osmolarity in the proximal tubule
(E) higher rate of H_2O reabsorption in the collecting ducts

17. Which of the following would cause an increase in both glomerular filtration rate (GFR) and renal plasma flow (RPF)?

(A) Hyperproteinemia
(B) A ureteral stone
(C) Dilation of the afferent arteriole
(D) Dilation of the efferent arteriole
(E) Constriction of the efferent arteriole

18. A patient has the following arterial blood values:

pH = 7.52
Pco_2 = 20 mm Hg
$[HCO_3^-]$ = 16 mEq/L
Which of the following statements about this patient is most likely to be correct?

(A) He is hypoventilating
(B) He has decreased ionized $[Ca^{2+}]$ in blood
(C) He has almost complete respiratory compensation
(D) He has an acid–base disorder caused by overproduction of fixed acid
(E) Appropriate renal compensation would cause his arterial $[HCO_3^-]$ to increase

19. Which of the following would best distinguish an otherwise healthy person with severe water deprivation from a person with the syndrome of inappropriate antidiuretic hormone (SIADH)?

(A) Free-water clearance (C_{H_2O})
(B) Urine osmolarity
(C) Plasma osmolarity
(D) Circulating levels of antidiuretic hormone (ADH)
(E) Corticopapillary osmotic gradient

20. Which of the following causes a decrease in renal Ca^{2+} clearance?

(A) Hypoparathyroidism
(B) Treatment with chlorothiazide
(C) Treatment with furosemide
(D) Extracellular fluid (ECF) volume expansion
(E) Hypermagnesemia

21. A patient arrives at the emergency room with low arterial pressure, reduced tissue turgor, and the following arterial blood values:

pH = 7.69
$[HCO_3^-]$ = 57 mEq/L
Pco_2 = 48 mm Hg
Which of the following responses would also be expected to occur in this patient?

(A) Hyperventilation
(B) Decreased K^+ secretion by the distal tubules
(C) Increased ratio of $H_2PO_4^-$ to HPO_4^{-2} in urine
(D) Exchange of intracellular H^+ for extracellular K^+

22. A woman has a plasma osmolarity of 300 mOsm/L and a urine osmolarity of 1200 mOsm/L. The correct diagnosis is

(A) syndrome of inappropriate antidiuretic hormone (SIADH)
(B) water deprivation
(C) central diabetes insipidus
(D) nephrogenic diabetes insipidus
(E) drinking large volumes of distilled water

23. A patient is infused with para-aminohippuric acid (PAH) to measure renal blood flow (RBF). She has a urine flow rate of 1 mL/min, a plasma [PAH] of 1 mg/mL, a urine [PAH] of 600 mg/mL, and a hematocrit of 45%. What is her "effective" RBF?

(A) 600 mL/min
(B) 660 mL/min
(C) 1091 mL/min
(D) 1333 mL/min

24. Which of the following substances has the highest renal clearance?

(A) Para-aminohippuric acid (PAH)
(B) Inulin
(C) Glucose
(D) Na^+
(E) Cl^-

25. A woman runs a marathon in 90°F weather and replaces all volume lost in sweat by drinking distilled water. After the marathon, she will have

(A) decreased total body water (TBW)
(B) decreased hematocrit
(C) decreased intracellular fluid (ICF) volume
(D) decreased plasma osmolarity
(E) increased intracellular osmolarity

26. Which of the following causes hyperkalemia?

(A) Exercise
(B) Alkalosis
(C) Insulin injection
(D) Decreased serum osmolarity
(E) Treatment with β-agonists

27. Which of the following is a cause of metabolic alkalosis?

(A) Diarrhea
(B) Chronic renal failure
(C) Ethylene glycol ingestion
(D) Treatment with acetazolamide
(E) Hyperaldosteronism
(F) Salicylate poisoning

28. Which of the following is an action of parathyroid hormone (PTH) on the renal tubule?

(A) Stimulation of adenylate cyclase
(B) Inhibition of distal tubule K^+ secretion
(C) Inhibition of distal tubule Ca^{2+} reabsorption
(D) Stimulation of proximal tubule phosphate reabsorption
(E) Inhibition of production of 1,25-dihydroxycholecalciferol

29. A man presents with hypertension and hypokalemia. Measurement of his arterial blood gases reveals a pH of 7.5 and a calculated HCO_3^- of 32 mEq/L. His serum cortisol and urinary vanillylmandelic acid (VMA) are normal, his serum aldosterone is increased, and his plasma renin activity is decreased. Which of the following is the most likely cause of his hypertension?

(A) Cushing syndrome
(B) Cushing disease
(C) Conn syndrome
(D) Renal artery stenosis
(E) Pheochromocytoma

30. Which set of arterial blood values describes a heavy smoker with a history of emphysema and chronic bronchitis who is becoming increasingly somnolent?

	pH	HCO_3^- (mEq/L)	Pco_2 (mm Hg)
(A)	7.65	48	45
(B)	7.50	15	20
(C)	7.40	24	40
(D)	7.32	30	60
(E)	7.31	16	33

31. Which set of arterial blood values describes a patient with partially compensated respiratory alkalosis after 1 month on a mechanical ventilator?

	pH	HCO_3^- (mEq/L)	Pco_2 (mm Hg)
(A)	7.65	48	45
(B)	7.50	15	20
(C)	7.40	24	40
(D)	7.32	30	60
(E)	7.31	16	33

32. Which set of arterial blood values describes a patient with chronic renal failure (eating a normal protein diet) and decreased urinary excretion of NH_4^+?

	pH	HCO_3^- (mEq/L)	Pco_2 (mm Hg)
(A)	7.65	48	45
(B)	7.50	15	20
(C)	7.40	24	40
(D)	7.32	30	60
(E)	7.31	16	33

33. Which set of arterial blood values describes a patient with untreated diabetes mellitus and increased urinary excretion of NH_4^+?

	pH	HCO_3^- (mEq/L)	Pco_2 (mm Hg)
(A)	7.65	48	45
(B)	7.50	15	20
(C)	7.40	24	40
(D)	7.32	30	60
(E)	7.31	16	33

34. Which set of arterial blood values describes a patient with a 5-day history of vomiting?

	pH	HCO_3^- (mEq/L)	Pco_2 (mm Hg)
(A)	7.65	48	45
(B)	7.50	15	20
(C)	7.40	24	40
(D)	7.32	30	60
(E)	7.31	16	33

The following figure applies to Questions 35–39.

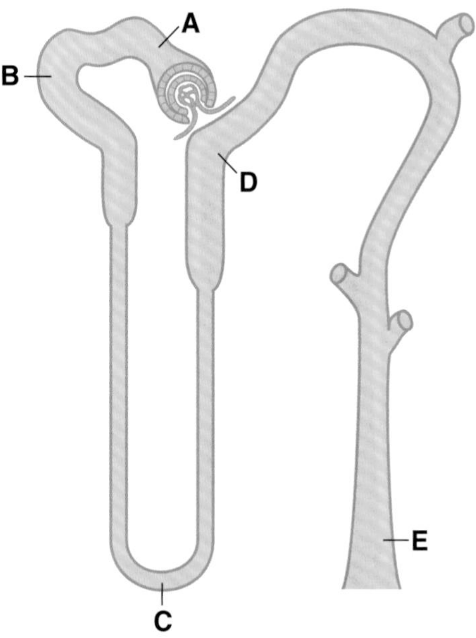

35. At which nephron site does the amount of K^+ in tubular fluid exceed the amount of filtered K^+ in a person on a high-K^+ diet?

(A) Site A
(B) Site B
(C) Site C
(D) Site D
(E) Site E

36. At which nephron site is the tubular fluid/plasma (TF/P) osmolarity lowest in a person who has been deprived of water?

(A) Site A
(B) Site B
(C) Site C
(D) Site D
(E) Site E

37. At which nephron site is the tubular fluid inulin concentration highest during antidiuresis?

(A) Site A
(B) Site B
(C) Site C
(D) Site D
(E) Site E

38. At which nephron site is the tubular fluid inulin concentration lowest?

(A) Site A
(B) Site B
(C) Site C
(D) Site D
(E) Site E

39. At which nephron site is the tubular fluid glucose concentration highest?

(A) Site A
(B) Site B
(C) Site C
(D) Site D
(E) Site E

The following graph applies to Questions 40–42. The curves show the percentage of the filtered load remaining in the tubular fluid at various sites along the nephron.

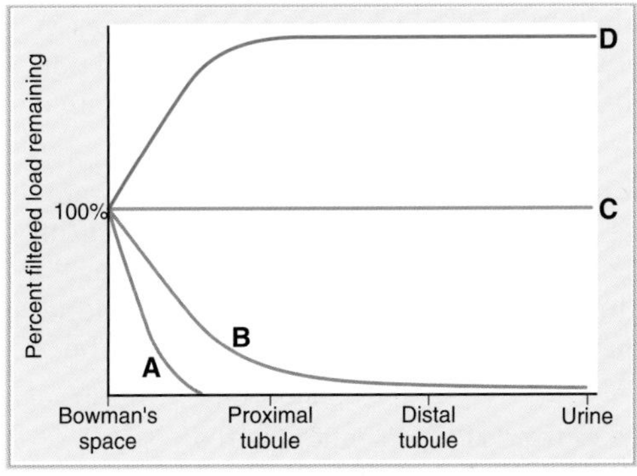

40. Which curve describes the inulin profile along the nephron?

(A) Curve A
(B) Curve B
(C) Curve C
(D) Curve D

41. Which curve describes the alanine profile along the nephron?

(A) Curve A
(B) Curve B
(C) Curve C
(D) Curve D

42. Which curve describes the para-aminohippuric acid (PAH) profile along the nephron?

(A) Curve A
(B) Curve B
(C) Curve C
(D) Curve D

43. A 5-year-old boy swallows a bottle of aspirin (salicylic acid) and is treated in the emergency room. The treatment produces a change in urine pH that increases the excretion of salicylic acid. What was the change in urine pH, and what is the mechanism of increased salicylic acid excretion?

(A) Acidification, which converts salicylic acid to its HA form
(B) Alkalinization, which converts salicylic acid to its A^- form
(C) Acidification, which converts salicylic acid to its A^- form
(D) Alkalinization, which converts salicylic acid to its HA form

44. A female graduate student is hyperventilating prior to her oral comprehensive examination. She is light-headed, and her feet and hands are numb and tingling. Which of the following set of blood values would be observed in the emergency room?

	pH	P_{CO_2}, mm Hg	P_{O_2}, mm Hg	Ionized Ca^{2+}
(A)	7.3	30	100	Decreased
(B)	7.3	50	90	Increased
(C)	7.4	40	100	Normal
(D)	7.5	30	110	Decreased
(E)	7.5	50	90	Increased

Answers and Explanations

1. **The answer is D** [V B 4 b]. Distal K^+ secretion is decreased by factors that decrease the driving force for passive diffusion of K^+ across the luminal membrane. Because spironolactone is an aldosterone antagonist, it reduces K^+ secretion. Alkalosis, a diet high in K^+, and hyperaldosteronism all increase $[K^+]$ in the distal cells and thereby increase K^+ secretion. Thiazide diuretics increase flow through the distal tubule and dilute the luminal $[K^+]$ so that the driving force for K^+ secretion is increased.
2. **The answer is D** [I C 2 a; VII C; Figure 5.15; Table 5.6]. After drinking distilled water, Jared will have an increase in intracellular fluid (ICF) and extracellular fluid (ECF) volumes, a decrease in plasma osmolarity, a suppression of antidiuretic hormone (ADH) secretion, and a positive free-water clearance (C_{H_2O}) and will produce *dilute* urine with a high flow rate. Adam, after drinking the same volume of isotonic NaCl, will have an increase in ECF volume only and no change in plasma osmolarity. Because Adam's ADH will not be suppressed, he will have a higher urine osmolarity, a lower urine flow rate, and a lower C_{H_2O} than Jared.
3. **The answer is A** [IX D 1 a–c; Tables 5.8 and 5.9]. An acid pH, together with decreased HCO_3^- and decreased P_{CO_2}, is consistent with metabolic acidosis with respiratory compensation (hyperventilation). Diarrhea causes gastrointestinal (GI) loss of HCO_3^-, creating a metabolic acidosis.
4. **The answer is D** [IX D 1 a–c; Tables 5.8 and 5.9]. The decreased arterial $[HCO_3^-]$ is caused by gastrointestinal (GI) loss of HCO_3^- from diarrhea, not by buffering of excess H^+ by HCO_3^-. The woman is hyperventilating as respiratory compensation for metabolic acidosis. Her hypokalemia cannot be the result of the exchange of intracellular H^+ for extracellular K^+, because she has an increase in extracellular H^+, which would drive the exchange in the other direction. Her circulating levels of aldosterone would be increased as a result of extracellular fluid (ECF) volume contraction, which leads to increased K^+ secretion by the distal tubule and hypokalemia.
5. **The answer is C** [II C 4, 5]. Glomerular filtration will stop when the net ultrafiltration pressure across the glomerular capillary is zero; that is, when the force that favors filtration (47 mm Hg) exactly equals the forces that oppose filtration (10 mm Hg + 37 mm Hg).
6. **The answer is D** [IX C 1 a, b]. Decreases in arterial P_{CO_2} cause a decrease in the reabsorption of filtered HCO_3^- by diminishing the supply of H^+ in the cell for secretion into the lumen. Reabsorption of filtered HCO_3^- is nearly 100% of the filtered load and requires carbonic anhydrase in the brush border to convert filtered HCO_3^- to CO_2 to proceed normally. This process causes little acidification of the urine and is not linked to net excretion of H^+ as titratable acid or NH_4^+.
7. **The answer is B** [II C 1]. To answer this question, calculate the glomerular filtration rate (GFR) and C_X. GFR = 150 mg/mL × 1 mL/min ÷ 1 mg/mL = 150 mL/min. C_X = 100 mg/mL × 1 mL/min ÷ 2 mg/mL = 50 mL/min. Because the clearance of X is less than the clearance of inulin (or GFR), *net reabsorption of X* must have occurred. Clearance data alone cannot determine whether there has also been secretion of X. Because GFR cannot be measured with a substance that is reabsorbed, X would not be suitable.
8. **The answer is A** [IX C 2]. Total daily production of fixed H^+ from catabolism of proteins and phospholipids (plus any additional fixed H^+ that is ingested) must be matched by the sum of excretion of H^+ as titratable acid plus NH_4^+ to maintain acid–base balance.

9. **The answer is C** [I B 1 a]. Mannitol is a marker substance for the extracellular fluid (ECF) volume. ECF volume = amount of mannitol/concentration of mannitol = 1 g − 0.2 g/0.08 g/L = 10 L.

10. **The answer is D** [III B; Figure 5.5]. At concentrations greater than at the transport maximum (T_m) for glucose, the carriers are saturated so that the reabsorption rate no longer matches the filtration rate. The difference is excreted in the urine. As the plasma glucose concentration increases, the excretion of glucose increases. When it is greater than the T_m, the renal vein glucose concentration will be less than the renal artery concentration because some glucose is being excreted in urine and therefore is not returned to the blood. The clearance of glucose is zero at concentrations lower than at T_m (or lower than threshold) when all of the filtered glucose is reabsorbed but is greater than zero at concentrations greater than T_m.

11. **The answer is D** [VII D; Table 5.6]. A person who produces hyperosmotic urine (1000 mOsm/L) will have a negative free-water clearance ($-C_{H_2O}$) [$C_{H_2O} = V - C_{osm}$]. All of the others will have a positive C_{H_2O} because they are producing hyposmotic urine as a result of the suppression of antidiuretic hormone (ADH) by water drinking, central diabetes insipidus, or nephrogenic diabetes insipidus.

12. **The answer is A** [IX B 3]. The Henderson-Hasselbalch equation can be used to calculate the ratio of HA/A⁻:

$$pH = pK + \log A^-/HA$$
$$7.4 = 5.4 + \log A^-/HA$$
$$2.0 = \log A^-/HA$$
$$100 = A^-/HA \text{ or } HA/A^- \text{ is } 1/100$$

13. **The answer is A** [II C 3; IV C 1 d (2)]. Increasing filtration fraction means that a larger portion of the renal plasma flow (RPF) is filtered across the glomerular capillaries. This increased flow causes an increase in the protein concentration and oncotic pressure of the blood leaving the glomerular capillaries. This blood becomes the peritubular capillary blood supply. The increased oncotic pressure in the peritubular capillary blood is a driving force *favoring reabsorption* in the proximal tubule. Extracellular fluid (ECF) volume expansion, decreased peritubular capillary protein concentration, and increased peritubular capillary hydrostatic pressure all inhibit proximal reabsorption. Oxygen deprivation would also inhibit reabsorption by stopping the Na^+–K^+ pump in the basolateral membranes.

14. **The answer is E** [I B 2 b–d]. Interstitial fluid volume is measured indirectly by determining the difference between extracellular fluid (ECF) volume and plasma volume. Inulin, a large fructose polymer that is restricted to the extracellular space, is a marker for ECF volume. Radioactive albumin is a marker for plasma volume.

15. **The answer is D** [III C; Figure 5.6]. At plasma concentrations that are lower than at the transport maximum (T_m) for para-aminohippuric acid (PAH) secretion, PAH concentration in the renal vein is nearly zero because the sum of filtration plus secretion removes virtually all PAH from the renal plasma. Thus, the PAH concentration in the renal vein is less than that in the renal artery because most of the PAH entering the kidney is excreted in urine. PAH clearance is greater than inulin clearance because PAH is filtered and secreted; inulin is only filtered.

16. **The answer is E** [VII D; Figures 5.14 and 5.15]. The person with water deprivation will have a higher plasma osmolarity and higher circulating levels of antidiuretic hormone (ADH). These effects will increase the rate of H_2O reabsorption in the collecting ducts and create a *negative* free-water clearance ($-C_{H_2O}$). Tubular fluid/plasma (TF/P) osmolarity in the proximal tubule is not affected by ADH.

17. **The answer is C** [II C 4; Table 5.3]. Dilation of the afferent arteriole will increase both renal plasma flow (RPF) (because renal vascular resistance is decreased) and glomerular filtration rate (GFR) (because glomerular capillary hydrostatic pressure is increased). Dilation

of the efferent arteriole will increase RPF but decrease GFR. Constriction of the efferent arteriole will decrease RPF (due to increased renal vascular resistance) and increase GFR. Both hyperproteinemia (↑ π in the glomerular capillaries) and a ureteral stone (↑ hydrostatic pressure in Bowman space) will oppose filtration and decrease GFR.

18. The answer is B [IX D 4; Table 5.8]. First, the acid–base disorder must be diagnosed. Alkaline pH, low Pco_2, and low HCO_3^- are consistent with respiratory alkalosis. In *respiratory alkalosis,* the [H^+] is decreased and less H^+ is bound to negatively charged sites on plasma proteins. As a result, more Ca^{2+} is bound to proteins and, therefore, the *ionized* [Ca^{2+}] decreases. There is no respiratory compensation for primary respiratory disorders. The patient is hyperventilating, which is the cause of the respiratory alkalosis. Appropriate renal compensation would be decreased reabsorption of HCO_3^-, which would cause his arterial [HCO_3^-] to decrease and his blood pH to decrease (become more normal).

19. The answer is C [VII B, D 4; Table 5.6]. Both individuals will have hyperosmotic urine, a negative free-water clearance ($-C_{H_2O}$), a normal corticopapillary gradient, and high circulating levels of antidiuretic hormone (ADH). The person with water deprivation will have a high plasma osmolarity, and the person with syndrome of inappropriate antidiuretic hormone (SIADH) will have a low plasma osmolarity (because of dilution by the inappropriate water reabsorption).

20. The answer is B [Table 5.11]. Thiazide diuretics have a unique effect on the distal tubule; they increase Ca^{2+} reabsorption, thereby decreasing Ca^{2+} excretion and clearance. Because parathyroid hormone (PTH) increases Ca^{2+} reabsorption, the lack of PTH will cause an increase in Ca^{2+} clearance. Furosemide inhibits Na^+ reabsorption in the thick ascending limb, and extracellular fluid (ECF) volume expansion inhibits Na^+ reabsorption in the proximal tubule. At these sites, Ca^{2+} reabsorption is linked to Na^+ reabsorption, and Ca^{2+} clearance would be increased. Because Mg^{2+} competes with Ca^{2+} for reabsorption in the thick ascending limb, hypermagnesemia will cause increased Ca^{2+} clearance.

21. The answer is D [IX D 2; Table 5.8]. First, the acid–base disorder must be diagnosed. Alkaline pH, with increased HCO_3^- and increased Pco_2, is consistent with metabolic alkalosis with respiratory compensation. The low blood pressure and decreased turgor suggest extracellular fluid (ECF) volume contraction. The reduced [H^+] in blood will cause intracellular H^+ to leave cells in exchange for extracellular K^+. The appropriate respiratory compensation is *hypoventilation,* which is responsible for the elevated Pco_2. H^+ excretion in urine will be decreased, so less titratable acid will be excreted. K^+ secretion by the distal tubules will be increased because aldosterone levels will be increased secondary to ECF volume contraction.

22. The answer is B [VII B; Figure 5.14]. This patient's plasma and urine osmolarity, taken together, are consistent with water deprivation. The plasma osmolarity is on the high side of normal, stimulating the posterior pituitary to secrete antidiuretic hormone (ADH). Secretion of ADH, in turn, acts on the collecting ducts to increase water reabsorption and produce hyperosmotic urine. Syndrome of inappropriate antidiuretic hormone (SIADH) would also produce hyperosmotic urine, but the plasma osmolarity would be lower than normal because of the excessive water retention. Central and nephrogenic diabetes insipidus and excessive water intake would all result in hyposmotic urine.

23. The answer is C [II B 2, 3]. Effective renal plasma flow (RPF) is calculated from the clearance of paraaminohippuric acid (PAH) [$C_{PAH} = U_{PAH} \times V/P_{PAH} = 600$ mL/min]. Renal blood flow (RBF) = RPF/1 – hematocrit = 1091 mL/min.

24. The answer is A [III D]. Para-aminohippuric acid (PAH) has the greatest clearance of all of the substances because it is both filtered and secreted. Inulin is only filtered. The other substances are filtered and subsequently reabsorbed; therefore, they will have clearances that are lower than the inulin clearance.

25. The answer is D [I C 2 f; Table 5.2]. By sweating and then replacing all volume by drinking H_2O, the woman has a *net loss of NaCl without a net loss of H_2O.* Therefore, her extracellular and plasma osmolarity will be decreased, and as a result, water will flow from extracellular fluid (ECF) to intracellular fluid (ICF). The intracellular osmolarity will also be decreased after the shift of water. Total body water (TBW) will be unchanged because the woman replaced all volume lost in sweat by drinking water. Hematocrit will be increased because of the shift of water from ECF to ICF and the shift of water into the red blood cells (RBCs), which causes their volume to increase.

26. **The answer is A** [Table 5.4]. Exercise causes a shift of K^+ from cells into blood. The result is hyperkalemia. Hyposmolarity, insulin, β-agonists, and alkalosis cause a shift of K^+ from blood into cells. The result is hypokalemia.

27. **The answer is E** [Table 5.9]. A cause of metabolic alkalosis is hyperaldosteronism; increased aldosterone levels cause increased H^+ secretion by the distal tubule and increased reabsorption of "new" HCO_3^-. Diarrhea causes loss of HCO_3^- from the gastrointestinal (GI) tract and acetazolamide causes loss of HCO_3^- in the urine, both resulting in hyperchloremic metabolic acidosis with normal anion gap. Ingestion of ethylene glycol and salicylate poisoning leads to metabolic acidosis with increased anion gap.

28. **The answer is A** [VI B; Table 5.7]. Parathyroid hormone (PTH) acts on the renal tubule by stimulating adenyl cyclase and generating cyclic adenosine monophosphate (cAMP). The major actions of the hormone are inhibition of phosphate reabsorption in the proximal tubule, stimulation of Ca^{2+} reabsorption in the distal tubule, and stimulation of 1,25-dihydroxycholecalciferol production. PTH does not alter the renal handling of K^+.

29. **The answer is C** [IV C 3 b; V B 4 b]. Hypertension, hypokalemia, metabolic alkalosis, elevated serum aldosterone, and decreased plasma renin activity are all consistent with a primary hyperaldosteronism (e.g., Conn syndrome). High levels of aldosterone cause increased Na^+ reabsorption (leading to increased blood pressure), increased K^+ secretion (leading to hypokalemia), and increased H^+ secretion (leading to metabolic alkalosis). In Conn syndrome, the increased blood pressure causes an increase in renal perfusion pressure, which inhibits renin secretion. Neither Cushing syndrome nor Cushing disease is a possible cause of this patient's hypertension because serum cortisol and adrenocorticotropic hormone (ACTH) levels are normal. Renal artery stenosis causes hypertension that is characterized by increased plasma renin activity. Pheochromocytoma is ruled out by the normal urinary excretion of vanillylmandelic acid (VMA).

30. **The answer is D** [IX D 3; Tables 5.8 and 5.9]. The history strongly suggests chronic obstructive pulmonary disease (COPD) as a cause of respiratory acidosis. Because of the COPD, the ventilation rate is decreased and CO_2 is retained. The $[H^+]$ and $[HCO_3^-]$ are increased by mass action. The $[HCO_3^-]$ is further increased by renal compensation for respiratory acidosis (increased HCO_3^- reabsorption by the kidney is facilitated by the high Pco_2).

31. **The answer is B** [IX D 4; Table 5.8]. The blood values in respiratory alkalosis show decreased Pco_2 (the cause) and decreased $[H^+]$ and $[HCO_3^-]$ by mass action. The $[HCO_3^-]$ is further decreased by renal compensation for chronic respiratory alkalosis (decreased HCO_3^- reabsorption).

32. **The answer is E** [IX D 1; Tables 5.8 and 5.9]. In patients who have chronic renal failure and ingest normal amounts of protein, fixed acids will be produced from the catabolism of protein. Because the failing kidney does not produce enough NH_4^+ to excrete all of the fixed acid, metabolic acidosis (with respiratory compensation) results.

33. **The answer is E** [IX D 1; Tables 5.8 and 5.9]. Untreated diabetes mellitus results in the production of keto acids, which are fixed acids that cause metabolic acidosis. Urinary excretion of NH_4^+ is increased in this patient because an adaptive increase in renal NH_3 synthesis has occurred in response to the metabolic acidosis.

34. **The answer is A** [IX D 2; Tables 5.8 and 5.9]. The history of vomiting (in the absence of any other information) indicates loss of gastric H^+ and, as a result, metabolic alkalosis (with respiratory compensation).

35. **The answer is E** [V B 4]. K^+ is secreted by the late distal tubule and collecting ducts. Because this secretion is affected by dietary K^+, a person who is on a high-K^+ diet can secrete more K^+ into the urine than was originally filtered. At all of the other nephron

sites, the amount of K^+ in the tubular fluid is either equal to the amount filtered (site *A*) or less than the amount filtered (because K^+ is reabsorbed in the proximal tubule and the loop of Henle).

36. **The answer is D** [VII B 3; Figure 5.16]. A person who is deprived of water will have high circulating levels of antidiuretic hormone (ADH). The tubular fluid/plasma (TF/P) osmolarity is 1.0 throughout the proximal tubule, regardless of ADH status. In antidiuresis, TF/P osmolarity is greater than 1.0 at site *C* because of equilibration of the tubular fluid with the large corticopapillary osmotic gradient. At site *E*, TF/P osmolarity is greater than 1.0 because of water reabsorption out of the collecting ducts and equilibration with the corticopapillary gradient. At site *D*, the tubular fluid is diluted because NaCl is reabsorbed in the thick ascending limb without water, making TF/P osmolarity less than 1.0.

37. **The answer is E** [IV A 2]. Because inulin, once filtered, is neither reabsorbed nor secreted, its concentration in tubular fluid reflects the amount of water remaining in the tubule. In antidiuresis, water is reabsorbed throughout the nephron (except in the thick ascending limb and cortical diluting segment). Thus, inulin concentration in the tubular fluid progressively rises along the nephron as water is reabsorbed, and will be highest in the final urine.

38. **The answer is A** [IV A 2]. The tubular fluid inulin concentration depends on the amount of water present. As water reabsorption occurs along the nephron, the inulin concentration progressively increases. Thus, the tubular fluid inulin concentration is lowest in Bowman space, prior to any water reabsorption.

39. **The answer is A** [IV C 1 a]. Glucose is extensively reabsorbed in the early proximal tubule by the Na^+–glucose cotransporter. The glucose concentration in tubular fluid is highest in Bowman space before any reabsorption has occurred.

40. **The answer is C** [IV A 2]. Once inulin is filtered, it is neither reabsorbed nor secreted. Thus, 100% of the filtered inulin remains in tubular fluid at each nephron site and in the final urine.

41. **The answer is A** [IV C 1 a]. Alanine, like glucose, is avidly reabsorbed in the early proximal tubule by a Na^+–amino acid cotransporter. Thus, the percentage of the filtered load of alanine remaining in the tubular fluid declines rapidly along the proximal tubule as alanine is reabsorbed into the blood.

42. **The answer is D** [III C; IVA 3]. Para-aminohippuric acid (PAH) is an organic acid that is filtered and subsequently secreted by the proximal tubule. The secretion process adds PAH to the tubular fluid; therefore, the amount that is present at the end of the proximal tubule is greater than the amount that was present in Bowman space.

43. **The answer is B** [III E]. Alkalinization of the urine converts more salicylic acid to its A^- form. The A^- form is charged and cannot back-diffuse from urine to blood. Therefore, it is trapped in the urine and excreted.

44. **The answer is D** [IX D 3; Tables 5.8 and 5.9]. The student is hyperventilating, which causes decreased P_{CO_2} and increased P_{O_2}. The decreased P_{CO_2} leads to increased pH. The increased pH causes a decrease in ionized Ca^{2+} concentration because H^+ and Ca^{2+} compete for binding to plasma albumin; with increased pH (decreased H^+ concentration), there is less binding of H^+ to hemoglobin, more binding of Ca^{2+}, and decreased free ionized Ca^{2+} concentration. The decreased free ionized Ca^{2+} concentration is responsible for her symptoms of tingling and numbness.

Chapter 6

Gastrointestinal Physiology

I. STRUCTURE AND INNERVATION OF THE GASTROINTESTINAL TRACT

A. Structure of the gastrointestinal (GI) tract (Figure 6.1)

1. Epithelial cells

- are specialized in different parts of the GI tract for **secretion** or **absorption.**

2. Muscularis mucosa

- Contraction causes a change in the surface area for secretion or absorption.

3. Circular muscle

- Contraction causes a **decrease in diameter** of the lumen of the GI tract.

4. Longitudinal muscle

- Contraction causes **shortening** of a segment of the GI tract.

5. Submucosal plexus (Meissner plexus) and myenteric plexus

- comprise the **enteric nervous system** of the GI tract.
- integrate and coordinate the motility, secretory, and endocrine functions of the GI tract.

B. Innervation of the GI tract

- The autonomic nervous system (ANS) of the GI tract comprises both extrinsic and intrinsic nervous systems.

1. Extrinsic innervation (parasympathetic and sympathetic nervous systems)

- **Efferent fibers** carry information from the brainstem and spinal cord to the GI tract.
- **Afferent fibers** carry sensory information from chemoreceptors and mechanoreceptors in the GI tract to the brainstem and spinal cord.

a. Parasympathetic nervous system

- is **usually excitatory** on the functions of the GI tract.
- is carried via the vagus and pelvic nerves.
- Preganglionic parasympathetic fibers synapse in the myenteric and submucosal plexuses.
- Cell bodies in the ganglia of the plexuses then send information to the smooth muscle, secretory cells, and endocrine cells of the GI tract.

(1) The **vagus nerve** innervates the esophagus, stomach, pancreas, and upper large intestine.

- Reflexes in which both afferent and efferent pathways are contained in the vagus nerve are called **vagovagal reflexes.**

(2) The **pelvic nerve** innervates the lower large intestine, rectum, and anus.

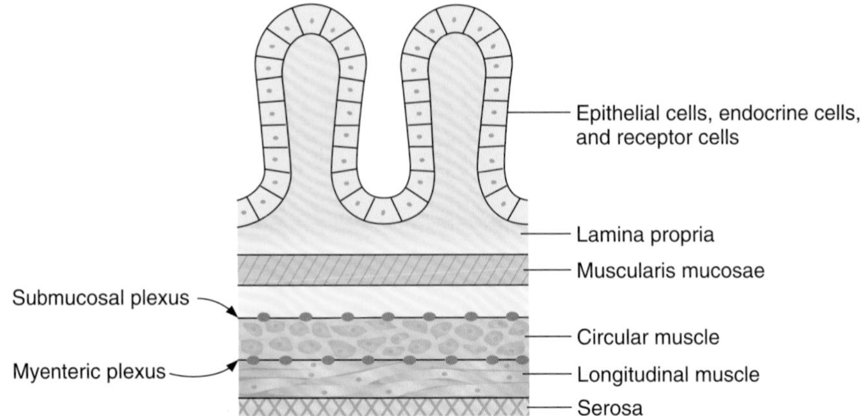

FIGURE 6.1. Structure of the gastrointestinal tract.

b. Sympathetic nervous system

- is **usually inhibitory** on the functions of the GI tract.
- Fibers originate in the spinal cord between T8 and L2.
- Preganglionic sympathetic cholinergic fibers synapse in the prevertebral ganglia.
- Postganglionic sympathetic adrenergic fibers leave the prevertebral ganglia and synapse in the myenteric and submucosal plexuses. Direct postganglionic adrenergic innervation of blood vessels and some smooth muscle cells also occurs.
- Cell bodies in the ganglia of the plexuses then send information to the smooth muscle, secretory cells, and endocrine cells of the GI tract.

2. Intrinsic innervation (enteric nervous system)

- coordinates and relays information from the parasympathetic and sympathetic nervous systems to the GI tract.
- uses **local reflexes** to relay information **within the GI tract.**
- controls most functions of the GI tract, especially motility and secretion, even in the absence of extrinsic innervation.

a. Myenteric plexus (Auerbach plexus)

- primarily controls the **motility** of the GI smooth muscle.

b. Submucosal plexus (Meissner plexus)

- primarily controls **secretion and blood flow.**
- receives sensory information from chemoreceptors and mechanoreceptors in the GI tract.

II. REGULATORY SUBSTANCES IN THE GASTROINTESTINAL TRACT (FIGURE 6.2)

A. GI hormones (Table 6.1)

- are released from endocrine cells in the GI mucosa into the portal circulation, enter the general circulation, and have physiologic actions on target cells.
- Four substances meet the requirements to be considered "official" GI hormones; others are considered "candidate" hormones. The four official GI hormones are **gastrin, cholecystokinin (CCK), secretin,** and **glucose-dependent insulinotropic peptide (GIP).**

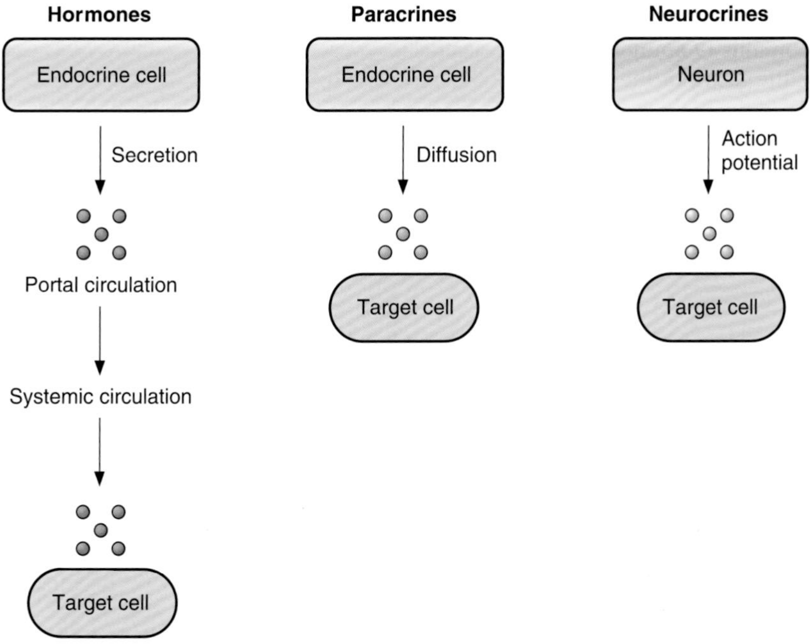

FIGURE 6.2. Gastrointestinal hormones, paracrines, and neurocrines.

table 6.1 Summary of Gastrointestinal (GI) Hormones

Hormones	Homology (Family)	Site of Secretion	Stimulus for Secretion	Actions
Gastrin	Gastrin–CCK	G cells of stomach	Small peptides and amino acids Distention of stomach Vagus (via GRP) Inhibited by H^+ in stomach Inhibited by somatostatin	↑ Gastric H^+ secretion Stimulates growth of gastric mucosa
CCK	Gastrin–CCK	I cells of duodenum and jejunum	Small peptides and amino acids Fatty acids	Stimulates contraction of gallbladder and relaxation of sphincter of Oddi ↑ Pancreatic enzyme and HCO_3^- secretion ↑ Growth of exocrine pancreas/gallbladder Inhibits gastric emptying
Secretin	Secretin–glucagon	S cells of duodenum	H^+ in duodenum Fatty acids in duodenum	↑ Pancreatic HCO_3^- secretion ↑ Biliary HCO_3^- secretion ↓ Gastric H^+ secretion
GIP	Secretin–glucagon	Duodenum and jejunum	Fatty acids, amino acids, and oral glucose	↑ Insulin secretion ↓ Gastric H^+ secretion

CCK = cholecystokinin; GIP = glucose-dependent insulinotropic peptide; GRP = gastrin-releasing peptide.

1. Gastrin

- contains 17 amino acids **("little gastrin").**
- Little gastrin is the form secreted in response to a meal.
- All of the biologic activity of gastrin resides in the **four C-terminal amino acids.**
- **"Big gastrin"** contains 34 amino acids, although it is not a dimer of little gastrin.

a. Actions of gastrin

(1) *increases H^+ secretion* by the gastric **parietal cells.**

(2) *stimulates growth of gastric mucosa* by stimulating the synthesis of RNA and new protein. Patients with **gastrin-secreting tumors** have hypertrophy and hyperplasia of the gastric mucosa.

b. Stimuli for secretion of gastrin

- Gastrin is secreted from the **G cells** of the **gastric antrum** in response to a meal.
- Gastrin is secreted in response to the following:

(1) *small peptides and amino acids* in the lumen of the stomach.

- The most potent stimuli for gastrin secretion are **phenylalanine** and **tryptophan.**

(2) *distention of the stomach.*

(3) vagal stimulation, mediated by **gastrin-releasing peptide (GRP).**

- Atropine does not block vagally mediated gastrin secretion because the mediator of the vagal effect is GRP, not acetylcholine (ACh).

c. Inhibition of gastrin secretion

- **H^+ in the lumen of the stomach** inhibits gastrin release. This negative feedback control ensures that gastrin secretion is inhibited if the stomach contents are sufficiently acidified.
- **Somatostatin** inhibits gastrin release.

d. Zollinger–Ellison syndrome (gastrinoma)

- occurs when gastrin is secreted by non–β-cell tumors of the pancreas.

2. CCK

- contains 33 amino acids.
- is **homologous to gastrin.**
- The five C-terminal amino acids are the same in CCK and gastrin.
- The biologic activity of CCK resides in the **C-terminal heptapeptide.** Thus, the heptapeptide contains the sequence that is homologous to gastrin and has both CCK and gastrin activity.

a. Actions of CCK

(1) stimulates **contraction of the gallbladder** and simultaneously causes **relaxation of the sphincter of Oddi** for secretion of bile.

(2) stimulates **pancreatic enzyme secretion.**

(3) potentiates secretin-induced stimulation of pancreatic HCO_3^- secretion.

(4) stimulates **growth of the exocrine pancreas.**

(5) **inhibits gastric emptying.** Thus, meals containing fat stimulate the secretion of CCK, which slows gastric emptying to allow more time for intestinal digestion and absorption.

b. Stimuli for the release of CCK

- CCK is released from the **I cells** of the **duodenal and jejunal mucosa** by

(1) **small peptides and amino acids.**

(2) **fatty acids and monoglycerides.**

- Triglycerides do not stimulate the release of CCK because they cannot cross intestinal cell membranes.

3. Secretin

- contains 27 amino acids.
- is **homologous to glucagon;** 14 of the 27 amino acids in secretin are the same as those in glucagon.
- All of the amino acids are required for biologic activity.

a. Actions of secretin

- are coordinated to reduce the amount of H^+ in the lumen of the small intestine.

(1) stimulates pancreatic HCO_3^- secretion and increases **growth of the exocrine pancreas.** Pancreatic HCO_3^- neutralizes H^+ in the intestinal lumen.
(2) stimulates HCO_3^- and H_2O secretion by the liver and increases **bile production.**
(3) inhibits H^+ secretion by gastric parietal cells.

b. Stimuli for the release of secretin

- Secretin is released by the **S cells** of the **duodenum** in response to

(1) H^+ in the lumen of the duodenum.
(2) fatty acids in the lumen of the duodenum.

4. GIP

- contain 42 amino acids.
- is **homologous to secretin and glucagon.**

a. Actions of GIP

(1) stimulates insulin release. In the presence of an oral glucose load, GIP causes the release of insulin from the pancreas. Thus, **oral glucose is more effective than intravenous glucose in causing insulin release** and, therefore, glucose utilization.
(2) inhibits H^+ secretion by gastric parietal cells.

b. Stimuli for the release of GIP

- GIP is secreted by the duodenum and jejunum.
- GIP is the only GI hormone that is released in response to fat, protein, and carbohydrate. GIP secretion is stimulated by **fatty acids, amino acids, and orally administered glucose.**

5. Candidate hormones

- are secreted by cells of the GI tract.
- **Motilin** increases GI motility and is involved in **interdigestive myoelectric complexes**.
- **Pancreatic polypeptide** inhibits pancreatic secretions.
- **Glucagon-like peptide-1 (GLP-1)** binds to pancreatic β-cells and **stimulates insulin secretion**. Analogues of GLP-1 may be helpful in the treatment of type 2 diabetes mellitus.
- **Leptin** decreases appetite.
- **Ghrelin** increases appetite.

B. Paracrines

- are released from endocrine cells in the GI mucosa.
- diffuse over short distances to act on target cells located in the GI tract.
- The GI paracrines are **somatostatin** and **histamine.**

1. Somatostatin

- is secreted by cells throughout the GI tract in response to H^+ in the lumen. Its secretion is inhibited by vagal stimulation.
- **inhibits the release of all GI hormones.**
- inhibits gastric H^+ secretion.

2. Histamine

- is secreted by mast cells of the gastric mucosa.
- **increases gastric H^+ secretion** directly and by potentiating the effects of gastrin and vagal stimulation.

C. Neurocrines

- are synthesized in neurons of the GI tract, moved by axonal transport down the axons, and released by action potentials in the nerves.
- Neurocrines then diffuse across the synaptic cleft to a target cell.
- The GI neurocrines are **vasoactive intestinal peptide (VIP), neuropeptide Y, nitric oxide (NO), GRP (bombesin), and enkephalins.**

1. **VIP**
   - contains 28 amino acids and is **homologous to secretin.**
   - is released from neurons in the mucosa and smooth muscle of the GI tract.
   - produces **relaxation** of GI smooth muscle, including the **lower esophageal sphincter.**
   - **stimulates pancreatic HCO_3^- secretion** and **inhibits gastric H^+ secretion.** In these actions, it resembles secretin.
   - is secreted by pancreatic islet cell tumors and is presumed to mediate **pancreatic cholera.**
2. **GRP (bombesin)**
   - is released from vagus nerves that innervate the G cells.
   - **stimulates gastrin release** from G cells.
3. **Enkephalins (met-enkephalin and leu-enkephalin)**
   - are secreted from nerves in the mucosa and smooth muscle of the GI tract.
   - **stimulate contraction of GI smooth muscle,** particularly the lower esophageal, pyloric, and ileocecal sphincters.
   - **inhibit intestinal secretion** of fluid and electrolytes. This action forms the basis for the usefulness of **opiates in the treatment of diarrhea.**

D. Satiety

- **Hypothalamic centers**

1. **Satiety center** (inhibits appetite) is located in the ventromedial nucleus of the hypothalamus.
2. **Feeding center** (stimulates appetite) is located in the lateral hypothalamic area of the hypothalamus.

- **Anorexigenic neurons** release proopiomelanocortin (POMC) in the hypothalamic centers and cause decreased appetite.
- **Orexigenic neurons** release neuropeptide Y in the hypothalamic centers and stimulate appetite.
- **Leptin** is secreted by fat cells. It stimulates anorexigenic neurons and inhibits orexigenic neurons, thus decreasing appetite.
- Insulin and GLP-1 inhibit appetite.
- **Ghrelin** is secreted by gastric cells. It stimulates orexigenic neurons and inhibits anorexigenic neurons, thus increasing appetite.

III. GASTROINTESTINAL MOTILITY

- Contractile tissue of the GI tract is almost exclusively **unitary smooth muscle,** with the exception of the pharynx, upper one-third of the esophagus, and external anal sphincter, all of which are **striated muscle.**
- Depolarization of **circular muscle** leads to contraction of a ring of smooth muscle and a **decrease in diameter** of that segment of the GI tract.
- Depolarization of **longitudinal muscle** leads to contraction in the longitudinal direction and a **decrease in length** of that segment of the GI tract.
- **Phasic contractions** occur in the esophagus, gastric antrum, and small intestine, which contract and relax periodically.
- **Tonic contractions** occur in the lower esophageal sphincter, orad stomach, and ileocecal and internal anal sphincters.

A. Slow waves (Figure 6.3)

- are **oscillating membrane potentials** inherent to the smooth muscle cells of some parts of the GI tract.
- occur spontaneously.

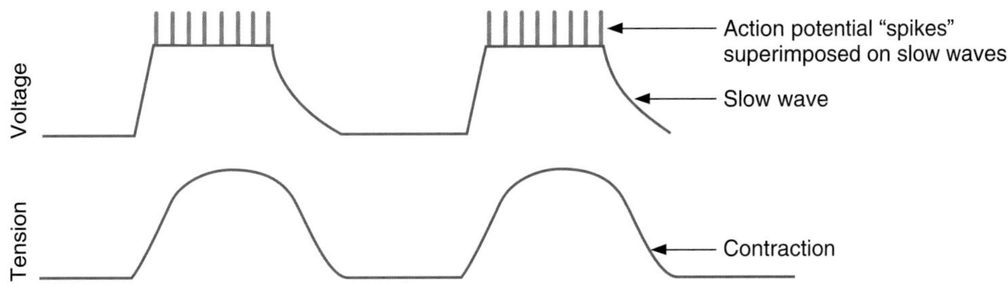

FIGURE 6.3. Gastrointestinal slow waves superimposed by action potentials. Action potentials produce subsequent contraction.

- originate in the **interstitial cells of Cajal,** which serve as the **pacemaker** for GI smooth muscle.
- are ***not*** **action potentials,** although they **determine the pattern of action potentials** and, therefore, the pattern of contraction.

1. Mechanism of slow wave production

- is the cyclic opening of Ca^{2+} channels (depolarization) followed by opening of K^+ channels (repolarization).
- **Depolarization during each slow wave** brings the membrane potential of smooth muscle cells closer to threshold and, therefore, **increases the probability that action potentials will occur.**
- Action potentials, produced on top of the background of slow waves, then initiate phasic contractions of the smooth muscle cells (see Chapter 1, VII B).

2. Frequency of slow waves

- varies along the GI tract, but is constant and characteristic for each part of the GI tract.
- is *not* influenced by neural or hormonal input. In contrast, the frequency of the action potentials that occur on top of the slow waves is modified by neural and hormonal influences.
- **sets the maximum frequency of contractions** for each part of the GI tract.
- **is lowest in the stomach** (3 slow waves/min) and **highest in the duodenum** (12 slow waves/min).

B. Chewing, swallowing, and esophageal peristalsis

1. Chewing

- lubricates food by mixing it with saliva.
- decreases the size of food particles to facilitate swallowing and to begin the digestive process.

2. Swallowing

- The swallowing reflex is **coordinated in the medulla.** Fibers in the vagus and glossopharyngeal nerves carry information between the GI tract and the medulla.
- The following sequence of events is involved in swallowing:

a. The nasopharynx closes and, at the same time, **breathing is inhibited.**

b. The laryngeal muscles contract to close the glottis and elevate the larynx.

c. **Peristalsis begins in the pharynx** to propel the food bolus toward the esophagus. Simultaneously, the **upper esophageal sphincter relaxes** to permit the food bolus to enter the esophagus.

3. Esophageal motility

- The esophagus propels the swallowed food into the stomach.
- Sphincters at either end of the esophagus prevent air from entering the upper esophagus and gastric acid from entering the lower esophagus.

- Because the esophagus is located in the thorax, intraesophageal pressure equals thoracic pressure, which is **lower than atmospheric pressure.** In fact, a balloon catheter placed in the esophagus can be used to measure intrathoracic pressure.
- The following sequence of events occurs as food moves into and down the esophagus:

a. As part of the swallowing reflex, the **upper esophageal sphincter relaxes** to permit swallowed food to enter the esophagus.

b. The upper esophageal sphincter then contracts so that food will not reflux into the pharynx.

c. A **primary peristaltic contraction** creates an area of high pressure behind the food bolus. The peristaltic contraction moves down the esophagus and propels the food bolus along. **Gravity** accelerates the movement.

d. A **secondary peristaltic contraction** clears the esophagus of any remaining food.

e. As the food bolus approaches the lower end of the esophagus, the **lower esophageal sphincter relaxes.** This relaxation is vagally mediated, and the neurotransmitter is **VIP.**

f. The orad region of the stomach relaxes **("receptive relaxation")** to allow the food bolus to enter the stomach.

4. Clinical correlations of esophageal motility

a. Gastroesophageal reflux (heartburn) may occur if the tone of the lower esophageal sphincter is decreased and gastric contents reflux into the esophagus.

b. Achalasia may occur if the lower esophageal sphincter does not relax during swallowing, with impaired esophageal peristalsis. Food accumulates in the esophagus, and there is dilation of the esophagus above the sphincter.

C. Gastric motility

- The stomach has three layers of smooth muscle—the usual longitudinal and circular layers and a third oblique layer.
- The stomach has three anatomic divisions—the **fundus, body,** and **antrum.**
- The **orad region** of the stomach includes the fundus and the proximal body. This region contains oxyntic glands and is responsible for receiving the ingested meal.
- The **caudad region** of the stomach includes the antrum and the distal body. This region is responsible for the contractions that mix food and propel it into the duodenum.

1. "Receptive relaxation"

- is a **vagovagal reflex** that is initiated by distention of the stomach and is abolished by vagotomy.
- The **orad region of the stomach relaxes** to accommodate the ingested meal.
- **CCK** participates in "receptive relaxation" by increasing the distensibility of the orad stomach.

2. Mixing and digestion

- The caudad region of the stomach contracts to mix the food with gastric secretions and begins the process of digestion. The size of food particles is reduced.

a. Slow waves in the caudad stomach occur at a frequency of 3 to 5 waves/min. They depolarize the smooth muscle cells.

b. If threshold is reached during the slow waves, action potentials are fired, followed by contraction. Thus, the frequency of slow waves sets the maximal frequency of contraction.

c. A **wave of contraction** closes the distal antrum. Thus, as the caudad stomach contracts, food is propelled back into the stomach to be mixed **(retropulsion).**

d. Gastric contractions are **increased by vagal stimulation and decreased by sympathetic stimulation.**

e. Even during fasting, contractions (the **"migrating myoelectric complex"**) occur at 90-minute intervals and clear the stomach of residual food. **Motilin** is the mediator of these contractions.

3. **Gastric emptying**

- The caudad region of the stomach contracts to propel food into the duodenum.

a. The rate of **gastric emptying is fastest** when the stomach contents are **isotonic.** If the stomach contents are hypertonic or hypotonic, gastric emptying is slowed.

b. **Fat inhibits gastric emptying** (i.e., increases gastric emptying time) by stimulating the release of **CCK.**

c. **H^+ in the duodenum inhibits gastric emptying** via direct neural reflexes. H^+ receptors in the duodenum relay information to the gastric smooth muscle via interneurons in the GI plexuses.

D. Small intestinal motility

- The small intestine functions in the **digestion and absorption** of nutrients. The small intestine mixes nutrients with digestive enzymes, exposes the digested nutrients to the absorptive mucosa, and then propels any nonabsorbed material to the large intestine.
- As in the stomach, **slow waves** set the basic electrical rhythm, which occurs at a frequency of 12 waves/min. Action potentials occur on top of the slow waves and lead to contractions.
- **Parasympathetic stimulation** increases intestinal smooth muscle contraction; **sympathetic stimulation** decreases it.

1. **Segmentation contractions**

- **mix the intestinal contents.**
- A section of small intestine contracts, sending the intestinal contents (chyme) in both orad and caudad directions. That section of small intestine then relaxes, and the contents move back into the segment.
- This **back-and-forth movement** produced by segmentation contractions causes mixing without any net forward movement of the chyme.

2. **Peristaltic contractions**

- are highly coordinated and **propel the chyme** through the small intestine toward the large intestine. Ideally, peristalsis occurs after digestion and absorption have taken place.
- **Contraction behind the bolus** and, simultaneously, **relaxation in front of the bolus** cause the chyme to be propelled caudally.
- The peristaltic reflex is **coordinated by the enteric nervous system.**

a. Food in the intestinal lumen is sensed by enterochromaffin cells, which release serotonin (**5-hydroxytryptamine, 5-HT**).

b. 5-HT binds to receptors on intrinsic primary afferent neurons (**IPANs**), which initiate the peristaltic reflex.

c. **Behind the food bolus**, excitatory transmitters cause contraction of circular muscle and inhibitory transmitters cause relaxation of longitudinal muscle. **In front of the bolus**, inhibitory transmitters cause relaxation of circular muscle and excitatory transmitters cause contraction of longitudinal muscle.

3. **Gastroileal reflex**

- is mediated by the extrinsic ANS and possibly by gastrin.
- The presence of food in the stomach triggers increased peristalsis in the ileum and relaxation of the ileocecal sphincter. As a result, the intestinal contents are delivered to the large intestine.

E. Large intestinal motility

- Fecal material moves from the cecum to the colon (i.e., through the ascending, transverse, descending, and sigmoid colons), to the rectum, and then to the anal canal.
- **Haustra**, or saclike segments, appear after contractions of the large intestine.

1. Cecum and proximal colon

- When the proximal colon is distended with fecal material, the ileocecal sphincter contracts to prevent reflux into the ileum.

a. Segmentation contractions in the proximal colon mix the contents and are responsible for the appearance of haustra.

b. Mass movements occur 1 to 3 times/day and cause the colonic contents to move distally for long distances (e.g., from the transverse colon to the sigmoid colon).

2. Distal colon

- Because most colonic water absorption occurs in the proximal colon, fecal material in the distal colon becomes semisolid and moves slowly. Mass movements propel it into the rectum.

3. Rectum, anal canal, and defecation

- The sequence of events for defecation is as follows:

a. As the rectum fills with fecal material, it contracts and the internal anal sphincter relaxes **(rectosphincteric reflex).**

b. Once the rectum is filled to about 25% of its capacity, there is an **urge to defecate.** However, defecation is prevented because the external anal sphincter is tonically contracted.

c. When it is convenient to defecate, the external anal sphincter is relaxed voluntarily. The smooth muscle of the rectum contracts, forcing the feces out of the body.

- Intra-abdominal pressure is increased by expiring against a closed glottis **(Valsalva maneuver).**

4. Gastrocolic reflex

- The presence of **food in the stomach** increases the motility of the colon and **increases the frequency of mass movements.**

a. The gastrocolic reflex has a rapid **parasympathetic** component that is initiated when the stomach is stretched by food.

b. A slower, hormonal component is mediated by CCK and gastrin.

5. Disorders of large intestinal motility

a. Emotional factors strongly influence large intestinal motility via the extrinsic ANS. **Irritable bowel syndrome** may occur during periods of stress and may result in **constipation** (increased segmentation contractions) or **diarrhea** (decreased segmentation contractions).

b. Megacolon (Hirschsprung disease), the **absence of the colonic enteric nervous system,** results in constriction of the involved segment, marked dilation and accumulation of intestinal contents proximal to the constriction, and severe constipation.

F. Vomiting

- A wave of reverse peristalsis begins in the small intestine, moving the GI contents in the orad direction.
- The gastric contents are eventually pushed into the esophagus. If the upper esophageal sphincter remains closed, **retching** occurs. If the pressure in the esophagus becomes high enough to open the upper esophageal sphincter, **vomiting** occurs.
- The **vomiting center** in the **medulla** is stimulated by tickling the back of the throat, gastric distention, and vestibular stimulation (motion sickness).
- The **chemoreceptor trigger zone** in the fourth ventricle is activated by emetics, radiation, and vestibular stimulation.

IV. GASTROINTESTINAL SECRETION (TABLE 6.2)

A. Salivary secretion

1. Functions of saliva

a. **initial starch digestion** by α-amylase (ptyalin) and **initial triglyceride digestion** by lingual lipase.

b. **lubrication** of ingested food by mucus.

c. **protection** of the mouth and esophagus by dilution and buffering of ingested foods.

2. Composition of saliva

a. Saliva is characterized by

(1) high volume (relative to the small size of the salivary glands).
(2) *high* K^+ and HCO_3^- concentrations.
(3) low Na^+ and Cl^- concentrations.
(4) *hypotonicity.*
(5) presence of α-amylase, lingual lipase, and kallikrein.

b. The composition of saliva varies with the salivary flow rate (Figure 6.4).

(1) *At the lowest flow rates,* saliva has the lowest osmolarity and lowest Na^+, Cl^-, and HCO_3^- concentrations but has the highest K^+ concentration.
(2) *At the highest flow rates* (up to 4 mL/min), the composition of saliva is closest to that of plasma.

table 6.2 Summary of Gastrointestinal (GI) Secretions

GI Secretion	Major Characteristics	Stimulated By	Inhibited By
Saliva	High HCO_3^- High K^+ Hypotonic α-Amylase Lingual lipase	Parasympathetic nervous system Sympathetic nervous system	Sleep Dehydration Atropine
Gastric secretion	HCl	Gastrin Parasympathetic nervous system Histamine	↓ Stomach pH Chyme in duodenum (via secretin and GIP) Somatostatin Atropine Cimetidine Omeprazole
	Pepsinogen	Parasympathetic nervous system	
	Intrinsic factor		
Pancreatic secretion	High HCO_3^- Isotonic	Secretin CCK (potentiates secretin) Parasympathetic nervous system	
	Pancreatic lipase, amylase, proteases	CCK Parasympathetic nervous system	
Bile	Bile salts Bilirubin Phospholipids Cholesterol	CCK (causes contraction of gallbladder and relaxation of sphincter of Oddi) Parasympathetic nervous system (causes contraction of gallbladder)	Ileal resection

CCK = cholecystokinin; GIP = glucose-dependent insulinotropic peptide.

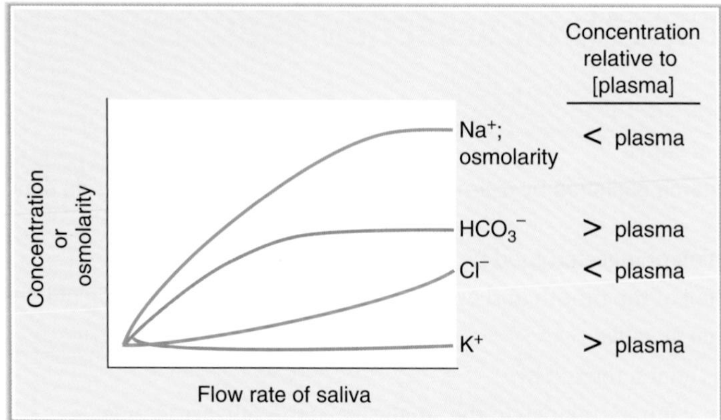

FIGURE 6.4. Composition of saliva as a function of salivary flow rate.

3. Formation of saliva (Figure 6.5)

- Saliva is formed by three major glands—the **parotid, submandibular, and sublingual glands.**
- The **structure** of each gland is similar to a bunch of grapes. The **acinus** (the blind end of each duct) is lined with acinar cells and secretes an initial saliva. A **branching duct system** is lined with columnar epithelial cells, which modify the initial saliva.
- When saliva production is stimulated, **myoepithelial cells,** which line the acinus and initial ducts, contract and eject saliva into the mouth.

a. The acinus

- **produces an initial saliva** with a composition **similar to plasma.**
- This initial saliva is **isotonic** and has the same Na^+, K^+, Cl^-, and HCO_3^- concentrations as plasma.

b. The ducts

- **modify the initial saliva** by the following processes:

(1) The ducts **reabsorb Na^+ and Cl^-**; therefore, the concentrations of these ions are lower than their plasma concentrations.

(2) The ducts **secrete K^+** and **HCO_3^-**; therefore, the concentrations of these ions are higher than their plasma concentrations.

(3) **Aldosterone** acts on the ductal cells to increase the reabsorption of Na^+ and the secretion of K^+ (analogous to its actions on the renal distal tubule).

(4) **Saliva becomes hypotonic** in the ducts because the ducts are relatively impermeable to water. Because more solute than water is reabsorbed by the ducts, the saliva becomes dilute relative to plasma.

(5) The **effect of flow rate** on saliva composition is explained primarily by changes in the contact time available for reabsorption and secretion processes to occur in the ducts.

- Thus, at **high flow rates,** saliva is most like the initial secretion from the acinus; it has the highest Na^+ and Cl^- concentrations and the lowest K^+ concentration.

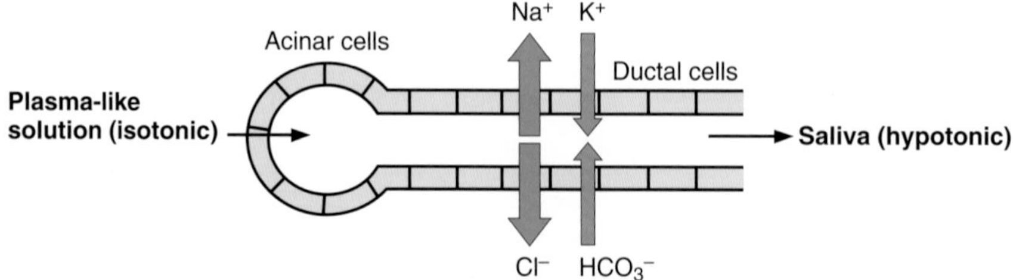

FIGURE 6.5. Modification of saliva by ductal cells.

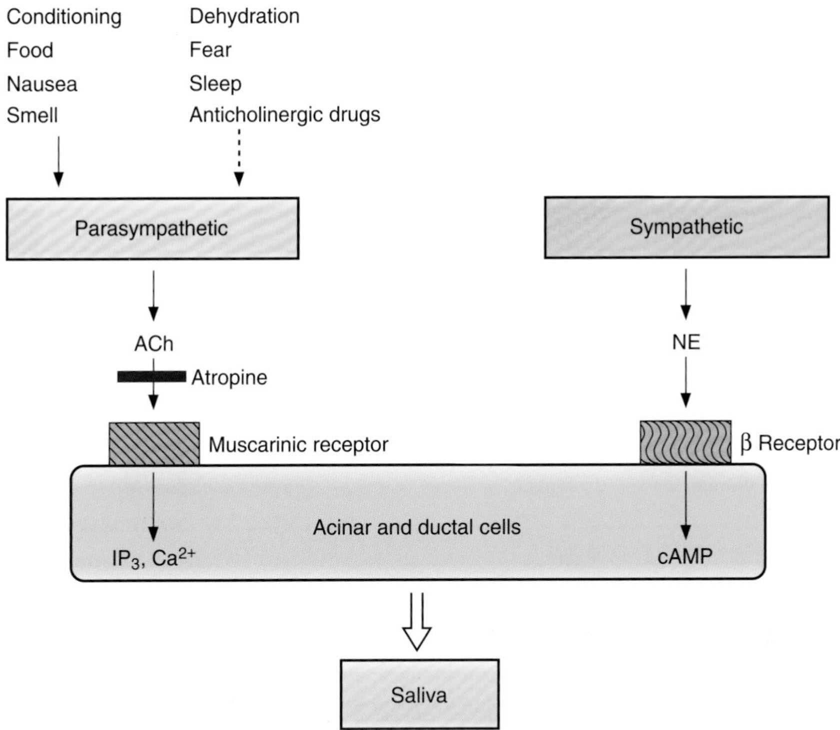

FIGURE 6.6. Regulation of salivary secretion. ACh = acetylcholine; cAMP = cyclic adenosine monophosphate; IP_3 = inositol 1,4,5-triphosphate; NE = norepinephrine.

- At **low flow rates,** saliva is least like the initial secretion from the acinus; it has the lowest Na^+ and Cl^- concentrations and the highest K^+ concentration.
- The only ion that does not "fit" this contact time explanation is HCO_3^-; HCO_3^- secretion is selectively stimulated when saliva secretion is stimulated.

4. Regulation of saliva production (Figure 6.6)

- Saliva production is controlled by the parasympathetic and sympathetic nervous systems (not by GI hormones).
- Saliva production is unique in that it is **increased by *both* parasympathetic and sympathetic activity.** Parasympathetic activity is more important, however.

a. Parasympathetic stimulation (cranial nerves VII and IX)

- **increases saliva production** by increasing transport processes in the acinar and ductal cells and by causing vasodilation.
- Cholinergic receptors on acinar and ductal cells are **muscarinic.**
- The second messenger is **inositol 1,4,5-triphosphate (IP_3)** and **increased intracellular [Ca^{2+}].**
- Anticholinergic drugs (e.g., **atropine**) inhibit the production of saliva and cause **dry mouth.**

b. Sympathetic stimulation

- **increases the production of saliva** and the growth of salivary glands, although the effects are smaller than those of parasympathetic stimulation.
- Receptors on acinar and ductal cells are **β-adrenergic.**
- The second messenger is **cyclic adenosine monophosphate (cAMP).**

c. Saliva production

- **is increased** (via activation of the parasympathetic nervous system) by **food** in the mouth, **smells, conditioned reflexes,** and **nausea.**

table 6.3 Gastric Cell Types and Their Secretions

Cell Type	Part of Stomach	Secretion Products	Stimulus for Secretion
Parietal cells	Body (fundus)	HCl	Gastrin Vagal stimulation (ACh) Histamine
		Intrinsic factor (essential)	
Chief cells	Body (fundus)	Pepsinogen (converted to pepsin at low pH)	Vagal stimulation (ACh)
G cells	Antrum	Gastrin	Vagal stimulation (via GRP) Small peptides Inhibited by somatostatin Inhibited by H^+ in stomach (via stimulation of somatostatin release)
Mucous cells	Antrum	Mucus Pepsinogen	Vagal stimulation (ACh)

ACh = acetylcholine; GRP = gastrin-releasing peptide.

- **is decreased** (via inhibition of the parasympathetic nervous system) by **sleep, dehydration, fear,** and **anticholinergic drugs.**

B. Gastric secretion

1. Gastric cell types and their secretions (Table 6.3 and Figure 6.7)

- **Parietal cells,** located in the body, secrete **HCl** and **intrinsic factor.**
- **Chief cells,** located in the body, secrete **pepsinogen.**
- **G cells,** located in the antrum, secrete **gastrin.**

2. Mechanism of gastric H^+ secretion (Figure 6.8)

- Parietal cells **secrete HCl into the lumen of the stomach** and, concurrently, **absorb HCO_3^-** into the bloodstream as follows:

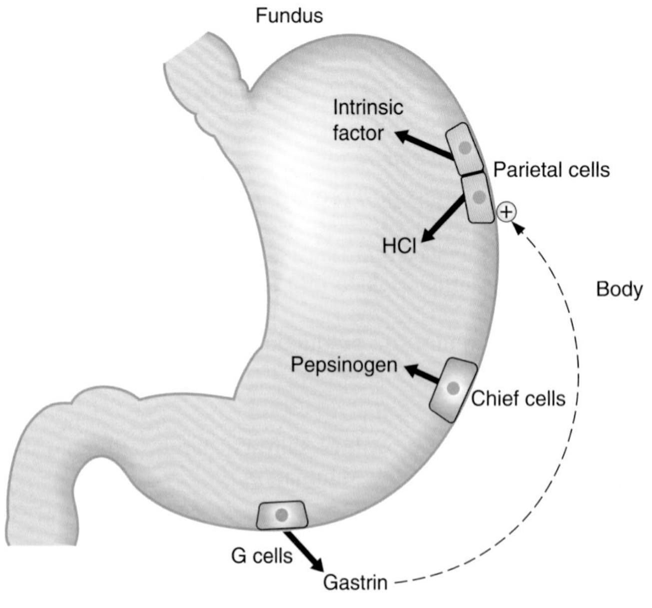

FIGURE 6.7. Gastric cell types and their functions.

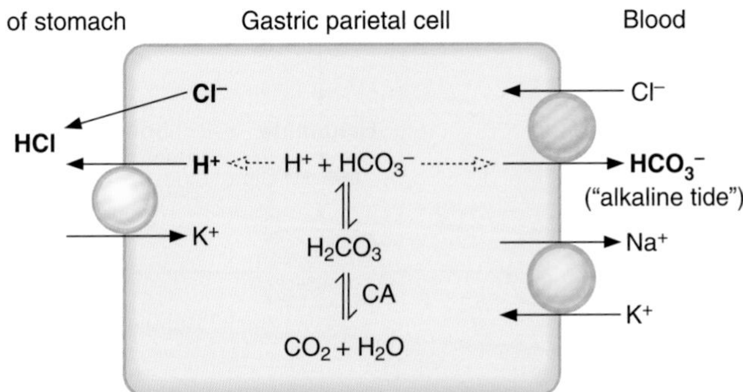

FIGURE 6.8. Simplified mechanism of H^+ secretion by gastric parietal cells. CA = carbonic anhydrase.

a. In the parietal cells, CO_2 and H_2O are converted to H^+ and HCO_3^-, catalyzed by **carbonic anhydrase.**

b. **H^+ is secreted into the lumen of the stomach** by the H^+–K^+ pump **(H^+, K^+-ATPase).** Cl^- is secreted along with H^+; thus, the secretion product of the parietal cells is HCl.

- The drug **omeprazole** (a "proton pump inhibitor") inhibits the H^+, K^+-ATPase and blocks H^+ secretion.

c. The **HCO_3^-** produced in the cells is absorbed into the bloodstream in exchange for Cl^- (**Cl^-–HCO_3^- exchange**). As HCO_3^- is added to the venous blood, the pH of the blood increases ("**alkaline tide**"). (Eventually, this HCO_3^- will be secreted in pancreatic secretions to neutralize H^+ in the small intestine.)

- If **vomiting** occurs, gastric H^+ never arrives in the small intestine, there is no stimulus for pancreatic HCO_3^- secretion, and the arterial blood becomes alkaline **(metabolic alkalosis).**

3. **Stimulation of gastric H^+ secretion** (Figure 6.9)

a. **Vagal stimulation**

- increases H^+ secretion by a direct pathway and an indirect pathway.
- In the direct path, the **vagus nerve innervates parietal cells** and stimulates H^+ secretion directly. The neurotransmitter at these synapses is **ACh,** the receptor on the parietal cells is **muscarinic** (M_3), and the second messengers for CCK are **IP_3** and **increased intracellular [Ca^{2+}].**
- In the indirect path, the **vagus nerve innervates G cells** and stimulates gastrin secretion, which then stimulates H^+ secretion by an endocrine action. The neurotransmitter at these synapses is **GRP** (not ACh).
- **Atropine,** a cholinergic muscarinic antagonist, inhibits H^+ secretion by blocking the direct pathway, which uses ACh as a neurotransmitter. However, atropine does not block H^+ secretion completely because it does not inhibit the indirect pathway, which uses GRP as a neurotransmitter.
- **Vagotomy** eliminates both direct and indirect pathways.

b. **Gastrin**

- is released in response to eating a meal (small peptides, distention of the stomach, vagal stimulation).
- stimulates H^+ secretion by interacting with the cholecystokinin$_B$ (CCK_B) receptor on the parietal cells.
- The second messenger for gastrin on the parietal cell is IP_3/Ca^{2+}.
- Gastrin also stimulates enterochromaffin-like (ECL) cells and histamine secretion, which stimulates H^+ secretion (not shown in figure).

c. **Histamine**

- is released from ECL cells in the gastric mucosa and diffuses to the nearby parietal cells.

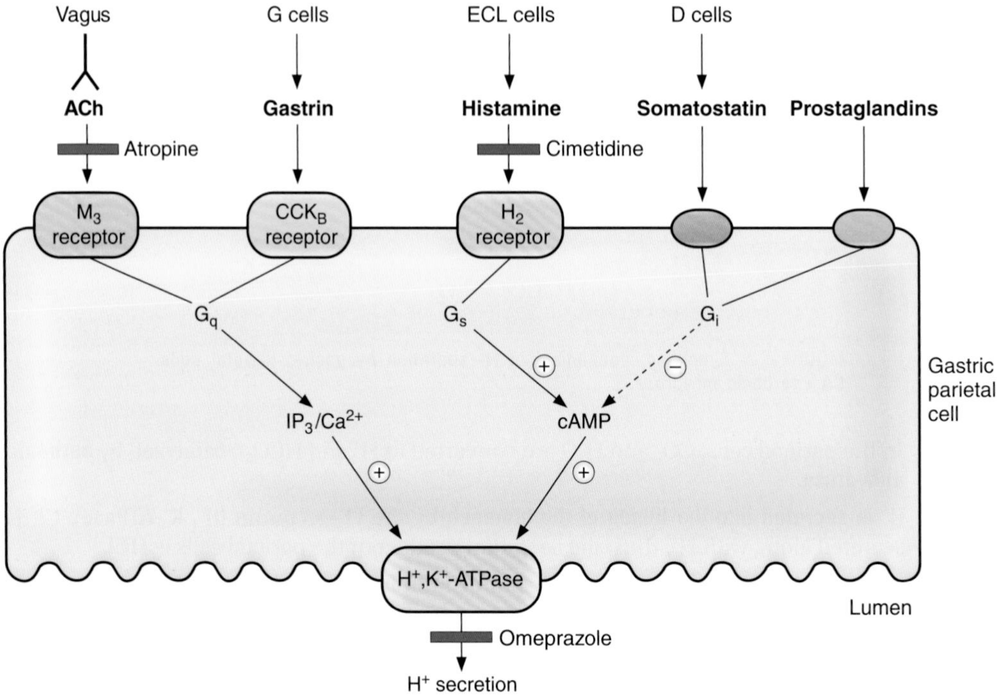

FIGURE 6.9. Agents that stimulate and inhibit H^+ secretion by gastric parietal cells. ACh = acetylcholine; cAMP = cyclic adenosine monophosphate; CCK = cholecystokinin; ECL = enterochromaffin-like; IP_3 = inositol 1,4,5-triphosphate; M = muscarinic.

- stimulates H^+ secretion by activating **H_2 receptors** on the parietal cell membrane.
- The H_2 receptor is coupled to adenylyl cyclase via a G_s protein.
- The second messenger for histamine is **cAMP.**
- H_2 receptor–blocking drugs, such as **cimetidine**, inhibit H^+ secretion by blocking the stimulatory effect of histamine.

d. Potentiating effects of ACh, histamine, and gastrin on H^+ secretion

- **Potentiation** occurs when the response to simultaneous administration of two stimulants is greater than the sum of responses to either agent given alone. As a result, low concentrations of stimulants given together can produce maximal effects.
- Potentiation of gastric H^+ secretion can be explained, in part, because **each agent has a different mechanism of action** on the parietal cell.

(1) *Histamine potentiates the actions of ACh and gastrin* in stimulating H^+ secretion.

- Thus, H_2 receptor blockers (e.g., **cimetidine**) are particularly effective in treating ulcers because they block both the direct action of histamine on parietal cells and the potentiating effects of histamine on ACh and gastrin.

(2) *ACh potentiates the actions of histamine and gastrin* in stimulating H^+ secretion.

- Thus, muscarinic receptor blockers, such as **atropine,** block both the direct action of ACh on parietal cells and the potentiating effects of ACh on histamine and gastrin.

4. Inhibition of gastric H^+ secretion

- **Negative feedback** mechanisms inhibit the secretion of H^+ by the parietal cells.

a. Low pH (<3.0) in the stomach

- **inhibits gastrin secretion** and thereby inhibits H^+ secretion.
- After a meal is ingested, H^+ secretion is stimulated by the mechanisms discussed previously (see IV B 2). After the meal is digested and the stomach emptied, further H^+ secretion decreases the pH of the stomach contents. When the pH of the stomach contents is less than 3.0, gastrin secretion is inhibited and, by negative feedback, inhibits further H^+ secretion.

b. Somatostatin (see Figure 6.9)

- inhibits gastric H^+ secretion by a direct pathway and an indirect pathway.
- In the **direct pathway,** somatostatin binds to receptors on the parietal cell that are coupled to adenylyl cyclase via a **G_i** protein, thus inhibiting adenylyl cyclase and **decreasing cAMP** levels. In this pathway, somatostatin antagonizes the stimulatory action of histamine on H^+ secretion.
- In the **indirect pathways** (not shown in Figure 6.9), somatostatin **inhibits release of histamine and gastrin,** thus decreasing H^+ secretion indirectly.

c. Prostaglandins (see Figure 6.9)

- inhibit gastric H^+ secretion by activating a **G_i** protein, inhibiting adenylyl cyclase and **decreasing cAMP levels.**
- Maintain the mucosal barrier and stimulate HCO_3^- secretion, thus protecting the gastric mucosa from the damaging effects of H^+.

5. Peptic ulcer disease

- is an ulcerative lesion of the gastric or duodenal mucosa.
- can occur when there is **loss of the protective mucous barrier** (of mucus and HCO_3^-) and/or **excessive secretion of H^+ and pepsin.**
- **Protective factors** are mucus, HCO_3^-, prostaglandins, mucosal blood flow, and growth factors.
- **Damaging factors** are H^+, pepsin, *Helicobacter pylori (H. pylori),* nonsteroidal anti-inflammatory drugs (NSAIDs), stress, smoking, and alcohol.

a. Gastric ulcers

- The gastric mucosa is damaged.
- Gastric **H^+ secretion is *decreased* because secreted H^+ leaks back through the damaged gastric mucosa.**
- **Gastrin levels are increased** because decreased H^+ secretion stimulates gastrin secretion.
- A major cause of gastric ulcer is the gram-negative bacterium ***Helicobacter pylori (H. pylori).***
- *H. pylori* colonizes the gastric mucus and releases cytotoxins that damage the gastric mucosa.
- *H. pylori* contains **urease,** which converts urea to NH_3, thus alkalinizing the local environment and permitting *H. pylori* to survive in the otherwise acidic gastric lumen.
- The **diagnostic test for *H. pylori*** involves drinking a solution of ^{13}C-urea, which is converted to $^{13}CO_2$ by urease and measured in the expired air.

b. Duodenal ulcers

- The duodenal mucosa is damaged.
- Gastric **H^+ secretion is *increased.*** Excess H^+ is delivered to the duodenum, damaging the duodenal mucosa.
- **Gastrin secretion in response to a meal is increased** (although baseline gastrin may be normal).
- *H. pylori* is also a major cause of duodenal ulcer. *H. pylori* inhibits somatostatin secretion (thus stimulating gastric H^+ secretion) and inhibits intestinal HCO_3^- secretion (so there is insufficient HCO_3^- to neutralize the H^+ load from the stomach).

c. Zollinger–Ellison syndrome

- occurs when a **gastrin-secreting tumor of the pancreas** causes increased H^+ secretion.
- H^+ secretion continues unabated because the gastrin secreted by pancreatic tumor cells is not subject to negative feedback inhibition by H^+.

6. Drugs that block gastric H^+ secretion (see Figure 6.9)

a. Atropine

- blocks H^+ secretion by inhibiting cholinergic muscarinic receptors on parietal cells, thereby inhibiting ACh stimulation of H^+ secretion.

b. Cimetidine

- blocks H_2 receptors and thereby inhibits histamine stimulation of H^+ secretion.
- is particularly effective in reducing H^+ secretion because it not only blocks the histamine stimulation of H^+ secretion but also blocks histamine's potentiation of ACh effects.

c. Omeprazole

- is a proton pump inhibitor.
- directly inhibits H^+, K^+-ATPase, and H^+ secretion.

C. Pancreatic secretion

- contains a high concentration of **HCO_3^-**, whose purpose is to neutralize the acidic chyme that reaches the duodenum.
- contains **enzymes** essential for the digestion of protein, carbohydrate, and fat.

1. Composition of pancreatic secretion

a. Pancreatic juice is characterized by

(1) high volume.
(2) virtually the same Na^+ and K^+ concentrations as plasma.
(3) much **higher HCO_3^- concentration than plasma.**
(4) much lower Cl^- concentration than plasma.
(5) *isotonicity.*
(6) pancreatic lipase, amylase, and proteases.

b. The composition of the aqueous component of pancreatic secretion varies with the flow rate (Figure 6.10).

- At **low flow rates**, the pancreas secretes an isotonic fluid that is composed mainly of Na^+ and **Cl^-**.
- At **high flow rates**, the pancreas secretes an isotonic fluid that is composed mainly of Na^+ and **HCO_3^-**.
- Regardless of the flow rate, pancreatic secretions are **isotonic.**

2. Formation of pancreatic secretion (Figure 6.11)

- Like the salivary glands, the exocrine pancreas resembles a bunch of grapes.
- The acinar cells of the exocrine pancreas make up most of its weight.

a. Acinar cells

- produce a small volume of initial pancreatic secretion, which is mainly Na^+ and Cl^-.

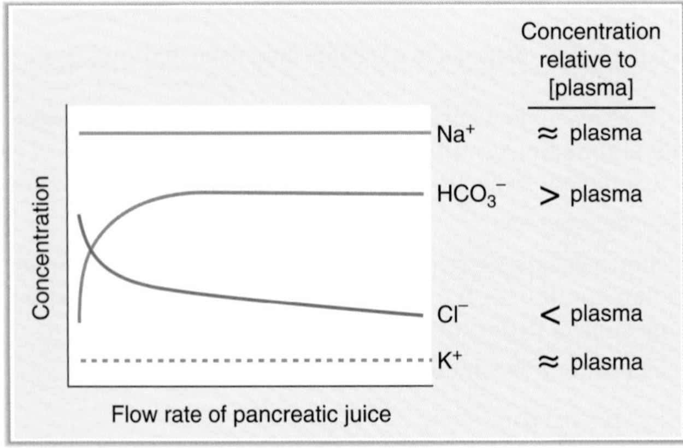

FIGURE 6.10. Composition of pancreatic secretion as a function of pancreatic flow rate.

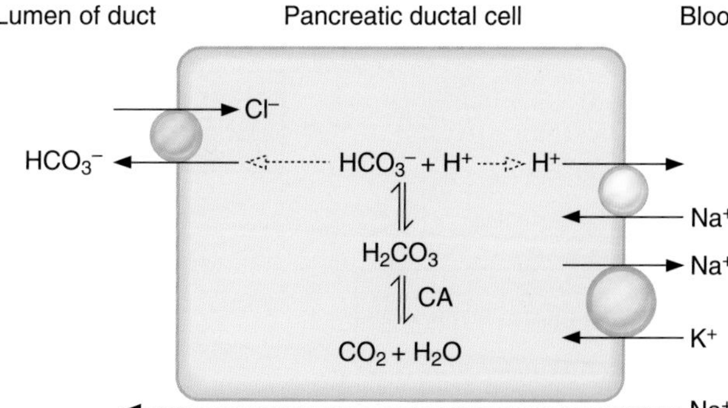

FIGURE 6.11. Modification of pancreatic secretion by ductal cells. CA = carbonic anhydrase.

b. **Ductal cells**

- modify the initial pancreatic secretion by **secreting HCO_3^-** and **absorbing Cl^-** via a **Cl^-–HCO_3^- exchange** mechanism in the luminal membrane.
- Because the pancreatic ducts are **permeable to water**, H_2O moves into the lumen to make the pancreatic secretion isosmotic.

3. **Stimulation of pancreatic secretion**

a. **Secretin**

- is secreted by the S cells of the duodenum in response to H^+ in the duodenal lumen.
- acts on the pancreatic **ductal cells** to increase **HCO_3^- secretion.**
- Thus, when H^+ is delivered from the stomach to the duodenum, secretin is released. As a result, HCO_3^- is secreted from the pancreas into the duodenal lumen to neutralize the H^+.
- The second messenger for secretin is **cAMP.**

b. **CCK**

- is secreted by the I cells of the duodenum in response to small peptides, amino acids, and fatty acids in the duodenal lumen.
- acts on the pancreatic **acinar cells** to increase **enzyme secretion** (amylase, lipases, proteases).
- potentiates the effect of secretin on ductal cells to stimulate HCO_3^- secretion.
- The second messengers for CCK are **IP_3** and **increased intracellular [Ca^{2+}].** The potentiating effects of CCK on secretin are explained by the different mechanisms of action for the two GI hormones (i.e., cAMP for secretin and IP_3/Ca^{2+} for CCK).

c. **ACh (via vagovagal reflexes)**

- is released in response to H^+, small peptides, amino acids, and fatty acids in the duodenal lumen.
- **stimulates enzyme secretion** by the acinar cells and, like CCK, potentiates the effect of secretin on HCO_3^- secretion.

4. **Cystic fibrosis**

- is a disorder of pancreatic secretion.
- results from a defect in Cl^- channels that is caused by a mutation in the **cystic fibrosis transmembrane conductance regulator (CFTR) gene.**
- is associated with a **deficiency of pancreatic enzymes** resulting in malabsorption and steatorrhea.

D. Bile secretion and gallbladder function (Figure 6.12)

1. **Composition and function of bile**

- Bile contains **bile salts,** phospholipids, cholesterol, and bile pigments (bilirubin).

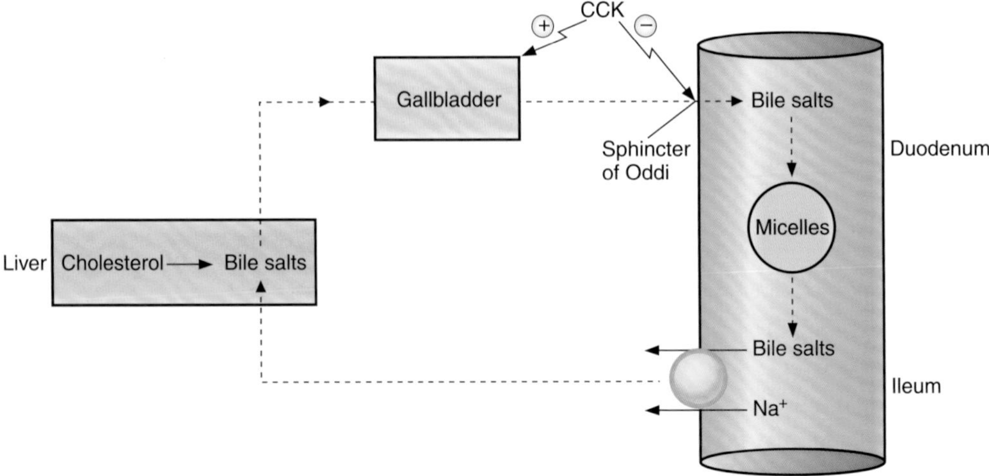

FIGURE 6.12. Recirculation of bile acids from the ileum to the liver. CCK = cholecystokinin.

a. **Bile salts**

- are **amphipathic** molecules because they have both hydrophilic and hydrophobic portions. In aqueous solution, bile salts orient themselves around droplets of lipid and keep the lipid droplets dispersed **(emulsification).**
- aid in the intestinal digestion and absorption of lipids by emulsifying and solubilizing them in **micelles.**

b. **Micelles**

- Above a **critical micellar concentration,** bile salts form micelles.
- Bile salts are positioned on the outside of the micelle, with their hydrophilic portions dissolved in the aqueous solution of the intestinal lumen and their hydrophobic portions dissolved in the micelle interior.
- Free fatty acids and monoglycerides are present in the inside of the micelle, essentially "solubilized" for subsequent absorption.

2. **Formation of bile**

- Bile is **produced continuously by hepatocytes.**
- Bile drains into the hepatic ducts and is stored in the gallbladder for subsequent release.
- **Choleretic agents** increase the formation of bile.
- Bile is formed by the following process:

a. **Primary bile acids** (cholic acid and chenodeoxycholic acid) are synthesized from cholesterol by hepatocytes.

- In the intestine, bacteria convert a portion of each of the primary bile acids to **secondary bile acids** (deoxycholic acid and lithocholic acid).
- Synthesis of new bile acids occurs, as needed, to replace bile acids that are excreted in the feces.

b. The bile acids are conjugated with glycine or taurine to form their respective **bile salts,** which are named for the parent bile acid (e.g., taurocholic acid is cholic acid conjugated with taurine).

c. Electrolytes and H_2O are added to the bile.

d. During the interdigestive period, the gallbladder is relaxed, the sphincter of Oddi is closed, and the gallbladder fills with bile.

e. Bile is **concentrated** in the gallbladder as a result of isosmotic absorption of solutes and H_2O.

3. **Contraction of the gallbladder**

   a. **CCK**

      - is released in response to **small peptides** and **fatty acids** in the duodenum.
      - tells the gallbladder that bile is needed to emulsify and absorb lipids in the duodenum.
      - causes **contraction of the gallbladder** and **relaxation of the sphincter of Oddi.**

   b. **ACh**

      - causes contraction of the gallbladder.

4. **Recirculation of bile acids to the liver**

   a. The **terminal ileum** contains a **Na^+–bile acid cotransporter**, which is a secondary active transporter that recirculates bile acids to the liver.

   b. Because bile acids are not recirculated until they reach the terminal ileum, bile acids are present for maximal absorption of lipids throughout the upper small intestine.

   c. After **ileal resection**, bile acids are not recirculated to the liver but are excreted in feces. The bile acid pool is thereby depleted, and fat absorption is impaired, resulting in **steatorrhea.**

V. DIGESTION AND ABSORPTION (TABLE 6.4)

- Carbohydrates, proteins, and lipids are digested and absorbed in the small intestine.
- The surface area for absorption in the small intestine is greatly increased by the presence of the **brush border.**

table 6.4 Summary of Digestion and Absorption

Nutrient	Digestion	Site of Absorption	Mechanism of Absorption
Carbohydrates	To monosaccharides (glucose, galactose, fructose)	Small intestine	Na^+-dependent cotransport (glucose, galactose) Facilitated diffusion (fructose)
Proteins	To amino acids, dipeptides, tripeptides	Small intestine	Na^+-dependent cotransport (amino acids) H^+-dependent cotransport (di- and tripeptides)
Lipids	To fatty acids, monoglycerides, cholesterol	Small intestine	Micelles form with bile salts in intestinal lumen Diffusion of fatty acids, monoglycerides, and cholesterol into cell Reesterification in cell to triglycerides and phospholipids Chylomicrons form in cell (requires apoprotein) and are transferred to lymph
Fat-soluble vitamins		Small intestine	Micelles with bile salts
Water-soluble vitamins Vitamin B_{12}		Small intestine Ileum of small intestine	Na^+-dependent cotransport Intrinsic factor–vitamin B_{12} complex
Bile acids		Ileum of small intestine	Na^+-dependent cotransport; recirculated to liver
Ca^{2+}		Small intestine	Vitamin D dependent (calbindin D-28K)
Fe^{2+}	Fe^{3+} is reduced to Fe^{2+}	Small intestine	Binds to apoferritin in cell Circulates in blood bound to transferrin

A. Carbohydrates

1. Digestion of carbohydrates

- **Only monosaccharides are absorbed.** Carbohydrates must be digested to glucose, galactose, and fructose for absorption to proceed.

a. α-Amylases (salivary and pancreatic) hydrolyze 1,4-glycosidic bonds in starch, yielding maltose, maltotriose, and α-limit dextrins.

b. Maltase, α-dextrinase, and sucrase in the intestinal brush border then hydrolyze the oligosaccharides to glucose.

c. Lactase, trehalase, and sucrase degrade their respective disaccharides to monosaccharides.

- **Lactase** degrades lactose to glucose and galactose.
- **Trehalase** degrades trehalose to glucose.
- **Sucrase** degrades sucrose to glucose and fructose.

2. Absorption of carbohydrates (Figure 6.13)

a. Glucose and galactose

- are transported from the intestinal lumen into the cells by a **Na^+-dependent cotransport (SGLT1)** in the luminal membrane. The sugar is transported "uphill" and Na^+ is transported "downhill."
- are then transported from cell to blood by facilitated diffusion (GLUT2).
- The Na^+–K^+ pump in the basolateral membrane keeps the intracellular [Na^+] low, thus maintaining the Na^+ gradient across the luminal membrane.
- Poisoning the Na^+–K^+ pump inhibits glucose and galactose absorption by dissipating the Na^+ gradient.

b. Fructose

- is transported exclusively by **facilitated diffusion;** therefore, it cannot be absorbed against a concentration gradient.

3. Clinical disorders of carbohydrate absorption

- **Lactose intolerance** results from the absence of brush border lactase and, thus, the inability to hydrolyze lactose to glucose and galactose for absorption. Nonabsorbed lactose and H_2O remain in the lumen of the GI tract and cause **osmotic diarrhea.**

B. Proteins

1. Digestion of proteins

a. Endopeptidases

- degrade proteins by hydrolyzing interior peptide bonds.

b. Exopeptidases

- hydrolyze one amino acid at a time from the C-terminus of proteins and peptides.

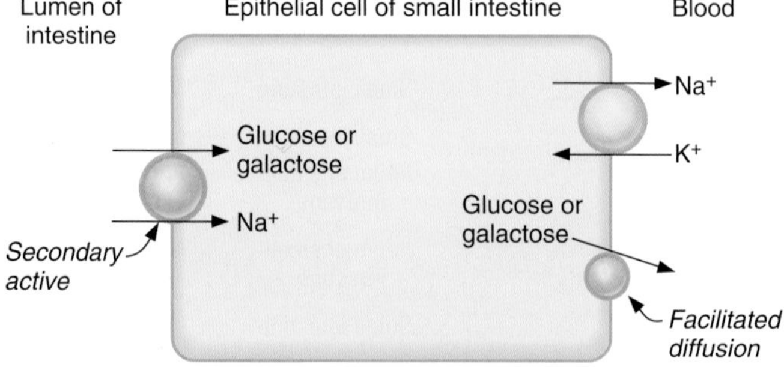

FIGURE 6.13. Mechanism of absorption of monosaccharides by intestinal epithelial cells. Glucose and galactose are absorbed by Na^+-dependent cotransport (secondary active), and fructose (not shown) is absorbed by facilitated diffusion.

c. Pepsin

- is not essential for protein digestion.
- is secreted as pepsinogen by the chief cells of the stomach.
- Pepsinogen is activated to pepsin by gastric H^+.
- The **optimum pH for pepsin is between 1 and 3.**
- When the pH is >5, pepsin is denatured. Thus, in the intestine, as HCO_3^- is secreted in pancreatic fluids, duodenal pH increases and pepsin is inactivated.

d. Pancreatic proteases

- include trypsin, chymotrypsin, elastase, carboxypeptidase A, and carboxypeptidase B.
- are secreted in inactive forms that are activated in the small intestine as follows:

(1) Trypsinogen is activated to **trypsin** by a brush border enzyme, enterokinase.

(2) Trypsin then converts chymotrypsinogen, proelastase, and procarboxypeptidase A and B to their active forms. (Even trypsinogen is converted to more trypsin by trypsin!)

(3) After their digestive work is complete, the pancreatic proteases degrade each other and are absorbed along with dietary proteins.

2. Absorption of proteins (Figure 6.14)

- Digestive products of protein can be **absorbed as amino acids, dipeptides, and tripeptides** (in contrast to carbohydrates, which can only be absorbed as monosaccharides).

a. Free amino acids

- **Na^+-dependent amino acid cotransport** occurs in the luminal membrane. It is analogous to the cotransporter for glucose and galactose.
- The amino acids are then transported from cell to blood by facilitated diffusion.
- There are **four separate carriers** for neutral, acidic, basic, and imino amino acids, respectively.

b. Dipeptides and tripeptides

- are absorbed faster than free amino acids.
- **H^+-dependent cotransport of dipeptides and tripeptides** also occurs in the luminal membrane.
- After the dipeptides and tripeptides are transported into the intestinal cells, cytoplasmic peptidases hydrolyze them to amino acids.
- The amino acids are then transported from cell to blood by facilitated diffusion.

C. Lipids

1. Digestion of lipids

a. Stomach

(1) In the stomach, **mixing** breaks lipids into droplets to increase the surface area for digestion by pancreatic enzymes.

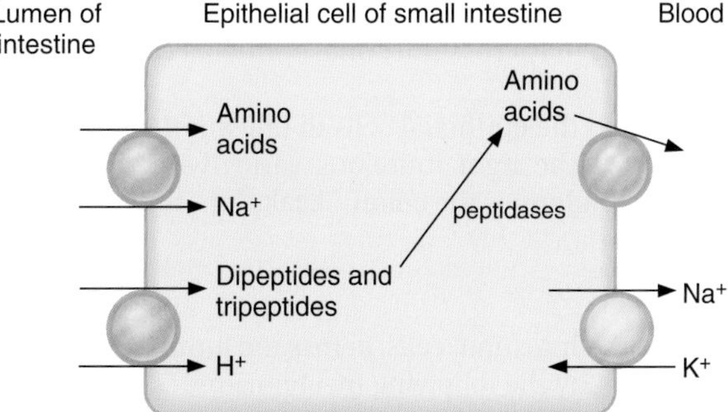

FIGURE 6.14. Mechanism of absorption of amino acids, dipeptides, and tripeptides by intestinal epithelial cells.

(2) **Lingual lipases** digest some of the ingested triglycerides to monoglycerides and fatty acids. However, most of the ingested lipids are digested in the intestine by pancreatic lipases.

(3) **CCK slows gastric emptying.** Thus, delivery of lipids from the stomach to the duodenum is slowed to allow adequate time for digestion and absorption in the intestine.

b. Small intestine

(1) **Bile acids** emulsify lipids in the small intestine, increasing the surface area for digestion.

(2) **Pancreatic lipases** hydrolyze lipids to fatty acids, monoglycerides, cholesterol, and lysolecithin. The enzymes are pancreatic lipase, cholesterol ester hydrolase, and phospholipase A_2.

(3) The hydrophobic products of lipid digestion are solubilized in **micelles** by **bile acids.**

2. Absorption of lipids

a. Micelles bring the products of lipid digestion into contact with the absorptive surface of the intestinal cells. Then, **fatty acids, monoglycerides, and cholesterol diffuse across the luminal membrane into the cells.** Glycerol is hydrophilic and is not contained in the micelles.

b. In the intestinal cells, the products of lipid digestion are **reesterified** to triglycerides, cholesterol ester, and phospholipids and, with apoproteins, form **chylomicrons.**

- Lack of apoprotein B results in the inability to transport chylomicrons out of the intestinal cells and causes **abetalipoproteinemia.**

c. Chylomicrons are transported out of the intestinal cells by **exocytosis.** Because chylomicrons are too large to enter the capillaries, they are transferred to **lymph vessels** and are added to the bloodstream via the thoracic duct.

3. Malabsorption of lipids—steatorrhea

- can be caused by any of the following:

a. pancreatic disease (e.g., pancreatitis, cystic fibrosis), in which the pancreas cannot synthesize adequate amounts of the enzymes (e.g., pancreatic lipase) needed for lipid digestion.

b. hypersecretion of gastrin, in which gastric H^+ secretion is increased and the duodenal pH is decreased. Low duodenal pH inactivates pancreatic lipase.

c. ileal resection, which leads to a depletion of the bile acid pool because the bile acids do not recirculate to the liver.

d. bacterial overgrowth, which may lead to deconjugation of bile acids and their "early" absorption in the upper small intestine. In this case, bile acids are not present throughout the small intestine to aid in lipid absorption.

e. decreased number of intestinal cells for lipid absorption (tropical sprue).

f. failure to synthesize apoprotein B, which leads to the inability to form chylomicrons.

D. Absorption and secretion of electrolytes and H_2O

- Electrolytes and H_2O may cross intestinal epithelial cells by either cellular or paracellular (between cells) routes.
- **Tight junctions** attach the epithelial cells to one another at the luminal membrane.
- The permeability of the tight junctions varies with the type of epithelium. A **"tight"** (impermeable) epithelium is the colon. **"Leaky"** (permeable) epithelia are the small intestine and gallbladder.

1. Absorption of NaCl

a. Na^+ moves into the intestinal cells, across the luminal membrane, and down its electrochemical gradient by the following mechanisms:

(1) passive diffusion (through Na^+ channels).

(2) Na^+–glucose or Na^+–amino acid cotransport.
(3) Na^+–Cl^- cotransport.
(4) Na^+–H^+ exchange.

- In the **small intestine,** Na^+–glucose cotransport, Na^+–amino acid cotransport, and Na^+–H^+ exchange mechanisms are most important. These cotransport and exchange mechanisms are similar to those in the renal proximal tubule.
- In the **colon,** passive diffusion via Na^+ channels is most important. The Na^+ channels of the colon are similar to those in the renal distal tubule and collecting ducts and are stimulated by **aldosterone.**

b. Na^+ is pumped out of the cell against its electrochemical gradient by the Na^+–K^+ pump in the basolateral membranes.

c. Cl^- absorption accompanies Na^+ absorption throughout the GI tract by the following mechanisms:

(1) passive diffusion by a paracellular route.
(2) Na^+–Cl^- cotransport.
(3) Cl^-–HCO_3^- exchange.

2. Absorption and secretion of K^+

a. Dietary K^+ is **absorbed in the small intestine** by passive diffusion via a paracellular route.

b. K^+ is actively **secreted in the colon** by a mechanism similar to that for K^+ secretion in the renal distal tubule.

- As in the distal tubule, K^+ secretion in the colon is stimulated by **aldosterone.**
- In **diarrhea,** K^+ secretion by the colon is increased because of a flow rate–dependent mechanism similar to that in the renal distal tubule. Excessive loss of K^+ in diarrheal fluid causes **hypokalemia.**

3. Absorption of H_2O

- is secondary to solute absorption.
- is **isosmotic in the small intestine and gallbladder.** The mechanism for coupling solute and water absorption in these epithelia is the same as that in the **renal proximal tubule.**
- In the **colon,** H_2O permeability is much lower than in the small intestine, and feces may be hypertonic.

4. Secretion of electrolytes and H_2O by the intestine

- The GI tract also secretes electrolytes from blood to lumen.
- The secretory mechanisms are located in the **crypts.** The absorptive mechanisms are located in the villi.

a. Cl^- is the primary ion secreted into the intestinal lumen. It is transported through Cl^- channels in the luminal membrane that are regulated by **cAMP.**

b. Na^+ is secreted into the lumen by passively following Cl^-. H_2O follows NaCl to maintain isosmotic conditions.

c. ***Vibrio cholerae*** **(cholera toxin)** causes diarrhea by stimulating Cl^- secretion.

- Cholera toxin catalyzes **adenosine diphosphate (ADP) ribosylation** of the α_s subunit of the G_s protein coupled to adenylyl cyclase, permanently activating it.
- Intracellular cAMP increases; as a result, **Cl^- channels** in the luminal membrane open.
- Na^+ and H_2O follow Cl^- into the lumen and lead to **secretory diarrhea.**
- Some strains of ***Escherichia coli*** cause diarrhea by a similar mechanism.
- Oral rehydration solutions contain Na^+, Cl^-, HCO_3^-, and glucose. The inclusion of glucose stimulates absorption via Na^+–glucose cotransport to offset secretory losses.

E. Absorption of other substances

1. Vitamins

a. Fat-soluble vitamins (A, D, E, and K) are incorporated into micelles and absorbed along with other lipids.

b. Most **water-soluble vitamins** are absorbed by Na^+-dependent cotransport mechanisms.

c. Vitamin B_{12} is absorbed in the ileum and requires **intrinsic factor.**

- The vitamin B_{12}–intrinsic factor complex binds to a receptor on the ileal cells and is absorbed.
- **Gastrectomy** results in the loss of gastric parietal cells, which are the source of intrinsic factor. Injection of vitamin B_{12} is required to prevent **pernicious anemia.**
- **Ileectomy** results in loss of absorption of the vitamin B_{12}–intrinsic factor complex and thus requires injection of vitamin B_{12}.

2. Calcium

- absorption in the small intestine depends on the presence of adequate amounts of the active form of vitamin D, **1,25-dihydroxycholecalciferol,** which is produced in the kidney. 1,25-Dihydroxycholecalciferol induces the synthesis of an intestinal Ca^{2+}-binding protein, **calbindin D-28K.**
- Vitamin D deficiency or chronic renal failure results in inadequate intestinal Ca^{2+} absorption, causing **rickets** in children and **osteomalacia** in adults.

3. Iron

- is absorbed as **heme iron** (iron bound to hemoglobin or myoglobin) or as **free Fe^{2+}**. In the intestinal cells, "heme iron" is degraded and free Fe^{2+} is released. The free Fe^{2+} binds to apoferritin and is transported into the blood.
- Free Fe^{2+} circulates in the **blood bound to transferrin,** which transports it from the small intestine to its storage sites in the liver and from the liver to the bone marrow for the synthesis of hemoglobin.
- Iron deficiency is the most common cause of anemia.

VI. LIVER PHYSIOLOGY

A. Bile formation and secretion (see IV D)

B. Bilirubin production and excretion (Figure 6.15)

- Hemoglobin is degraded to **bilirubin** by the reticuloendothelial system.
- Bilirubin is carried in the circulation bound to albumin.
- In the liver, bilirubin is conjugated with glucuronic acid via the enzyme **UDP glucuronyl transferase.**
- A portion of **conjugated bilirubin** is excreted in the urine, and a portion is secreted into bile.
- In the intestine, conjugated bilirubin is converted to **urobilinogen,** which is returned to the liver via the enterohepatic circulation, and **urobilin** and **stercobilin,** which are excreted in feces.

C. Metabolic functions of the liver

1. Carbohydrate metabolism

- performs gluconeogenesis, stores glucose as glycogen, and releases stored glucose into the circulation.

2. Protein metabolism

- synthesizes nonessential amino acids.
- synthesizes plasma proteins.

3. Lipid metabolism

- participates in fatty acid oxidation.
- synthesizes lipoproteins, cholesterol, and phospholipids.

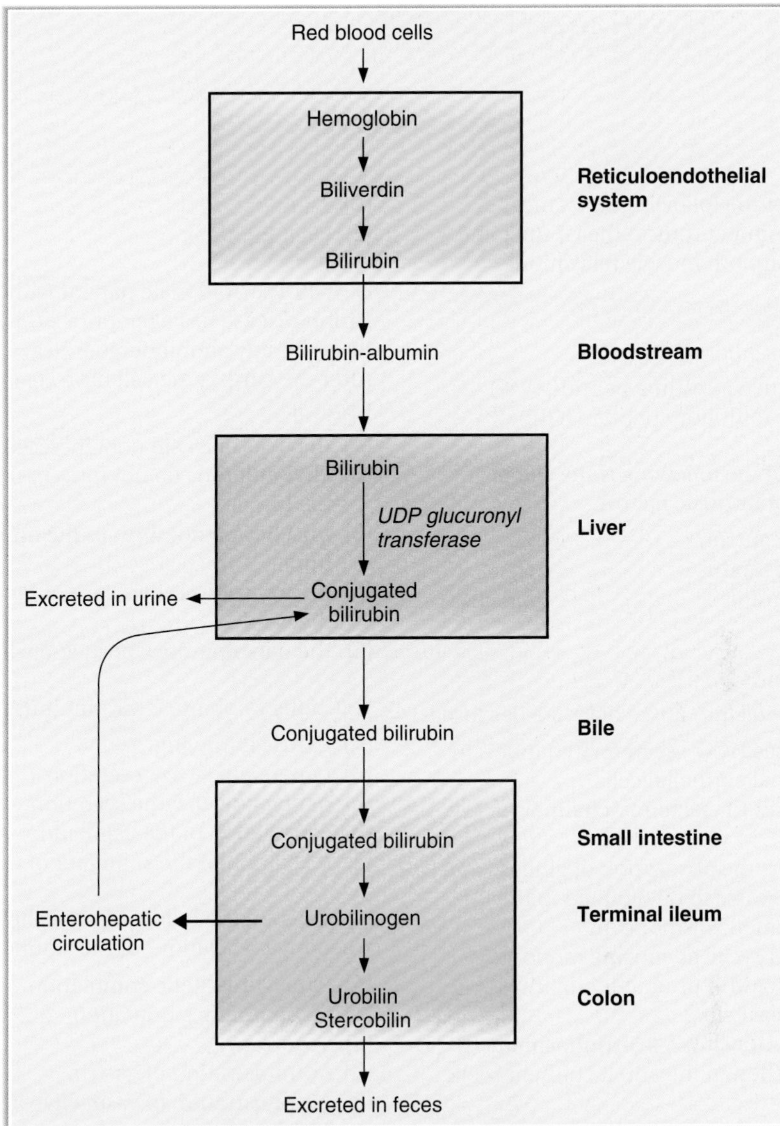

FIGURE 6.15. Bilirubin metabolism. UDP = uridine diphosphate.

D. Detoxification

- Potentially toxic substances are presented to the liver via the portal circulation.
- The liver modifies these substances in "first-pass metabolism."
- **Phase I reactions** are catalyzed by cytochrome P-450 enzymes, which are followed by **phase II reactions** that conjugate the substances.

Review Test

1. Which of the following substances is released from neurons in the GI tract and produces smooth muscle relaxation?

(A) Secretin
(B) Gastrin
(C) Cholecystokinin (CCK)
(D) Vasoactive intestinal peptide (VIP)
(E) Gastric inhibitory peptide (GIP)

2. Which of the following is the site of secretion of intrinsic factor?

(A) Gastric antrum
(B) Gastric fundus
(C) Duodenum
(D) Ileum
(E) Colon

3. *Vibrio cholerae* causes diarrhea because it

(A) increases HCO_3^- secretory channels in intestinal epithelial cells
(B) increases Cl^- secretory channels in crypt cells
(C) prevents the absorption of glucose and causes water to be retained in the intestinal lumen isosmotically
(D) inhibits cyclic adenosine monophosphate (cAMP) production in intestinal epithelial cells
(E) inhibits inositol 1,4,5-triphosphate (IP_3) production in intestinal epithelial cells

4. Cholecystokinin (CCK) has some gastrin-like properties because both CCK and gastrin

(A) are released from G cells in the stomach
(B) are released from I cells in the duodenum
(C) are members of the secretin-homologous family
(D) have five identical C-terminal amino acids
(E) have 90% homology of their amino acids

5. Which of the following is transported in intestinal epithelial cells by a Na^+-dependent cotransport process?

(A) Fatty acids
(B) Triglycerides
(C) Fructose
(D) Alanine
(E) Oligopeptides

6. A 49-year-old male patient with severe Crohn disease has been unresponsive to drug therapy and undergoes ileal resection. After the surgery, he will have steatorrhea because

(A) the liver bile acid pool increases
(B) chylomicrons do not form in the intestinal lumen
(C) micelles do not form in the intestinal lumen
(D) dietary triglycerides cannot be digested
(E) the pancreas does not secrete lipase

7. Cholecystokinin (CCK) inhibits

(A) gastric emptying
(B) pancreatic HCO_3^- secretion
(C) pancreatic enzyme secretion
(D) contraction of the gallbladder
(E) relaxation of the sphincter of Oddi

8. Which of the following abolishes "receptive relaxation" of the stomach?

(A) Parasympathetic stimulation
(B) Sympathetic stimulation
(C) Vagotomy
(D) Administration of gastrin
(E) Administration of vasoactive intestinal peptide (VIP)
(F) Administration of cholecystokinin (CCK)

9. Secretion of which of the following substances is inhibited by low pH?

(A) Secretin
(B) Gastrin
(C) Cholecystokinin (CCK)
(D) Vasoactive intestinal peptide (VIP)
(E) Gastric inhibitory peptide (GIP)

10. Which of the following is the site of secretion of gastrin?

(A) Gastric antrum
(B) Gastric fundus
(C) Duodenum
(D) Ileum
(E) Colon

11. Micelle formation is necessary for the intestinal absorption of

(A) glycerol
(B) galactose
(C) leucine
(D) bile acids
(E) vitamin B_{12}
(F) vitamin D

12. Which of the following changes occurs during defecation?

(A) Internal anal sphincter is relaxed
(B) External anal sphincter is contracted
(C) Rectal smooth muscle is relaxed
(D) Intra-abdominal pressure is lower than when at rest
(E) Segmentation contractions predominate

13. Which of the following is characteristic of saliva?

(A) Hypotonicity relative to plasma
(B) A lower HCO_3^- concentration than plasma
(C) The presence of proteases
(D) Secretion rate that is increased by vagotomy
(E) Modification by the salivary ductal cells involves reabsorption of K^+ and HCO_3^-

14. Which of the following substances is secreted in response to an oral glucose load?

(A) Secretin
(B) Gastrin
(C) Cholecystokinin (CCK)
(D) Vasoactive intestinal peptide (VIP)
(E) Glucose-dependent insulinotropic peptide (GIP)

15. Which of the following is true about the secretion from the exocrine pancreas?

(A) It has a higher Cl^- concentration than does plasma
(B) It is stimulated by the presence of HCO_3^- in the duodenum
(C) Pancreatic HCO_3^- secretion is increased by gastrin
(D) Pancreatic enzyme secretion is increased by cholecystokinin (CCK)
(E) It is hypotonic

16. Which of the following substances must be further digested before it can be absorbed by specific carriers in intestinal cells?

(A) Fructose
(B) Sucrose
(C) Alanine
(D) Dipeptides
(E) Tripeptides

17. Slow waves in small intestinal smooth muscle cells are

(A) action potentials
(B) phasic contractions
(C) tonic contractions
(D) oscillating resting membrane potentials
(E) oscillating release of cholecystokinin (CCK)

18. A 24-year-old male graduate student participates in a clinical research study on intestinal motility. Peristalsis of the small intestine

(A) mixes the food bolus
(B) is coordinated by the central nervous system (CNS)
(C) involves contraction of circular smooth muscle behind and in front of the food bolus
(D) involves contraction of circular smooth muscle behind the food bolus and relaxation of circular smooth muscle in front of the bolus
(E) involves relaxation of circular and longitudinal smooth muscle simultaneously throughout the small intestine

19. A 38-year-old male patient with a duodenal ulcer is treated successfully with the drug cimetidine. The basis for cimetidine's inhibition of gastric H^+ secretion is that it

(A) blocks muscarinic receptors on parietal cells
(B) blocks H_2 receptors on parietal cells
(C) increases intracellular cyclic adenosine monophosphate (cAMP) levels
(D) blocks H^+,K^+-adenosine triphosphatase (ATPase)
(E) enhances the action of acetylcholine (ACh) on parietal cells

20. Which of the following substances inhibits gastric emptying?

(A) Secretin
(B) Gastrin
(C) Cholecystokinin (CCK)
(D) Vasoactive intestinal peptide (VIP)
(E) Gastric inhibitory peptide (GIP)

21. When parietal cells are stimulated, they secrete

(A) HCl and intrinsic factor
(B) HCl and pepsinogen
(C) HCl and HCO_3^-
(D) HCO_3^- and intrinsic factor
(E) mucus and pepsinogen

22. A 44-year-old woman is diagnosed with Zollinger–Ellison syndrome. Which of the following findings is consistent with the diagnosis?

(A) Decreased serum gastrin levels
(B) Increased serum insulin levels
(C) Increased absorption of dietary lipids
(D) Decreased parietal cell mass
(E) Peptic ulcer disease

23. Which of the following is the site of Na^+–bile acid cotransport?

(A) Gastric antrum
(B) Gastric fundus
(C) Duodenum
(D) Ileum
(E) Colon

Answers and Explanations

1. **The answer is D** [II C 1]. Vasoactive intestinal peptide (VIP) is a gastrointestinal (GI) neurocrine that causes relaxation of GI smooth muscle. For example, VIP mediates the relaxation response of the lower esophageal sphincter when a bolus of food approaches it, allowing passage of the bolus into the stomach.
2. **The answer is B** [IV B 1; Table 6.3; Figure 6.7]. Intrinsic factor is secreted by the parietal cells of the gastric fundus (as is HCl). It is absorbed, with vitamin B_{12}, in the ileum.
3. **The answer is B** [V D 4 c]. Cholera toxin activates adenylate cyclase and increases cyclic adenosine monophosphate (cAMP) in the intestinal crypt cells. In the crypt cells, cAMP activates the Cl^--secretory channels and produces a primary secretion of Cl^- with Na^+ and H_2O following.
4. **The answer is D** [II A 2]. The two hormones have five identical amino acids at the C-terminus. Biologic activity of cholecystokinin (CCK) is associated with the seven C-terminal amino acids, and biologic activity of gastrin is associated with the four C-terminal amino acids. Because this CCK heptapeptide contains the five common amino acids, it is logical that CCK should have some gastrin-like properties. G cells secrete gastrin. I cells secrete CCK. The secretin family includes glucagon.
5. **The answer is D** [V A–C; Table 6.4]. Fructose is the only monosaccharide that is not absorbed by Na^+-dependent cotransport; it is transported by facilitated diffusion. Amino acids are absorbed by Na^+-dependent cotransport, but oligopeptides (larger peptide units) are not. Triglycerides are not absorbed without further digestion. The products of lipid digestion, such as fatty acids, are absorbed by simple diffusion.
6. **The answer is C** [IV D 4]. Ileal resection removes the portion of the small intestine that normally transports bile acids from the lumen of the gut and recirculates them to the liver. Because this process maintains the bile acid pool, new synthesis of bile acids is needed only to replace those bile acids that are lost in the feces. With ileal resection, most of the bile acids secreted are excreted in the feces, and the liver pool is significantly diminished. Bile acids are needed for micelle formation in the intestinal lumen to solubilize the products of lipid digestion so that they can be absorbed. Chylomicrons are formed *within* the intestinal epithelial cells and are transported to lymph vessels.
7. **The answer is A** [II A 2 a; Table 6.1]. Cholecystokinin (CCK) inhibits gastric emptying and therefore helps to slow the delivery of food from the stomach to the intestine during periods of high digestive activity. CCK stimulates both functions of the exocrine pancreas—HCO_3^- secretion and digestive enzyme secretion. It also stimulates the delivery of bile from the gallbladder to the small intestinal lumen by causing contraction of the gallbladder while relaxing the sphincter of Oddi.
8. **The answer is C** [III C 1]. "Receptive relaxation" of the orad region of the stomach is initiated when food enters the stomach from the esophagus. This parasympathetic (vagovagal) reflex is abolished by vagotomy.
9. **The answer is B** [II A 1; Table 6.1]. Gastrin's principal physiologic action is to increase H^+ secretion. H^+ secretion decreases the pH of the stomach contents. The decreased pH, in turn, inhibits further secretion of gastrin—a classic example of negative feedback.
10. **The answer is A** [II A 1 b; Table 6.3; Figure 6.7]. Gastrin is secreted by the G cells of the gastric antrum. HCl and intrinsic factor are secreted by the fundus.

11. **The answer is F** [V E 1; Table 6.4]. Micelles provide a mechanism for solubilizing fat-soluble nutrients in the aqueous solution of the intestinal lumen until the nutrients can be brought into contact with and absorbed by the intestinal epithelial cells. Because vitamin D is fat soluble, it is absorbed in the same way as other dietary lipids. Glycerol is one product of lipid digestion that is water soluble and is not included in micelles. Galactose and leucine are absorbed by Na^+-dependent cotransport. Although bile acids are a key ingredient of micelles, they are absorbed by a specific Na^+-dependent cotransporter in the ileum. Vitamin B_{12} is water soluble; thus, its absorption does not require micelles.

12. **The answer is A** [III E 3]. Both the internal and external anal sphincters must be relaxed to allow feces to be expelled from the body. Rectal smooth muscle contracts and intra-abdominal pressure is elevated by expiring against a closed glottis (Valsalva maneuver). Segmentation contractions are prominent in the small intestine during digestion and absorption.

13. **The answer is A** [IV A 2 a; Table 6.2]. Saliva is characterized by hypotonicity and a high HCO_3^- concentration (relative to plasma) and by the presence of α-amylase and lingual lipase (not proteases). The high HCO_3^- concentration is achieved by secretion of HCO_3^- into saliva by the ductal cells (not reabsorption of HCO_3^-). Because control of saliva production is parasympathetic, it is abolished by vagotomy.

14. **The answer is E** [II A 4; Table 6.4]. Glucose-dependent insulinotropic peptide (GIP) is the only gastrointestinal (GI) hormone that is released in response to all three categories of nutrients—fat, protein, and carbohydrate. Oral glucose releases GIP, which, in turn, causes the release of insulin from the endocrine pancreas. This action of GIP explains why oral glucose is more effective than intravenous glucose in releasing insulin.

15. **The answer is D** [II A 2, 3; Table 6.2]. The major anion in pancreatic secretions is HCO_3^- (which is found in higher concentration than in plasma), and the Cl^- concentration is lower than in plasma. Pancreatic secretion is stimulated by the presence of fatty acids in the duodenum. Secretin (not gastrin) stimulates pancreatic HCO_3^- secretion, and cholecystokinin (CCK) stimulates pancreatic enzyme secretion. Pancreatic secretions are always isotonic, regardless of flow rate.

16. **The answer is B** [V A, B; Table 6.4]. Only monosaccharides can be absorbed by intestinal epithelial cells. Disaccharides, such as sucrose, must be digested to monosaccharides before they are absorbed. On the other hand, proteins are hydrolyzed to amino acids, dipeptides, or tripeptides, and all three forms are transported into intestinal cells for absorption.

17. **The answer is D** [III A; Figure 6.3]. Slow waves are oscillating resting membrane potentials of the gastrointestinal (GI) smooth muscle. The slow waves bring the membrane potential toward or to threshold, but *are not themselves action potentials.* If the membrane potential is brought to threshold by a slow wave, then action potentials occur, followed by contraction.

18. **The answer is D** [III D 2]. Peristalsis is contractile activity that is coordinated by the enteric nervous system (not the central nervous system [CNS]) and propels the intestinal contents forward. Normally, it takes place after sufficient mixing, digestion, and absorption have occurred. To propel the food bolus forward, the circular smooth muscle must simultaneously contract behind the bolus and relax in front of the bolus; at the same time, longitudinal smooth muscle relaxes (lengthens) behind the bolus and contracts (shortens) in front of the bolus.

19. **The answer is B** [IV B 3 c, d (1), 6]. Cimetidine is a reversible inhibitor of H_2 receptors on parietal cells and blocks H^+ secretion. Cyclic adenosine monophosphate (cAMP) (the second messenger for histamine) levels would be expected to decrease, not increase. Cimetidine also blocks the action of acetylcholine (ACh) to stimulate H^+ secretion. Omeprazole blocks H^+, K^+-adenosine triphosphatase (ATPase) directly.

20. **The answer is C** [II A 2 a; Table 6.1]. Cholecystokinin (CCK) is the most important hormone for digestion and absorption of dietary fat. In addition to causing contraction of the gallbladder, it inhibits gastric emptying. As a result, chyme moves more slowly from the stomach to the small intestine, thus allowing more time for fat digestion and absorption.

21. **The answer is A** [IV B I; Table 6.3]. The gastric parietal cells secrete HCl and intrinsic factor. The chief cells secrete pepsinogen.

22. **The answer is E** [II A 1 d; V C 3 b]. Zollinger–Ellison syndrome (gastrinoma) is a tumor of the non–β-cell pancreas. The tumor secretes gastrin, which then circulates to the gastric parietal cells to produce increased H^+ secretion, peptic ulcer, and parietal cell growth (trophic effect of gastrin). Because the tumor does not involve the pancreatic β-cells, insulin levels should be unaffected. Absorption of lipids is decreased (not increased) because increased H^+ secretion decreases the pH of the intestinal lumen and inactivates pancreatic lipases.

23. **The answer is D** [IV D 4]. Bile salts are recirculated to the liver in the enterohepatic circulation via a Na^+–bile acid cotransporter located in the ileum of the small intestine.

Chapter 7

Endocrine Physiology

I. OVERVIEW OF HORMONES

A. See Table 7.1 for a list of hormones, including abbreviations, glands of origin, and major actions.

B. Hormone synthesis

1. Protein and peptide hormone synthesis

- **Preprohormone** synthesis occurs in the **endoplasmic reticulum** and is directed by a specific mRNA.
- **Signal peptides are cleaved** from the preprohormone, producing a **prohormone**, which is transported to the Golgi apparatus.
- Additional peptide sequences are cleaved in the Golgi apparatus to form the **hormone**, which is packaged in secretory granules for later release.

2. Steroid hormone synthesis

- Steroid hormones are **derivatives of cholesterol** (the biosynthetic pathways are described in V A 1).

3. Amine hormone synthesis

- Amine hormones (thyroid hormones, epinephrine, norepinephrine) are **derivatives of tyrosine** (the biosynthetic pathway for thyroid hormones is described in IV A).

C. Regulation of hormone secretion

1. Negative feedback

- is the most commonly applied principle for regulating hormone secretion.
- is self-limiting.
- A hormone has biologic actions that, directly or indirectly, inhibit further secretion of the hormone.
- **For example**, insulin is secreted by the pancreatic beta cells in response to an increase in blood glucose. In turn, insulin causes an increase in glucose uptake into cells that results in decreased blood glucose concentration. The decrease in blood glucose concentration then decreases further secretion of insulin.
- **For example**, parathyroid hormone is secreted by the chief cells of the parathyroid gland in response to a decrease in serum Ca^{2+} concentration. In turn, parathyroid hormone's actions on bone, kidney, and intestine all act in concert to increase the serum Ca^{2+} concentration. The increased serum Ca^{2+} concentration then decreases further parathyroid hormone secretion.

2. Positive feedback

- is rare.
- is explosive and self-reinforcing.

table 7.1 Master List of Hormones

Hormone	Abbreviation	Gland of Origin	Major Actions*
Thyrotropin-releasing hormone	TRH	Hypothalamus	Stimulates secretion of TSH and prolactin
Corticotropin-releasing hormone	CRH	Hypothalamus	Stimulates secretion of ACTH
Gonadotropin-releasing hormone	GnRH	Hypothalamus	Stimulates secretion of LH and FSH
Growth hormone–releasing hormone	GHRH	Hypothalamus	Stimulates secretion of growth hormone
Somatotropin release–inhibiting hormone (somatostatin)	SRIF	Hypothalamus	Inhibits secretion of growth hormone
Prolactin-inhibiting factor (dopamine)	PIF	Hypothalamus	Inhibits secretion of prolactin
Thyroid-stimulating hormone	TSH	Anterior pituitary	Stimulates synthesis and secretion of thyroid hormones
Follicle-stimulating hormone	FSH	Anterior pituitary	Stimulates growth of ovarian follicles and estrogen secretion Promotes sperm maturation (testes)
Luteinizing hormone	LH	Anterior pituitary	Stimulates ovulation, formation of corpus luteum, and synthesis of estrogen and progesterone (ovary) Stimulates synthesis and secretion of testosterone (testes)
Growth hormone	GH	Anterior pituitary	Stimulates protein synthesis and overall growth
Prolactin		Anterior pituitary	Stimulates milk production and breast development
Adrenocorticotropic hormone	ACTH	Anterior pituitary	Stimulates synthesis and secretion of adrenal cortical hormones
Melanocyte-stimulating hormone	MSH	Anterior pituitary	Stimulates melanin synthesis (? humans)
Oxytocin		Posterior pituitary	Milk ejection; uterine contraction
Antidiuretic hormone (vasopressin)	ADH	Posterior pituitary	Stimulates H_2O reabsorption by renal collecting ducts and contraction of arterioles
L-thyroxine Triiodothyronine	T_4 T_3	Thyroid gland	Skeletal growth; ↑ O_2 consumption; heat production; ↑ protein, fat, and carbohydrate use; maturation of nervous system (perinatal)
Glucocorticoids (cortisol)		Adrenal cortex	Stimulates gluconeogenesis; anti-inflammatory; immunosuppression
Estradiol		Ovary	Growth and development of female reproductive organs; follicular phase of menstrual cycle
Progesterone		Ovary	Luteal phase of menstrual cycle
Testosterone		Testes	Spermatogenesis; male secondary sex characteristics
Parathyroid hormone	PTH	Parathyroid gland	↑ serum [Ca^{2+}]; ↓ serum [phosphate]
Calcitonin		Thyroid gland (parafollicular cells)	↓ serum [Ca^{2+}]
Aldosterone		Adrenal cortex	↑ renal Na^+ reabsorption; ↑ renal K^+ secretion; ↑ renal H^+ secretion
1,25-Dihydroxycholecalciferol		Kidney (activation)	↑ intestinal Ca^{2+} absorption; ↑ bone mineralization
Insulin		Pancreas (beta cells)	↓ blood [glucose]; ↓ blood [amino acid]; ↓ blood [fatty acid]
Glucagon		Pancreas (alpha cells)	↑ blood [glucose]; ↑ blood [fatty acid]
Human chorionic gonadotropin	HCG	Placenta	↓ estrogen and progesterone synthesis in corpus luteum of pregnancy
Human placental lactogen	HPL	Placenta	Same actions as growth hormone and prolactin during pregnancy

*See text for more complete description of each hormone.

- A hormone has biologic actions that, directly or indirectly, cause more secretion of the hormone.
- **For example**, the surge of luteinizing hormone (LH) that occurs just before ovulation is a result of positive feedback of estrogen on the anterior pituitary. LH then acts on the ovaries and causes more secretion of estrogen.

D. Regulation of receptors

- Hormones determine the sensitivity of the target tissue by **regulating the number or sensitivity of receptors.**

1. Down-regulation of receptors

- A hormone **decreases the number or affinity of receptors** for itself or for another hormone. For example, in the uterus, progesterone down-regulates its own receptor and the receptor for estrogen.

2. Up-regulation of receptors

- A hormone **increases the number or affinity of receptors** for itself or for another hormone.
- **For example**, in the ovary, estrogen up-regulates its own receptor and the receptor for LH.

II. CELL MECHANISMS AND SECOND MESSENGERS (TABLE 7.2)

A. G proteins

- are **guanosine triphosphate (GTP)-binding proteins** that couple hormone receptors to adjacent effector molecules. For example, in the cyclic adenosine monophosphate (cAMP) second messenger system, G proteins couple the hormone receptor to adenylate cyclase.
- are used in the adenylate cyclase and **inositol 1,4,5-triphosphate (IP_3) second messenger systems.**
- have intrinsic **GTPase activity.**
- have three subunits: α, β, and γ.
- The **α subunit** can bind either guanosine diphosphate (GDP) or GTP. When GDP is bound to the α subunit, the G protein is inactive. When GTP is bound, the G protein is active.
- G proteins can be either stimulatory (G_s) or inhibitory (G_i). Stimulatory or inhibitory activity resides in the α subunits, which are accordingly called α_s and α_i.

table 7.2 Mechanisms of Hormone Action

cAMP Mechanism	IP_3 Mechanism	Steroid Hormone Mechanism	Other Mechanisms
ACTH	GnRH	Glucocorticoids	**Tyrosine Kinase Mechanism**
LH and FSH	TRH	Estrogen	Insulin
TSH	GHRH	Testosterone	IGF-1
ADH (V_2 receptor)	Angiotensin II	Progesterone	Growth hormone
HCG	ADH (V_1 receptor)	Aldosterone	Prolactin
MSH	Oxytocin	Vitamin D	
CRH	α_1 Receptors	Thyroid hormone	**cGMP Mechanism**
β_1 and β_2 receptors			ANP
Calcitonin			Nitric oxide
PTH			
Glucagon			

ANP = atrial natriuretic peptide; cAMP = cyclic adenosine monophosphate; cGMP = cyclic guanosine monophosphate; IGF = insulin-like growth factor; IP_3 = inositol 1,4,5-triphosphate.

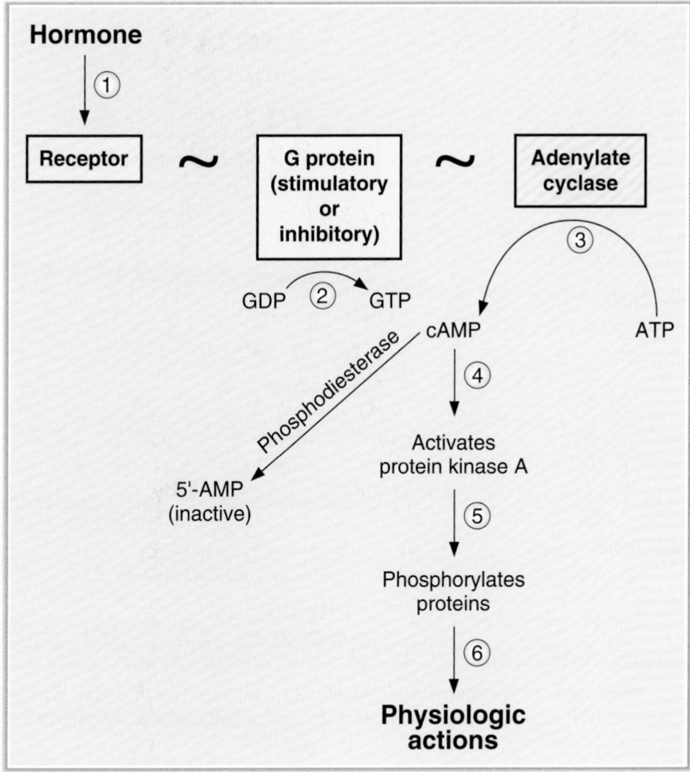

FIGURE 7.1. Mechanism of hormone action—adenylate cyclase. ATP = adenosine triphosphate; cAMP = cyclic adenosine monophosphate; GDP = guanosine diphosphate; GTP = guanosine triphosphate.

B. Adenylate cyclase mechanism (Figure 7.1)

1. **Hormone binds to a receptor in the cell membrane** (step 1).
2. **GDP is released from the G protein and replaced by GTP** (step 2), which activates the G protein. The G protein then activates or inhibits adenylate cyclase. If the G protein is stimulatory (G_s), then adenylate cyclase will be activated. If the G protein is inhibitory (G_i), then adenylate cyclase will be inhibited (not shown). Intrinsic GTPase activity in the G protein converts GTP back to GDP (not shown).
3. **Activated adenylate cyclase** then catalyzes the conversion of adenosine triphosphate (ATP) to cAMP (step 3).
4. **cAMP activates protein kinase A** (step 4), which phosphorylates specific proteins (step 5), producing specific physiologic actions (step 6).
5. **cAMP is degraded to 5′-AMP by phosphodiesterase**, which is inhibited by **caffeine**. Therefore, phosphodiesterase inhibitors would be expected to augment the physiologic actions of cAMP.

C. IP_3 mechanism (Figure 7.2)

1. **Hormone binds to a receptor** in the cell membrane (step 1) and, via a G protein (step 2), **activates phospholipase C** (step 3).
2. **Phospholipase C liberates diacylglycerol and IP_3** from membrane lipids (step 4).
3. **IP_3 mobilizes Ca^{2+} from the endoplasmic reticulum** (step 5). Together, Ca^{2+} and diacylglycerol **activate protein kinase C** (step 6), which phosphorylates proteins and causes specific **physiologic actions** (step 7).

D. Catalytic receptor mechanisms

- Hormone binds to extracellular receptors that have, or are associated with, enzymatic activity on the intracellular side of the membrane.

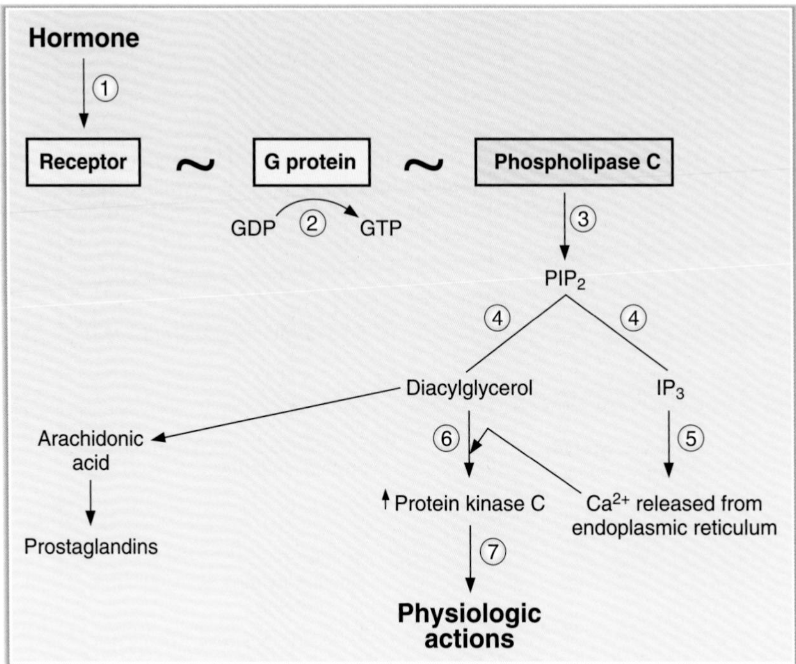

FIGURE 7.2. Mechanism of hormone action—inositol 1,4,5-triphosphate (IP_3)–Ca^{2+}. GDP = guanosine diphosphate; GTP = guanosine triphosphate; PIP_2 = phosphatidylinositol 4,5-diphosphate.

1. Guanylyl cyclase

a. Atrial natriuretic peptide (ANP) acts through *receptor* guanylyl cyclase, where the extracellular side of the receptor binds ANP and the intracellular side of the receptor has guanylyl cyclase activity. Activation of guanylyl cyclase converts GTP to cyclic GMP, which is the second messenger.

b. Nitric oxide (NO) acts through *cytosolic* guanylyl cyclase. Activation of guanylyl cyclase converts GTP to cyclic GMP, which is the second messenger.

2. Tyrosine kinases (Figure 7.3)

- Hormone binds to extracellular receptors that have, or are associated with, tyrosine kinase activity. When activated, tyrosine kinase phosphorylates tyrosine moieties on proteins, leading to the hormone's physiologic actions.

a. Receptor tyrosine kinase

- Hormone binds to the extracellular side of the receptor.
- The intracellular side of the receptor has intrinsic tyrosine kinase activity.
- One type of receptor tyrosine kinase is a **monomer** (e.g., receptor for nerve growth factor). Binding of hormone or ligand causes dimerization of the receptor, activation of intrinsic tyrosine kinase, and phosphorylation of tyrosine moieties.
- Another type of receptor tyrosine kinase is a **dimer** (e.g., receptors for **insulin** and **insulin-like growth factor [IGF]**). Binding of hormone activates intrinsic tyrosine kinase, leading to phosphorylation of tyrosine moieties.
- Insulin receptors are also discussed in Section VI C 2.

b. Tyrosine kinase–associated receptor

- is the mechanism of action of **growth hormone**.
- Growth hormone binds to the extracellular side of the receptor.
- The intracellular side of the receptor does *not* have tyrosine kinase activity but is noncovalently *associated with* tyrosine kinase (e.g., Janus family of receptor-associated tyrosine kinase, **JAK**).

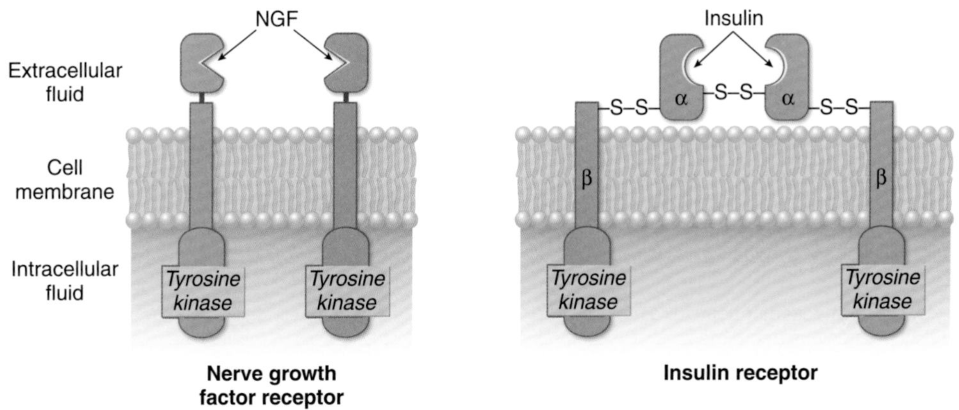

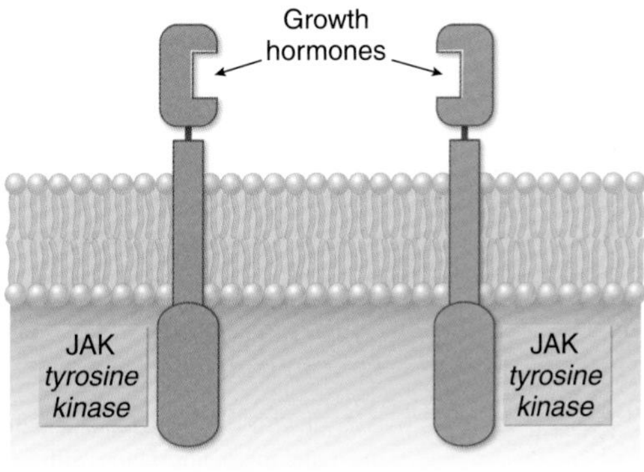

FIGURE 7.3. Tyrosine kinase receptors. Nerve growth factor and insulin utilize receptor tyrosine kinases. Growth hormone utilizes a tyrosine kinase–associated receptor. JAK = Janus family of receptor-associated tyrosine kinase; NGF = nerve growth factor.

- Binding of growth hormone causes dimerization of the receptor and activation of tyrosine kinase in the associated protein (e.g., JAK).
- Targets of JAK include signal transducers and activators of transcription (**STAT**), which cause transcription of new mRNAs and new protein synthesis.

E. Steroid hormone and thyroid hormone mechanism (Figure 7.4)

1. Steroid (or thyroid) hormone diffuses across the cell membrane and binds to its **receptor** (step 1).
2. The hormone–receptor complex enters the nucleus and dimerizes (step 2).
3. The **hormone–receptor dimers** are transcription factors that **bind to steroid-responsive elements (SREs)** of DNA (step 3) and initiate DNA transcription (step 4).
4. **New messenger RNA** is produced, leaves the nucleus, and is translated to synthesize new proteins (step 5).
5. The **new proteins** that are synthesized have specific physiologic actions. For example, 1,25-dihydroxycholecalciferol induces the synthesis of calbindin D-28K, a Ca^{2+}-binding protein in the intestine; aldosterone induces the synthesis of Na^+ channels in the renal principal cells.

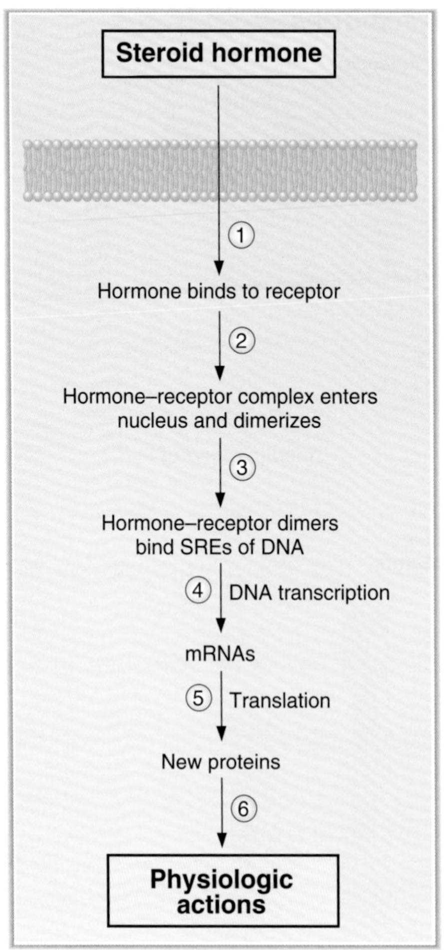

FIGURE 7.4. Mechanism of hormone action—steroid hormones and thyroid hormone. SREs = steroid-responsive elements.

III. PITUITARY GLAND (HYPOPHYSIS)

A. Hypothalamic–pituitary relationships

1. **The anterior lobe of the pituitary gland** is linked to the **hypothalamus by the hypothalamic–hypophysial portal system**. Thus, blood from the hypothalamus that contains high concentrations of hypothalamic hormones is delivered directly to the anterior pituitary. Hypothalamic hormones (e.g., growth hormone–releasing hormone [GHRH]) then stimulate or inhibit the release of anterior pituitary hormones (e.g., growth hormone).
2. **The posterior lobe of the pituitary gland** is derived from neural tissue. The nerve **cell bodies are located in hypothalamic nuclei**. Posterior pituitary hormones are synthesized in the nerve cell bodies, packaged in secretory granules, and transported down the axons to the posterior pituitary for release into the circulation.

B. Hormones of the anterior lobe of the pituitary

- are growth hormone, prolactin, thyroid-stimulating hormone (TSH), LH, follicle-stimulating hormone (FSH), and adrenocorticotropic hormone (ACTH).
- Growth hormone and prolactin are discussed in detail in this section. TSH, LH, FSH, and ACTH are discussed in context (e.g., TSH with thyroid hormone) in later sections of this chapter.

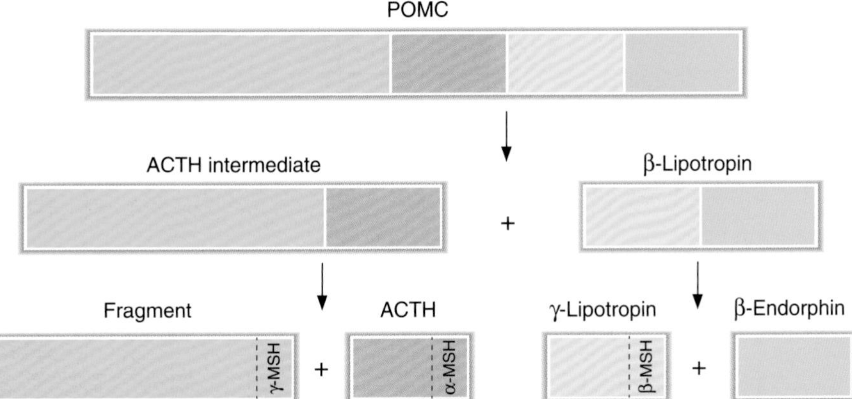

FIGURE 7.5. Proopiomelanocortin (POMC) is the precursor for adrenocorticotropic hormone (ACTH), β-lipotropin, and β-endorphin in the anterior pituitary. MSH = melanocyte-stimulating hormone.

1. **TSH, LH, and FSH**
   - belong to the same glycoprotein family. Each has an α subunit and a β subunit. The **α subunits are identical**. The β subunits are different and are responsible for the unique biologic activity of each hormone.
2. **ACTH, melanocyte-stimulating hormone (MSH), β-lipotropin, and β-endorphin** (Figure 7.5)
   - are derived from a single precursor, **proopiomelanocortin (POMC)**.
   - **α-MSH and β-MSH** are produced in the intermediary lobe, which is rudimentary in adult humans.
3. **Growth hormone (somatotropin)**
   - is the most important hormone for normal growth to adult size.
   - is a single-chain polypeptide that is **homologous with prolactin** and human placental lactogen.

   a. **Regulation of growth hormone secretion** (Figure 7.6)
   - Growth hormone is released in **pulsatile** fashion.
   - **Secretion is increased** by sleep, stress, hormones related to puberty, starvation, exercise, and hypoglycemia.
   - **Secretion is decreased** by somatostatin, somatomedins, obesity, hyperglycemia, and pregnancy.

   (1) *Hypothalamic control—GHRH and somatostatin*
   - **GHRH** stimulates the synthesis and secretion of growth hormone.
   - **Somatostatin** inhibits secretion of growth hormone by blocking the response of the anterior pituitary to GHRH.

   (2) *Negative feedback control by somatomedins*
   - Somatomedins are produced when growth hormone acts on target tissues. Somatomedins **inhibit the secretion of growth hormone** by acting directly on the anterior pituitary and by stimulating the secretion of somatostatin from the hypothalamus.

   (3) *Negative feedback control by GHRH and growth hormone*
   - **GHRH** inhibits its own secretion from the hypothalamus.
   - **Growth hormone** also inhibits its own secretion by stimulating the secretion of somatostatin from the hypothalamus.

   b. **Actions of growth hormone**
   - In the liver, growth hormone generates the production of **somatomedins (IGFs)**, which serve as the intermediaries of several physiologic actions.
   - The **IGF receptor** has **tyrosine kinase activity**, similar to the insulin receptor.

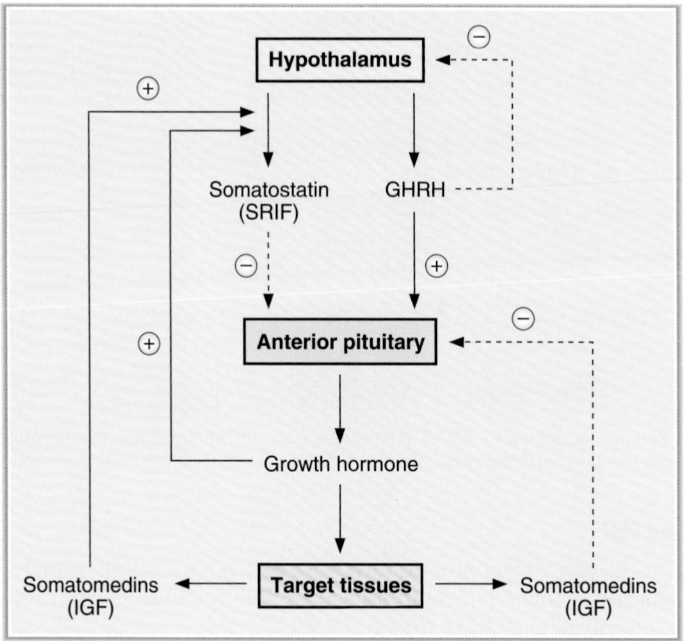

FIGURE 7.6. Control of growth hormone secretion. GHRH = growth hormone–releasing hormone; IGF = insulin-like growth factor; SRIF = somatotropin release–inhibiting factor.

(1) *Direct actions of growth hormone*

(a) ↓ glucose uptake into cells **(diabetogenic)**.
(b) ↑ lipolysis.
(c) ↑ protein synthesis in muscle and ↑ lean body mass.
(d) ↑ production of **IGF**.

(2) *Actions of growth hormone via IGF*

(a) ↑ protein synthesis in chondrocytes and ↑ **linear growth (pubertal growth spurt)**.
(b) ↑ protein synthesis in muscle and ↑ **lean body mass**.
(c) ↑ protein synthesis in most organs and ↑ **organ size**.

c. Pathophysiology of growth hormone

(1) *Growth hormone deficiency*

- in children causes **dwarfism**, failure to grow, short stature, mild obesity, and delayed puberty.
- can be caused by

(a) lack of anterior pituitary growth hormone.
(b) hypothalamic dysfunction (↓ GHRH).
(c) failure to generate IGF in the liver.
(d) growth hormone receptor deficiency.

(2) *Growth hormone excess*

- Hypersecretion of growth hormone causes **acromegaly**.

(a) **Before puberty**, excess growth hormone causes increased linear growth **(gigantism)**.
(b) **After puberty**, excess growth hormone causes increased periosteal bone growth, increased organ size, and glucose intolerance.

- can be treated with **somatostatin analogues (e.g., octreotide)**, which inhibit growth hormone secretion.

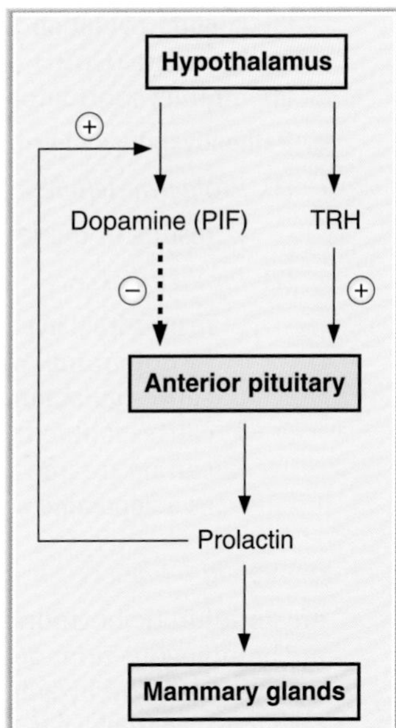

FIGURE 7.7. Control of prolactin secretion. PIF = prolactin-inhibiting factor; TRH = thyrotropin-releasing hormone.

4. Prolactin

- is the major hormone responsible for **lactogenesis.**
- participates, with estrogen, in breast development.
- is structurally **homologous to growth hormone.**

a. Regulation of prolactin secretion (Figure 7.7 and Table 7.3)

(1) *Hypothalamic control by dopamine and thyrotropin-releasing hormone (TRH)*

- Prolactin secretion is **tonically inhibited by dopamine** (prolactin-inhibiting factor [PIF]) secreted by the hypothalamus. Thus, interruption of the hypothalamic–pituitary tract causes increased secretion of prolactin and sustained lactation.
- TRH increases prolactin secretion.

(2) *Negative feedback control*

- Prolactin inhibits its own secretion by stimulating the hypothalamic release of dopamine.

b. Actions of prolactin

(1) stimulates **milk production** in the breast (casein, lactalbumin).
(2) stimulates **breast development** (in a supportive role with estrogen).

table 7.3 Regulation of Prolactin Secretion

Factors that Increase Prolactin Secretion	Factors that Decrease Prolactin Secretion
Estrogen (pregnancy)	Dopamine
Breast-feeding	Bromocriptine (dopamine agonist)
Sleep	Somatostatin
Stress	Prolactin (by negative feedback)
TRH	
Dopamine antagonists	

TRH = thyrotropin-releasing hormone.

(3) **inhibits ovulation** by decreasing synthesis and release of gonadotropin-releasing hormone (GnRH).
(4) inhibits spermatogenesis (by decreasing GnRH).

c. Pathophysiology of prolactin

(1) *Prolactin deficiency* (destruction of the anterior pituitary)
- results in the **failure to lactate**.

(2) *Prolactin excess*
- **results from hypothalamic destruction** (due to loss of the tonic "inhibitory" control by dopamine) or from prolactin-secreting tumors **(prolactinomas)**.
- causes **galactorrhea** and decreased libido.
- causes **failure to ovulate** and **amenorrhea** because it inhibits GnRH secretion.
- can be treated with **bromocriptine**, which reduces prolactin secretion by acting as a **dopamine agonist**.

C. Hormones of the posterior lobe of the pituitary

- are antidiuretic hormone (ADH) and oxytocin.
- are homologous nonapeptides.
- are **synthesized in hypothalamic nuclei** and are packaged in secretory granules with their respective **neurophysins**.
- travel down the nerve axons for secretion by the posterior pituitary.

1. ADH (see Chapter 5, VII)
- originates primarily in the **supraoptic nuclei** of the hypothalamus.
- regulates serum osmolarity by increasing the H_2O permeability of the late distal tubules and collecting ducts.

a. Regulation of ADH secretion (Table 7.4)

b. Actions of ADH

(1) **↑ H_2O permeability (aquaporin 2, AQP2)** of the principal cells of the late distal tubule and collecting duct (via a **V_2 receptor** and an adenylate cyclase–cAMP mechanism).
(2) **constriction of vascular smooth muscle** (via a **V_1 receptor** and an IP_3/Ca^{2+} mechanism).
(3) **increased Na^+–$2Cl^-$–K^+ cotransport** in thick ascending limb, leading to increased countercurrent multiplication and increased corticopapillary osmotic gradient.
(4) **increase urea recycling** in inner medullary collecting ducts, leading to increased corticopapillary osmotic gradient.

c. Pathophysiology of ADH (see Chapter 5, VII)

2. Oxytocin
- originates primarily in the **paraventricular nuclei** of the hypothalamus.
- causes **ejection of milk from the breast** when stimulated by suckling.

table 7.4 Regulation of ADH Secretion

Factors that Increase ADH Secretion	Factors that Decrease ADH Secretion
↑ serum osmolarity	↓ serum osmolarity
Volume contraction	Ethanol
Pain	α-Agonists
Nausea (powerful stimulant)	ANP
Hypoglycemia	
Nicotine, opiates, antineoplastic drugs	

ADH = antidiuretic hormone; ANP = atrial natriuretic peptide.

a. Regulation of oxytocin secretion

(1) *Suckling*

- is the major stimulus for oxytocin secretion.
- Afferent fibers carry impulses from the nipple to the spinal cord. Relays in the hypothalamus trigger the release of oxytocin from the posterior pituitary.
- The sight or sound of the infant may stimulate the hypothalamic neurons to secrete oxytocin, even in the absence of suckling.

(2) *Dilation of the cervix and orgasm*

- increases the secretion of oxytocin.

b. Actions of oxytocin

(1) *Contraction of myoepithelial cells in the breast*

- Milk is forced from the mammary alveoli into the ducts and ejected.

(2) *Contraction of the uterus*

- During pregnancy, oxytocin receptors in the uterus are up-regulated as parturition approaches, although the role of oxytocin in normal labor is uncertain.
- Oxytocin can be used to induce labor and **reduce postpartum bleeding.**

IV. THYROID GLAND

A. Synthesis of thyroid hormones (Figure 7.8)

- Each step in synthesis is **stimulated by TSH.**

1. Thyroglobulin is synthesized from tyrosine in the thyroid follicular cells, packaged in secretory vesicles, and extruded into the follicular lumen (step 1).

2. The iodide (I^-) pump, or Na^+–I^- cotransport

- is present in the thyroid follicular epithelial cells.
- actively transports I^- into the thyroid follicular cells for subsequent incorporation into thyroid hormones (step 2).
- is **inhibited by thiocyanate and perchlorate anions.**

3. Oxidation of I^- to I_2

- is catalyzed by a **peroxidase enzyme** in the follicular cell membrane (step 3).
- **I_2** is the reactive form, which will be "organified" by combination with tyrosine on thyroglobulin.
- The peroxidase enzyme is **inhibited by propylthiouracil,** which is used therapeutically to reduce thyroid hormone synthesis for the treatment of hyperthyroidism.
- The same peroxidase enzyme catalyzes the remaining organification and coupling reactions involved in the synthesis of thyroid hormones.

4. Organification of I_2

- At the junction of the follicular cells and the follicular lumen, tyrosine residues of thyroglobulin react with I_2 to form **monoiodotyrosine (MIT) and diiodotyrosine (DIT)** (step 4).
- High levels of I^- inhibit organification and, therefore, inhibit synthesis of thyroid hormone **(Wolff–Chaikoff effect).**

5. Coupling of MIT and DIT

- While MIT and DIT are attached to thyroglobulin, two coupling reactions occur (step 5).

a. When two molecules of DIT combine, **thyroxine (T_4)** is formed.

b. When one molecule of DIT combines with one molecule of MIT, **triiodothyronine (T_3)** is formed.

- More T_4 than T_3 is synthesized, although T_3 is more active.

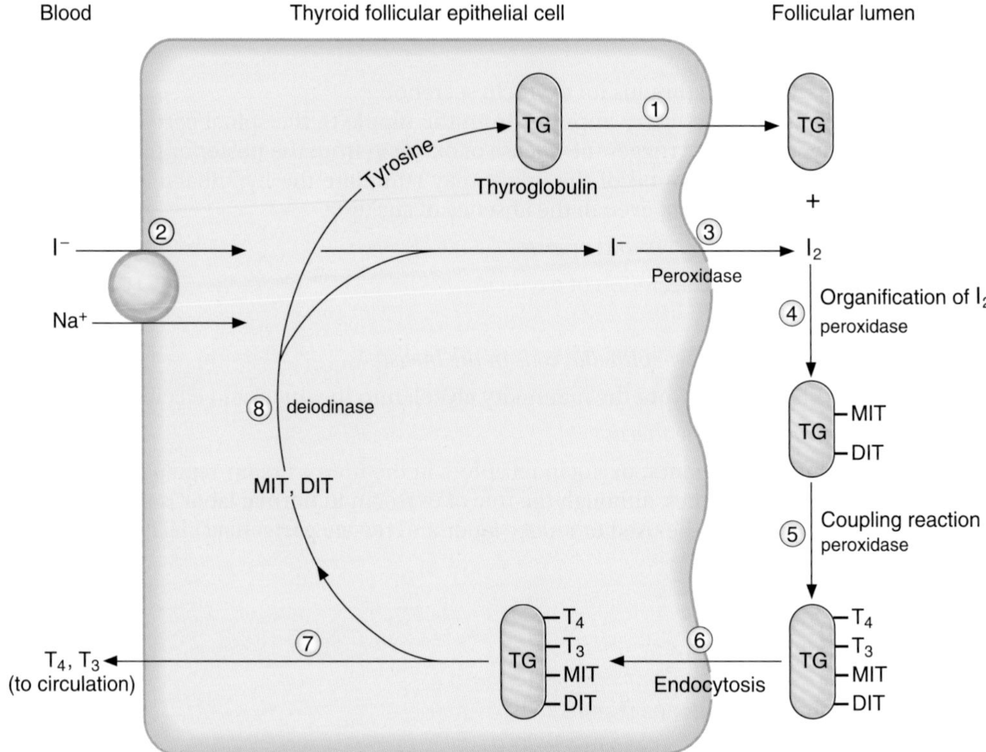

FIGURE 7.8. Steps in the synthesis of thyroid hormones. Each step is stimulated by thyroid-stimulating hormone. DIT = diiodotyrosine; I^- = iodide; MIT = monoiodotyrosine; T_3 = triiodothyronine; T_4 = thyroxine; TG = thyroglobulin.

c. Iodinated thyroglobulin is stored in the follicular lumen until the thyroid gland is stimulated to secrete thyroid hormones.

6. Stimulation of thyroid cells by TSH

- When the thyroid cells are stimulated, iodinated thyroglobulin is taken back into the follicular cells by endocytosis (step 6). Lysosomal enzymes then digest thyroglobulin, releasing T_4 and T_3 into the circulation (step 7).
- Leftover MIT and DIT are deiodinated by **thyroid deiodinase** (step 8). The I_2 that is released is reutilized to synthesize more thyroid hormones. Therefore, deficiency of thyroid deiodinase mimics I_2 deficiency.

7. Binding of T_3 and T_4

- In the circulation, most of the T_3 and T_4 is bound to thyroxine-binding globulin **(TBG)**.

a. In **hepatic failure**, TBG levels decrease, leading to a decrease in total thyroid hormone levels, but normal levels of free hormone.

b. In **pregnancy**, TBG levels increase, leading to an increase in total thyroid hormone levels, but normal levels of free hormone (i.e., clinically, euthyroid).

8. Conversion of T_4 to T_3 and reverse T_3 (rT_3)

- In the peripheral tissues, T_4 is converted to T_3 by **5′-iodinase** (or to rT_3).
- **T_3 is more biologically active than T_4.**
- rT_3 is inactive.

B. Regulation of thyroid hormone secretion (Figure 7.9)

1. Hypothalamic–pituitary control—TRH and TSH

a. **TRH** is secreted by the hypothalamus and stimulates the secretion of TSH by the anterior pituitary.

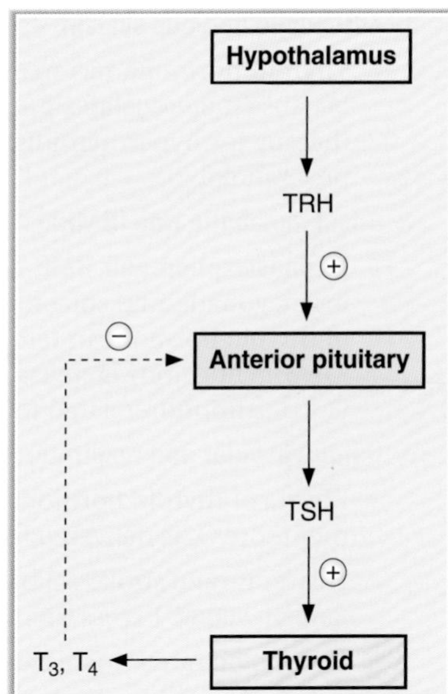

FIGURE 7.9. Control of thyroid hormone secretion. T_3 = triiodothyronine; T_4 = thyroxine; TRH = thyrotropin-releasing hormone; TSH = thyroid-stimulating hormone.

b. **TSH** increases both the synthesis and the secretion of thyroid hormones by the follicular cells via an **adenylate cyclase–cAMP** mechanism.

- Chronic elevation of TSH causes **hypertrophy** of the thyroid gland.

c. **T_3 down-regulates TRH receptors** in the anterior pituitary and thereby inhibits TSH secretion.

2. **Thyroid-stimulating immunoglobulins**

- are components of the immunoglobulin G (IgG) fraction of plasma proteins and are **antibodies to TSH receptors** on the thyroid gland.
- bind to TSH receptors and, like TSH, **stimulate the thyroid gland to secrete T_3 and T_4.**
- circulate in high concentrations in patients with **Graves disease**, which is characterized by high circulating levels of thyroid hormones and, accordingly, low concentrations of TSH (caused by feedback inhibition of thyroid hormones on the anterior pituitary).

C. Actions of thyroid hormone

- **T_3 is three to four times more potent than T_4.** The target tissues convert T_4 to T_3 (see IV A 8).

1. **Growth**

- Attainment of adult stature requires thyroid hormone.
- Thyroid hormones act synergistically with growth hormone and somatomedins to promote **bone formation.**
- Thyroid hormones stimulate **bone maturation** as a result of ossification and fusion of the growth plates. **In thyroid hormone deficiency, bone age is less than chronologic age.**

2. **Central nervous system (CNS)**

a. **Perinatal period**

- Maturation of the CNS **requires thyroid hormone in the perinatal period.**
- Thyroid hormone deficiency causes irreversible mental retardation. Because there is only a brief perinatal period when thyroid hormone replacement therapy is helpful, **screening for neonatal hypothyroidism is mandatory.**

b. **Adulthood**

- **Hyperthyroidism** causes hyperexcitability and irritability.
- **Hypothyroidism** causes listlessness, slowed speech, somnolence, impaired memory, and decreased mental capacity.

3. Autonomic nervous system

- Thyroid hormone has many of the same actions as the sympathetic nervous system because it **up-regulates β_1-adrenergic receptors in the heart**. Therefore, a useful adjunct therapy for hyperthyroidism is treatment with a β-adrenergic blocking agent, such as propranolol.

4. Basal metabolic rate (BMR)

- **O_2 consumption and BMR are increased** by thyroid hormone in all tissues except the brain, gonads, and spleen. The resulting increase in heat production underlies the role of thyroid hormone in temperature regulation.
- Thyroid hormone **increases the synthesis of Na^+, K^+-ATPase** and consequently increases O_2 consumption related to Na^+–K^+ pump activity.

5. Cardiovascular and respiratory systems

- Effects of thyroid hormone on cardiac output and ventilation rate combine to ensure that more O_2 is delivered to the tissues.

a. Heart rate and stroke volume are increased. These effects combine to produce **increased cardiac output**. Excess thyroid hormone can cause **high-output heart failure**.

b. Ventilation rate is increased.

6. Metabolic effects

- Overall, metabolism is increased to meet the demand for substrate associated with the increased rate of O_2 consumption.

a. Glucose absorption from the gastrointestinal tract is increased.

b. **Glycogenolysis**, **gluconeogenesis**, and **glucose oxidation** (driven by demand for ATP) are increased.

c. Lipolysis is increased.

d. Protein synthesis and degradation are increased. The overall effect of thyroid hormone is **catabolic**.

D. Pathophysiology of the thyroid gland (Table 7.5)

V. ADRENAL CORTEX AND ADRENAL MEDULLA (FIGURE 7.10)

A. Adrenal cortex

1. Synthesis of adrenocortical hormones (Figure 7.11)

- The **zona glomerulosa** produces **aldosterone**.
- The **zonae fasciculata and reticularis** produce glucocorticoids (**cortisol**) and androgens (**dehydroepiandrosterone** and **androstenedione**).

a. 21-carbon steroids

- include **progesterone, deoxycorticosterone, aldosterone,** and **cortisol**.
- Progesterone is the precursor for the others in the 21-carbon series.
- **Hydroxylation at C-21** leads to the production of deoxycorticosterone, which has mineralocorticoid (but not glucocorticoid) activity.
- **Hydroxylation at C-17** leads to the production of glucocorticoids (cortisol).

b. 19-carbon steroids

- have **androgenic activity** and are precursors to the estrogens.
- If the steroid has been previously hydroxylated at C-17, the $C_{20,21}$ side chain can be cleaved to yield the 19-carbon steroids **dehydroepiandrosterone** or **androstenedione** in the adrenal cortex.
- Adrenal androgens have a ketone group at C-17 and are excreted as **17-ketosteroids** in the urine.
- In the testes, androstenedione is converted to testosterone.

table 7.5 Pathophysiology of the Thyroid Gland

	Hyperthyroidism	Hypothyroidism
Symptoms	↑ metabolic rate Weight loss Negative nitrogen balance ↑ heat production (sweating) ↑ cardiac output Dyspnea Tremor, weakness Exophthalmos Goiter	↓ metabolic rate Weight gain Positive nitrogen balance ↓ heat production (cold sensitivity) ↓ cardiac output Hypoventilation Lethargy, mental slowness Drooping eyelids Myxedema Growth and mental retardation (perinatal) Goiter
Causes	Graves disease (antibodies to TSH receptor) Thyroid neoplasm	Thyroiditis (autoimmune thyroiditis; Hashimoto thyroiditis) Surgical removal of thyroid I^- deficiency Cretinism (congenital) ↓ TRH or TSH
TSH levels	↓ (because of feedback inhibition on anterior pituitary by high thyroid hormone levels)	↑ (because of decreased feedback inhibition on anterior pituitary by low thyroid hormone levels) ↓ (if primary defect is in hypothalamus or anterior pituitary)
Treatment	Propylthiouracil (inhibits thyroid hormone synthesis by blocking peroxidase) Thyroidectomy ^{131}I (destroys thyroid) β-blockers (adjunct therapy)	Thyroid hormone replacement

See Table 7.1 for abbreviations.

c. 18-carbon steroids

- have **estrogenic activity**.
- Oxidation of the A ring **(aromatization)** to produce estrogens occurs in the **ovaries** and **placenta**, but not in the adrenal cortex or testes.

2. Regulation of secretion of adrenocortical hormones

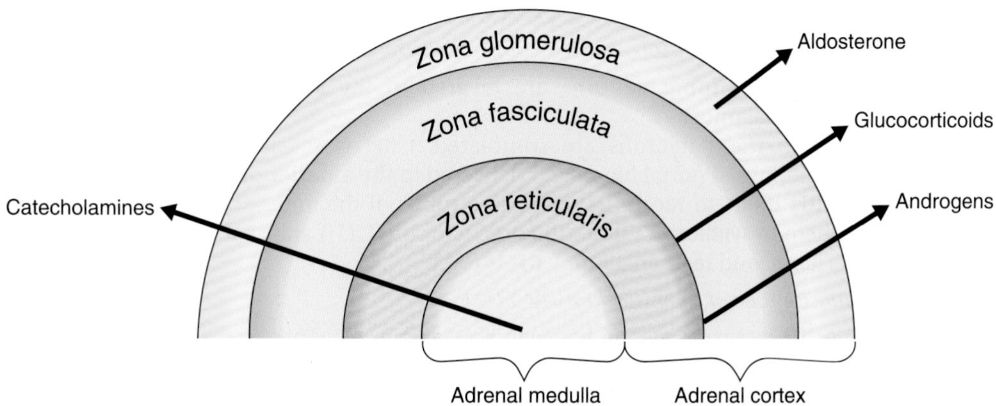

FIGURE 7.10. Secretory products of the adrenal cortex and medulla.

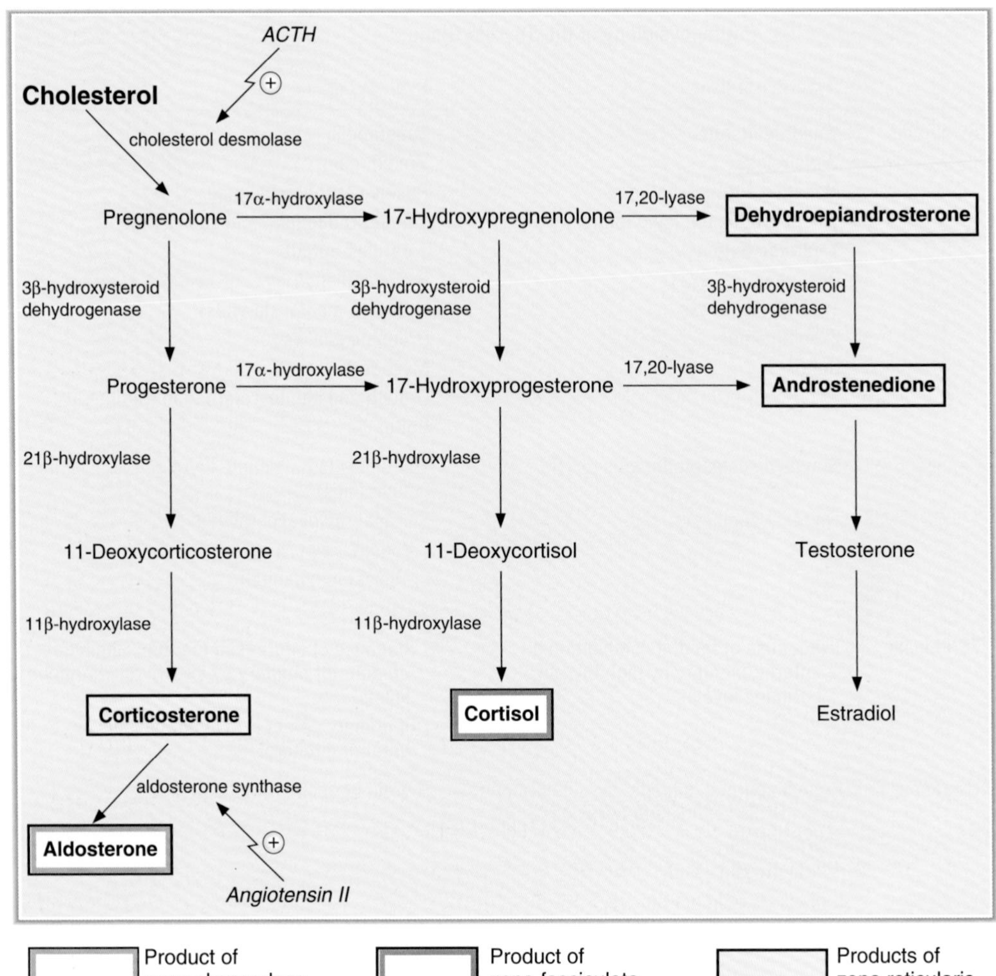

FIGURE 7.11. Synthetic pathways for glucocorticoids, androgens, and mineralocorticoids in the adrenal cortex. ACTH = adrenocorticotropic hormone.

a. Glucocorticoid secretion (Figure 7.12)

- oscillates with a 24-hour periodicity or **circadian rhythm.**
- For those who sleep at night, **cortisol levels are highest just before waking** (≈8 AM) and **lowest in the evening** (≈12 midnight).

(1) *Hypothalamic control—corticotropin-releasing hormone (CRH)*

- CRH-containing neurons are located in the **paraventricular nuclei** of the hypothalamus.
- When these neurons are stimulated, CRH is released into hypothalamic–hypophysial portal blood and delivered to the anterior pituitary.
- CRH binds to receptors on corticotrophs of the anterior pituitary and directs them to **synthesize POMC** (the precursor to ACTH) and **secrete ACTH.**
- The second messenger for CRH is **cAMP.**

(2) *Anterior lobe of the pituitary—ACTH*

- **ACTH increases steroid hormone synthesis** in all zones of the adrenal cortex by stimulating **cholesterol desmolase** and increasing the conversion of cholesterol to pregnenolone.

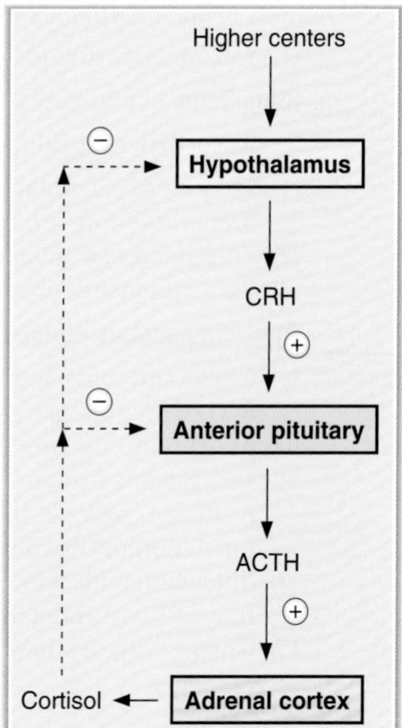

FIGURE 7.12. Control of glucocorticoid secretion. ACTH = adrenocorticotropic hormone; CRH = corticotropin-releasing hormone.

- **ACTH also up-regulates its own receptor** so that the sensitivity of the adrenal cortex to ACTH is increased.
- Chronically increased levels of ACTH cause hypertrophy of the adrenal cortex.
- The second messenger for ACTH is **cAMP**.

(3) *Negative feedback control—cortisol*

- **Cortisol inhibits the secretion of CRH** from the hypothalamus and the **secretion of ACTH** from the anterior pituitary.
- When cortisol (glucocorticoid) levels are chronically elevated, the secretion of CRH and ACTH is inhibited by negative feedback.
- The **dexamethasone suppression test** is based on the ability of dexamethasone (a synthetic glucocorticoid) to inhibit ACTH secretion. In normal persons, low-dose dexamethasone inhibits or "suppresses" ACTH secretion and, consequently, cortisol secretion. In persons with **ACTH-secreting tumors**, low-dose dexamethasone does not inhibit cortisol secretion but high-dose dexamethasone does. In persons with **adrenal cortical tumors**, neither low- nor high-dose dexamethasone inhibits cortisol secretion.

b. Aldosterone secretion (see Chapter 3, VI B)

- is under tonic control by ACTH, but is separately regulated by the renin–angiotensin system and by serum potassium.

(1) Renin–angiotensin–aldosterone system

(a) Decreases in blood volume cause a decrease in renal perfusion pressure, which in turn increases renin secretion. **Renin**, an enzyme, catalyzes the conversion of angiotensinogen to angiotensin I. Angiotensin I is converted to **angiotensin II** by **angiotensin-converting enzyme (ACE)**.

(b) Angiotensin II acts on the zona glomerulosa of the adrenal cortex to **increase the conversion of corticosterone to aldosterone**.

(c) Aldosterone increases renal Na^+ reabsorption, thereby restoring extracellular fluid (ECF) volume and blood volume to normal.

(2) Hyperkalemia increases aldosterone secretion. Aldosterone increases renal K^+ secretion, restoring serum $[K^+]$ to normal.

3. Actions of glucocorticoids (cortisol)

- Overall, glucocorticoids are essential for the **response to stress.**

a. Stimulation of gluconeogenesis

- Glucocorticoids increase gluconeogenesis by the following mechanisms:

(1) They **increase protein catabolism** in muscle and decrease protein synthesis, thereby providing more amino acids to the liver for gluconeogenesis.
(2) They **decrease glucose utilization** and insulin sensitivity of adipose tissue.
(3) They **increase lipolysis**, which provides more glycerol to the liver for gluconeogenesis.

b. Anti-inflammatory effects

(1) Glucocorticoids **induce the synthesis of lipocortin**, an **inhibitor of phospholipase A_2**. (Phospholipase A_2 is the enzyme that liberates arachidonate from membrane phospholipids, providing the precursor for prostaglandin and leukotriene synthesis.) Because prostaglandins and leukotrienes are involved in the inflammatory response, glucocorticoids have anti-inflammatory properties by inhibiting the formation of the precursor (arachidonate).
(2) Glucocorticoids inhibit the production of interleukin-2 (IL-2) and inhibit the proliferation of T lymphocytes.
(3) Glucocorticoids **inhibit the release of histamine and serotonin** from mast cells and platelets.

c. Suppression of the immune response

- Glucocorticoids **inhibit the production of IL-2** and T lymphocytes, both of which are critical for cellular immunity. In pharmacologic doses, glucocorticoids are used to **prevent rejection of transplanted organs.**

d. Maintenance of vascular responsiveness to catecholamines

- Cortisol **up-regulates α_1 receptors** on arterioles, increasing their sensitivity to the vasoconstrictor effect of norepinephrine. Thus, with cortisol excess, arterial pressure increases; with cortisol deficiency, arterial pressure decreases.

4. Actions of mineralocorticoids (aldosterone) (see Chapters 3 and 5)

a. ↑ renal Na^+ reabsorption (action on the principal cells of the late distal tubule and collecting duct).

b. ↑ renal K^+ secretion (action on the principal cells of the late distal tubule and collecting duct).

c. ↑ renal H^+ secretion (action on the α-intercalated cells of the late distal tubule and collecting duct).

5. Pathophysiology of the adrenal cortex (Table 7.6)

a. Adrenocortical insufficiency

(1) *Primary adrenocortical insufficiency—Addison disease*

- is most commonly caused by autoimmune **destruction of the adrenal cortex** and causes acute **adrenal crisis.**
- is characterized by the following:

(a) ↓ adrenal glucocorticoid, androgen, and mineralocorticoid.
(b) ↑ ACTH. (Low cortisol levels stimulate ACTH secretion by negative feedback.)
(c) Hypoglycemia (caused by cortisol deficiency).
(d) Weight loss, weakness, nausea, and vomiting.
(e) Hyperpigmentation. (Low cortisol levels stimulate ACTH secretion; ACTH contains the MSH fragment.)
(f) ↓ pubic and axillary hair in women (caused by the deficiency of adrenal androgens).
(g) ECF volume contraction, hypotension, hyperkalemia, and **metabolic acidosis** (caused by aldosterone deficiency).

table 7.6 Pathophysiology of the Adrenal Cortex

Disorder	Clinical Features	ACTH Levels	Treatment
Addison disease (e.g., primary adrenocortical insufficiency)	Hypoglycemia Anorexia, weight loss, nausea, vomiting Weakness Hypotension Hyperkalemia Metabolic acidosis Decreased pubic and axillary hair in women Hyperpigmentation	Increased (negative feedback effect of decreased cortisol)	Replacement of glucocorticoids and mineralocorticoids
Cushing syndrome (e.g., primary adrenal hyperplasia)	Hyperglycemia Muscle wasting Central obesity Round face, supraclavicular fat, buffalo hump Osteoporosis Striae Virilization and menstrual disorders in women Hypertension	Decreased (negative feedback effect of increased cortisol)	Ketoconazole Metyrapone
Cushing disease (excess ACTH)	Same as Cushing syndrome	Increased	Surgical removal of ACTH-secreting tumor
Conn syndrome (aldosterone-secreting tumor)	Hypertension Hypokalemia Metabolic alkalosis Decreased renin		Spironolactone (aldosterone antagonist) Surgical removal of aldosterone-secreting tumor
21β-Hydroxylase deficiency (↓ glucocorticoids and mineralocorticoids; ↑ adrenal androgens)	Virilization of women Early acceleration of linear growth Early appearance of pubic and axillary hair Symptoms of glucocorticoid and mineralocorticoid deficiency	Increased (negative feedback effect of decreased cortisol)	Replacement of glucocorticoids and mineralocorticoids
17α-Hydroxylase deficiency (↓ adrenal androgens and glucocorticoids; ↑ mineralocorticoids)	Lack of pubic and axillary hair in women Symptoms of glucocorticoid deficiency Symptoms of mineralocorticoid excess	Increased (negative feedback effect of decreased cortisol)	Replacement of glucocorticoids Aldosterone antagonist

See Table 7.1 for abbreviation.

(2) *Secondary adrenocortical insufficiency*

- is caused by primary **deficiency of ACTH.**
- **does *not* exhibit hyperpigmentation** (because there is a deficiency of ACTH).
- **does *not* exhibit volume contraction, hyperkalemia,** or **metabolic acidosis** (because aldosterone levels are normal).
- Symptoms are otherwise similar to those of Addison disease.

b. Adrenocortical excess—Cushing syndrome

- is most commonly caused by the administration of **pharmacologic doses of glucocorticoids.**
- is also caused by primary **hyperplasia of the adrenal glands.**
- is called **Cushing disease** when it is caused by overproduction of ACTH.

- is characterized by the following:

(1) ↑ **cortisol and androgen levels.**
(2) ↓ ACTH (if caused by primary adrenal hyperplasia or pharmacologic doses of glucocorticosteroids); ↑ ACTH (if caused by overproduction of ACTH, as in Cushing disease).
(3) **hyperglycemia** (caused by elevated cortisol levels).
(4) ↑ protein catabolism and muscle wasting.
(5) central obesity (round face, supraclavicular fat, buffalo hump).
(6) poor wound healing.
(7) **virilization of women** (caused by elevated levels of adrenal androgens).
(8) **hypertension** (caused by elevated levels of cortisol and aldosterone).
(9) osteoporosis (elevated cortisol levels cause increased bone resorption).
(10) striae.

- **Ketoconazole**, an inhibitor of steroid hormone synthesis, can be used to treat Cushing disease.

c. Hyperaldosteronism—Conn syndrome

- is caused by an aldosterone-secreting tumor.
- is characterized by the following:

(1) **hypertension** (because aldosterone increases Na^+ reabsorption, which leads to increases in ECF volume and blood volume).
(2) **hypokalemia** (because aldosterone increases K^+ secretion).
(3) metabolic alkalosis (because aldosterone increases H^+ secretion).
(4) ↓ **renin** secretion (because increased ECF volume and blood pressure inhibit renin secretion by negative feedback).

d. 21β-Hydroxylase deficiency

- is the most common biochemical abnormality of the steroidogenic pathway (see Figure 7.11).
- belongs to a group of disorders characterized by **adrenogenital syndrome**.
- is characterized by the following:

(1) ↓ **cortisol and aldosterone levels** (because the enzyme block prevents the production of 11-deoxycorticosterone and 11-deoxycortisol, the precursors for cortisol and aldosterone).
(2) ↑ 17-hydroxyprogesterone and progesterone levels (because of accumulation of intermediates above the enzyme block).
(3) ↑ **ACTH** (because of decreased feedback inhibition by cortisol).
(4) hyperplasia of zona fasciculata and zona reticularis (because of high levels of ACTH).
(5) ↑ **adrenal androgens** (because 17-hydroxyprogesterone is their major precursor) and ↑ **urinary 17-ketosteroids.**
(6) **virilization in women.**
(7) early acceleration of linear growth and early appearance of pubic and axillary hair.
(8) suppression of gonadal function in both men and women.

e. 17α-Hydroxylase deficiency is characterized by the following:

(1) ↓ **androgen and glucocorticoid levels** (because the enzyme block prevents the production of 17-hydroxypregnenolone and 17-hydroxyprogesterone).
(2) ↑ **mineralocorticoid levels** (because intermediates accumulate to the left of the enzyme block and are shunted toward the production of mineralocorticoids).
(3) lack of pubic and axillary hair (which depends on adrenal androgens) in women.
(4) hypoglycemia (because of decreased glucocorticoids).
(5) metabolic alkalosis, hypokalemia, and hypertension (because of increased mineralocorticoids).
(6) ↑ **ACTH** (because decreased cortisol levels stimulate ACTH secretion by negative feedback).

B. Adrenal medulla (see Chapter 2, I A 4)

VI. ENDOCRINE PANCREAS–GLUCAGON AND INSULIN (TABLE 7.7)

A. Organization of the endocrine pancreas

- The islets of Langerhans contain three major cell types (Table 7.8). Other cells secrete pancreatic polypeptide.
- **Gap junctions** link beta cells to each other, alpha cells to each other, and beta cells to alpha cells for rapid communication.
- The portal blood supply of the islets allows blood from the beta cells (containing insulin) to bathe the alpha and delta cells, again for rapid cell-to-cell communication.

B. Glucagon

1. Regulation of glucagon secretion (Table 7.9)

- The major factor that regulates glucagon secretion is the blood glucose concentration. **Decreased blood glucose stimulates glucagon secretion.**
- Increased blood amino acids stimulate glucagon secretion, which prevents hypoglycemia caused by unopposed insulin in response to a high protein meal.

2. Actions of glucagon

- Glucagon acts on the liver and adipose tissue.
- The second messenger for glucagon is **cAMP**.

a. Glucagon increases the blood glucose concentration.

(1) It **increases glycogenolysis** and prevents the recycling of glucose into glycogen.

(2) It **increases gluconeogenesis**. Glucagon decreases the production of fructose 2,6-bisphosphate, decreasing phosphofructokinase activity; in effect, substrate is directed toward glucose formation rather than toward glucose breakdown.

b. Glucagon increases blood fatty acid and ketoacid concentration.

- Glucagon **increases lipolysis**. The inhibition of fatty acid synthesis in effect "shunts" substrates toward gluconeogenesis.
- Ketoacids (β-hydroxybutyrate and acetoacetate) are produced from acetyl-coenzyme A (CoA), which results from fatty acid degradation.

c. Glucagon increases urea production.

- Amino acids are used for gluconeogenesis (stimulated by glucagon), and the resulting amino groups are incorporated into urea.

table 7.7 Comparison of Insulin and Glucagon

	Stimulus for Secretion	Major Actions	Overall Effect on Blood Levels
Insulin (tyrosine kinase receptor)	↑ blood glucose ↑ amino acids ↑ fatty acids Glucagon GIP Growth hormone Cortisol	Increases glucose uptake into cells and glycogen formation Decreases glycogenolysis and gluconeogenesis Increases protein synthesis Increases fat deposition and decreases lipolysis Increases K^+ uptake into cells	↓ [glucose] ↓ [amino acid] ↓ [fatty acid] ↓ [ketoacid] Hypokalemia
Glucagon (cAMP mechanism)	↓ blood glucose ↑ amino acids CCK Norepinephrine, epinephrine, ACh	Increases glycogenolysis and gluconeogenesis Increases lipolysis and ketoacid production	↑ [glucose] ↑ [fatty acid] ↑ [ketoacid]

ACh = acetylcholine; cAMP = cyclic adenosine monophosphate; CCK = cholecystokinin; GIP = glucose-dependent insulinotropic peptide.

table 7.8 Cell Types of the Islets of Langerhans

Type of Cell	Location	Function
Beta	Central islet	Secrete insulin
Alpha	Outer rim of islet	Secrete glucagon
Delta	Intermixed	Secrete somatostatin and gastrin

C. Insulin

- contains an A chain and a B chain, joined by two disulfide bridges.
- **Proinsulin is synthesized as a single-chain peptide.** Within storage granules, a connecting peptide (C peptide) is removed by proteases to yield insulin. The **C peptide** is packaged and secreted along with insulin, and its concentration is used to monitor beta cell function in diabetic patients who are receiving exogenous insulin.

1. Regulation of insulin secretion (Table 7.10)

a. Blood glucose concentration

- is the major factor that regulates insulin secretion.
- **Increased blood glucose stimulates insulin secretion.** An initial burst of insulin is followed by sustained secretion.

b. Mechanism of insulin secretion

- Glucose, the stimulant for insulin secretion, binds to the **Glut 2** receptor on the beta cells.
- Inside the beta cells, glucose is oxidized to **ATP**, which closes K^+ channels in the cell membrane and leads to **depolarization** of the beta cells. Similar to the action of ATP, **sulfonylurea drugs** (e.g., tolbutamide, glyburide) stimulate insulin secretion by closing these K^+ channels.
- Depolarization **opens Ca^{2+} channels**, which leads to an increase in intracellular [Ca^{2+}] and then to **secretion of insulin**.

2. Insulin receptor (see Figure 7.3)

- is found on target tissues for insulin.
- is a tetramer, with two α subunits and two β subunits.

a. The α subunits are located on the extracellular side of the cell membrane.

b. The **β subunits** span the cell membrane and have **intrinsic tyrosine kinase activity.** When insulin binds to the receptor, tyrosine kinase is activated and autophosphorylates the β subunits. The phosphorylated receptor then phosphorylates intracellular proteins.

table 7.9 Regulation of Glucagon Secretion

Factors that Increase Glucagon Secretion	Factors that Decrease Glucagon Secretion
↓ blood glucose	↑ blood glucose
↑ amino acids (especially arginine)	Insulin
CCK (alerts alpha cells to a protein meal)	Somatostatin
Norepinephrine, epinephrine	Fatty acids, ketoacids
ACh	

ACh = acetylcholine; CCK = cholecystokinin.

table 7.10 Regulation of Insulin Secretion

Factors that Increase Insulin Secretion	Factors that Decrease Insulin Secretion
↑ blood glucose	↓ blood glucose
↑ amino acids (arginine, lysine, leucine)	Somatostatin
↑ fatty acids	Norepinephrine, epinephrine
Glucagon	
GIP	
ACh	

ACh = acetylcholine; GIP = glucose-dependent insulinotropic peptide.

c. The insulin–receptor complexes enter the target cells.

d. Insulin **down-regulates** its own receptors in target tissues.

- Therefore, the number of insulin receptors is **increased in starvation** and **decreased in obesity** (e.g., type 2 diabetes mellitus).

3. Actions of insulin

- Insulin acts on the liver, adipose tissue, and muscle.

a. Insulin decreases blood glucose concentration by the following mechanisms:

(1) It **increases uptake of glucose** into target cells by directing the insertion of glucose transporters into cell membranes. As glucose enters the cells, the blood glucose concentration decreases.

(2) It **promotes formation of glycogen** from glucose in muscle and liver and simultaneously inhibits glycogenolysis.

(3) It **decreases gluconeogenesis**. Insulin increases the production of fructose 2,6-bisphosphate, increasing phosphofructokinase activity. In effect, substrate is directed away from glucose formation.

b. Insulin decreases blood fatty acid and ketoacid concentrations.

- In adipose tissue, insulin **stimulates fat deposition** and **inhibits lipolysis**.
- Insulin **inhibits ketoacid formation** in the liver because decreased fatty acid degradation provides less acetyl-CoA substrate for ketoacid formation.

c. Insulin decreases blood amino acid concentration.

- Insulin stimulates amino acid uptake into cells, increases protein synthesis, and inhibits protein degradation. Thus, insulin is **anabolic**.

d. Insulin decreases blood K^+ concentration.

- Insulin increases K^+ uptake into cells, thereby decreasing blood [K^+].

4. Insulin pathophysiology—diabetes mellitus

- **Case study:** A woman is brought to the emergency room. She is hypotensive and breathing rapidly; her breath has the odor of ketones. Analysis of her blood shows severe hyperglycemia, hyperkalemia, and blood gas values that are consistent with metabolic acidosis.
- **Explanation:**

a. Hyperglycemia

- is consistent with insulin deficiency.
- In the absence of insulin, glucose uptake into cells is decreased, as is storage of glucose as glycogen.
- If tests were performed, the woman's blood would have shown increased levels of both amino acids (because of increased protein catabolism) and fatty acids (because of increased lipolysis).

b. Hypotensionis

- is a result of ECF volume contraction.
- The high blood glucose concentration results in a high filtered load of glucose that exceeds the reabsorptive capacity (T_m) of the kidney.
- The unreabsorbed glucose acts as an osmotic diuretic in the urine and causes ECF volume contraction.

c. Metabolic acidosis

- is caused by overproduction of ketoacids (β-hydroxybutyrate and acetoacetate).
- The **increased ventilation rate**, or Kussmaul respiration, is the respiratory compensation for metabolic acidosis.

d. Hyperkalemia

- results from the lack of insulin; normally, insulin promotes K^+ uptake into cells.

D. Somatostatin

- is secreted by the delta cells of the pancreas.
- inhibits the secretion of insulin, glucagon, and gastrin.

VII. CALCIUM METABOLISM (PARATHYROID HORMONE, VITAMIN D, CALCITONIN) (TABLE 7.11)

A. Overall Ca^{2+} homeostasis (Figure 7.13)

- 40% of the total Ca^{2+} in blood is **bound to plasma proteins.**
- 60% of the total Ca^{2+} in blood is not bound to proteins and is ultrafilterable. **Ultrafilterable Ca^{2+}** includes Ca^{2+} that is complexed to anions such as phosphate and free, ionized Ca^{2+}.
- **Free, ionized Ca^{2+} is biologically active.**
- Serum [Ca^{2+}] is determined by the interplay of intestinal absorption, renal excretion, and bone remodeling (bone resorption and formation). Each component is hormonally regulated.
- To maintain Ca^{2+} balance, net intestinal absorption must be balanced by urinary excretion.

table 7.11 Summary of Hormones that Regulate Ca^{2+}

	PTH	Vitamin D	Calcitonin
Stimulus for secretion	↓ serum [Ca^{2+}]	↓ serum [Ca^{2+}] ↑ PTH ↓ serum [phosphate]	↑ serum [Ca^{2+}]
Action on			
Bone	↑ resorption	↑ resorption	↓ resorption
Kidney	↓ P reabsorption (↑ urinary cAMP) ↑ Ca^{2+} reabsorption	↑ P reabsorption ↑ Ca^{2+} reabsorption	
Intestine	↑ Ca^{2+} absorption (via activation of vitamin D)	↑ Ca^{2+} absorption (calbindin D-28K) ↑ P absorption	
Overall effect on			
Serum [Ca^{2+}]	↑	↑	↓
Serum [phosphate]	↓	↑	

cAMP = cyclic adenosine monophosphate. See Table 7.1 for other abbreviation.

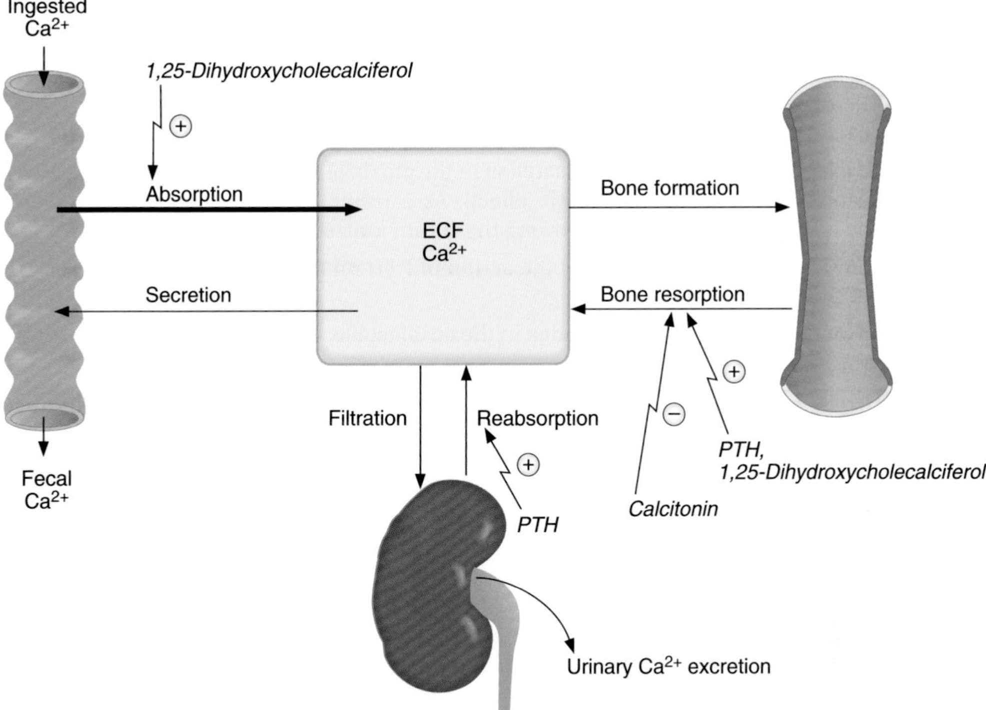

FIGURE 7.13. Hormonal regulation of Ca^{2+} metabolism. ECF = extracellular fluid; PTH = parathyroid hormone.

1. Positive Ca^{2+} balance

- is seen in growing children.
- Intestinal Ca^{2+} absorption exceeds urinary excretion, and the excess is deposited in the growing bones.

2. Negative Ca^{2+} balance

- is seen in women during pregnancy or lactation.
- Intestinal Ca^{2+} absorption is less than Ca^{2+} excretion, and the deficit comes from the maternal bones.

B. Parathyroid hormone (PTH)

- is the major hormone for the regulation of serum [Ca^{2+}].
- is synthesized and secreted by the **chief cells** of the parathyroid glands.

1. Secretion of PTH

- is controlled by the serum [Ca^{2+}] binding to **Ca^{2+}-sensing receptors** in the parathyroid chief cell membrane. **Decreased serum [Ca^{2+}] increases PTH secretion**, whereas increased serum Ca^{2+} decreases PTH secretion.
- Decreased serum Ca^{2+} causes decreased binding to the Ca^{2+}-sensing receptor, which stimulates PTH secretion.
- Mild decreases in serum [Mg^{2+}] stimulate PTH secretion.
- Severe decreases in serum [Mg^{2+}] inhibit PTH secretion and produce symptoms of hypoparathyroidism (e.g., hypocalcemia).
- The second messenger for PTH secretion by the parathyroid gland is cAMP.

2. Actions of PTH

- are coordinated to produce an **increase in serum [Ca^{2+}]** and a **decrease in serum [phosphate].**
- The second messenger for PTH actions on its target tissues is **cAMP**.

a. **PTH increases bone resorption**, which brings both Ca^{2+} and phosphate from bone mineral into the ECF. Alone, this effect on bone would not increase the serum ionized [Ca^{2+}] because phosphate complexes Ca^{2+}.

- Resorption of the organic matrix of bone is reflected in **increased hydroxyproline excretion.**

b. **PTH inhibits renal phosphate reabsorption** in the **proximal tubule** and, therefore, increases phosphate excretion **(phosphaturic effect)**. As a result, the phosphate resorbed from bone is excreted in the urine, allowing the serum ionized [Ca^{2+}] to increase.

- cAMP generated as a result of the action of PTH on the proximal tubule is excreted in the urine **(urinary cAMP)**.

c. **PTH increases renal Ca^{2+} reabsorption** in the **distal tubule**, which also increases the serum [Ca^{2+}].

d. **PTH increases intestinal Ca^{2+} absorption** indirectly by stimulating the production of 1,25-dihydroxycholecalciferol in the kidney (see VII C).

3. **Pathophysiology of PTH** (Table 7.12)

a. **Primary hyperparathyroidism**

- is most commonly caused by **parathyroid adenoma.**
- is characterized by the following:

(1) ↑ serum [Ca^{2+}] (hypercalcemia).
(2) ↓ serum [phosphate] (hypophosphatemia).
(3) ↑ urinary phosphate excretion (phosphaturic effect of PTH).
(4) ↑ urinary Ca^{2+} excretion (caused by the increased filtered load of Ca^{2+}).
(5) ↑ urinary cAMP.
(6) ↑ bone resorption.

table 7.12 Pathophysiology of PTH

Disorder	PTH	1,25-Dihydroxy-cholecalciferol	Bone	Urine	Serum [Ca^{2+}]	Serum [P]
Primary hyperparathyroidism	↑	↑ (PTH stimulates 1α-hydroxylase)	↑ resorption	↑ P excretion (phosphaturia) ↑ Ca^{2+} excretion (high filtered load of Ca^{2+}) ↑ urinary cAMP	↑	↓
Humoral hypercalcemia of malignancy	↓	—	↑ resorption	↑ P excretion	↑	↓
Surgical hypoparathyroidism	↓	↓	↓ resorption	↓ P excretion ↓ urinary cAMP	↓	↑
Pseudohypoparathyroidism	↑	↓	↓ resorption (defective G_s)	↓ P excretion ↓ urinary cAMP (defective G_s)	↓	↑
Chronic renal failure	↑ (2°)	↓ (caused by renal failure)	Osteomalacia (caused by ↓ 1,25-dihydroxy-cholecalciferol) ↑ resorption (caused by ↑ PTH)	↓ P excretion (caused by ↓ GFR)	↓ (caused by ↓ 1,25-dihydroxy-cholecalciferol)	↑ (caused by ↓ P excretion)

cAMP = cyclic adenosine monophosphate; GFR = glomerular filtration rate. See Table 7.1 for other abbreviation.

b. Humoral hypercalcemia of malignancy

- is caused by **PTH-related peptide (PTH-rp)** secreted by some malignant tumors (e.g., breast, lung). PTH-rp has all of the physiologic actions of PTH, including increased bone resorption, increased renal Ca^{2+} reabsorption, and decreased renal phosphate reabsorption.
- is characterized by the following:

(1) ↑ serum [Ca^{2+}] (hypercalcemia).
(2) ↓ serum [phosphate] (hypophosphatemia).
(3) ↑ urinary phosphate excretion (phosphaturic effect of PTH-rp).
(4) ↓ serum PTH levels (due to feedback inhibition from the high serum Ca^{2+}).

- Can be treated with an inhibitor of bone resorption (e.g., etidronate or pamidronate) and furosemide.

c. Hypoparathyroidism

- is most commonly a result of **thyroid surgery**, or it is **congenital**.
- is characterized by the following:

(1) ↓ serum [Ca^{2+}] (hypocalcemia) and **tetany**.
(2) ↑ serum [phosphate] (hyperphosphatemia).
(3) ↓ urinary phosphate excretion.

d. Pseudohypoparathyroidism type Ia—Albright hereditary osteodystrophy

- is the result of **defective G_s protein** in the kidney and bone, which causes end-organ **resistance to PTH**.
- **Hypocalcemia** and **hyperphosphatemia** occur (as in hypoparathyroidism), which are not correctable by the administration of exogenous PTH.
- Circulating **PTH levels are elevated** (stimulated by hypocalcemia).

e. Chronic renal failure

- Decreased glomerular filtration rate (GFR) leads to decreased filtration of phosphate, phosphate retention, and **increased serum [phosphate]**.
- Increased serum phosphate complexes Ca^{2+} and leads to **decreased ionized [Ca^{2+}]**.
- **Decreased production of 1,25-dihydroxycholecalciferol** by the diseased renal tissue also contributes to the decreased ionized [Ca^{2+}] (see VII C 1).
- Decreased [Ca^{2+}] causes **secondary hyperparathyroidism**.
- The combination of increased PTH levels and decreased 1,25-dihydroxycholecalciferol produces **renal osteodystrophy**, in which there is increased bone resorption and osteomalacia.

f. Familial hypocalciuric hypercalcemia (FHH)

- autosomal dominant disorder with decreased urinary Ca^{2+} excretion and increased serum Ca^{2+}.
- caused by **inactivating mutations** of the Ca^{2+}-sensing receptors that regulate PTH secretion.

C. Vitamin D

- provides Ca^{2+} and phosphate to ECF for bone mineralization.
- In children, vitamin D deficiency causes **rickets**.
- In adults, vitamin D deficiency causes **osteomalacia**.

1. Vitamin D metabolism (Figure 7.14)

- Cholecalciferol, 25-hydroxycholecalciferol, and 24,25-dihydroxycholecalciferol are inactive.
- The active form of vitamin D is **1,25-dihydroxycholecalciferol**.
- The production of 1,25-dihydroxycholecalciferol in the kidney is catalyzed by the enzyme 1α-hydroxylase.
- **1α-hydroxylase activity is increased** by the following:

a. ↓ serum [Ca^{2+}].

b. ↑ PTH levels.

c. ↓ serum [phosphate].

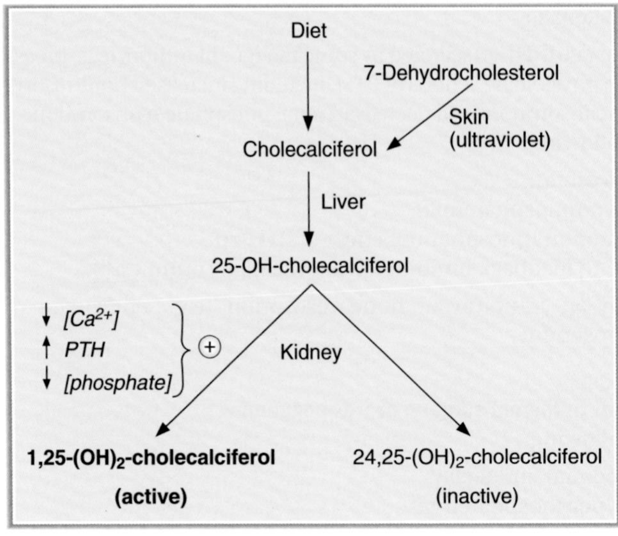

FIGURE 7.14. Steps and regulation in the synthesis of 1,25-dihydroxycholecalciferol. PTH = parathyroid hormone.

2. **Actions of 1,25-dihydroxycholecalciferol**
   - are coordinated to **increase both [Ca^{2+}] and [phosphate]** in ECF to **mineralize new bone.**

   a. **Increases intestinal Ca^{2+} absorption.** Vitamin D–dependent Ca^{2+}-binding protein **(calbindin D-28K)** is induced by 1,25-dihydroxycholecalciferol.
      - PTH increases intestinal Ca^{2+} absorption indirectly by stimulating 1α-hydroxylase and increasing production of the active form of vitamin D.

   b. **Increases intestinal phosphate absorption.**

   c. **Increases renal reabsorption of Ca^{2+} and phosphate**, analogous to its actions on the intestine.

   d. **Increases bone resorption**, which provides Ca^{2+} and phosphate from "old" bone to mineralize "new" bone.

D. Calcitonin

- is synthesized and secreted by the **parafollicular cells** of the thyroid.
- secretion is stimulated by an increase in serum [Ca^{2+}].
- acts primarily to **inhibit bone resorption.**
- can be used to **treat hypercalcemia.**

VIII. SEXUAL DIFFERENTIATION (FIGURE 7.15)

- **Genetic sex** is defined by the sex chromosomes, **XY** in males and **XX** in females.
- **Gonadal sex** is defined by the presence of **testes** in males and **ovaries** in females.
- **Phenotypic sex** is defined by the characteristics of the **internal genital tract** and the **external genitalia.**

A. Male phenotype

- The testes of gonadal males secrete **anti-müllerian hormone** and **testosterone.**
- Testosterone stimulates the growth and differentiation of the wolffian ducts, which develop into the male internal genital tract.
- Anti-müllerian hormone causes atrophy of the müllerian ducts (which would have become the female internal genital tract).

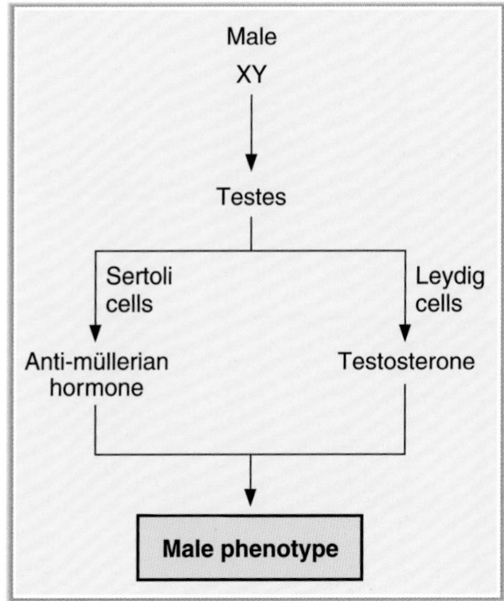

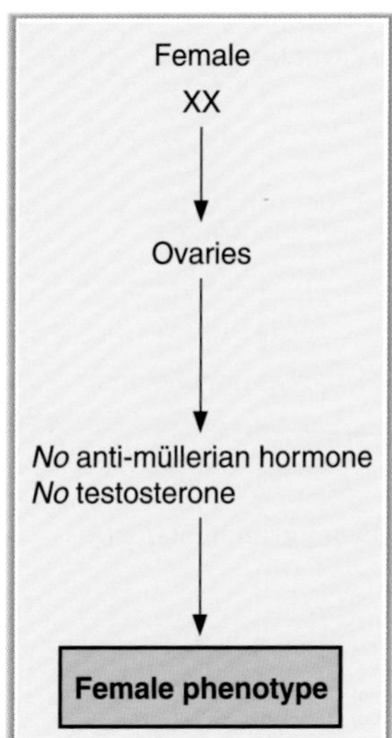

FIGURE 7.15. Sexual differentiation in males and females.

B. Female phenotype

- The ovaries of gonadal females secrete estrogen, but not anti-müllerian hormone or testosterone.
- Without testosterone, the wolffian ducts do not differentiate.
- Without anti-müllerian hormone, the müllerian ducts are not suppressed and therefore develop into the female internal genital tract.

IX. MALE REPRODUCTION

A. Synthesis of testosterone (Figure 7.16)

- Testosterone is the major androgen synthesized and secreted by the **Leydig cells.**
- Leydig cells do not contain 21β-hydroxylase or 11β-hydroxylase (in contrast to the adrenal cortex) and, therefore, do not synthesize glucocorticoids or mineralocorticoids.
- LH (in an analogous action to ACTH in the adrenal cortex) increases testosterone synthesis by stimulating cholesterol desmolase, the first step in the pathway.
- Accessory sex organs (e.g., **prostate**) contain **5α-reductase**, which converts testosterone to its active form, dihydrotestosterone.
- **5α-reductase inhibitors (finasteride)** may be used to treat **benign prostatic hyperplasia** because they block the activation of testosterone to dihydrotestosterone in the prostate.

B. Regulation of testes (Figure 7.17)

1. Hypothalamic control—GnRH

- Arcuate nuclei of the hypothalamus secrete GnRH into the hypothalamic–hypophysial portal blood. GnRH stimulates the anterior pituitary to secrete FSH and LH.

2. Anterior pituitary—FSH and LH

- **FSH acts on the Sertoli cells** to maintain **spermatogenesis.** The Sertoli cells also secrete **inhibin**, which is involved in negative feedback of FSH secretion.
- **LH acts on the Leydig cells** to promote **testosterone synthesis.** Testosterone acts via an intratesticular paracrine mechanism to reinforce the spermatogenic effects of FSH in the Sertoli cells.

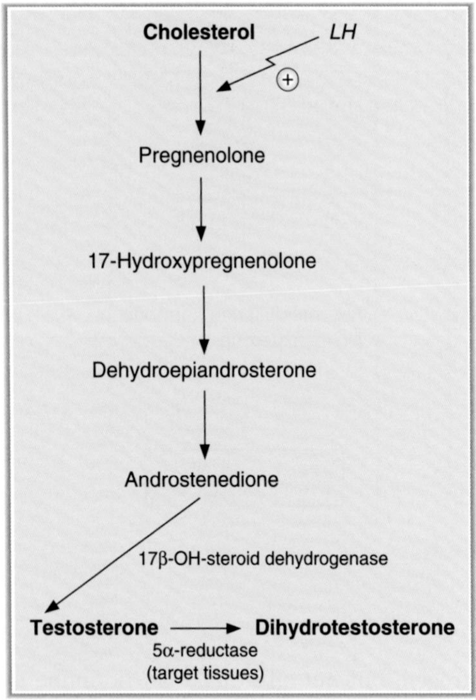

FIGURE 7.16. Synthesis of testosterone. LH = luteinizing hormone.

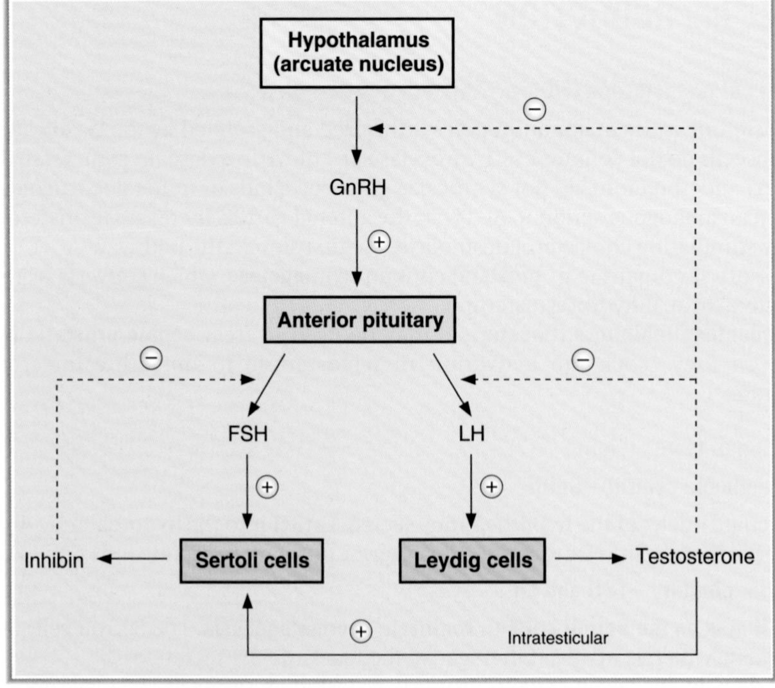

FIGURE 7.17. Control of male reproductive hormones. FSH = follicle-stimulating hormone; GnRH = gonadotropin-releasing hormone; LH = luteinizing hormone.

3. **Negative feedback control—testosterone and inhibin**

- **Testosterone inhibits the secretion of LH** by inhibiting the release of GnRH from the hypothalamus and by directly inhibiting the release of LH from the anterior pituitary.
- **Inhibin** (produced by the Sertoli cells) **inhibits the secretion of FSH** from the anterior pituitary.

C. Actions of testosterone or dihydrotestosterone

1. **Actions of testosterone**

- differentiation of epididymis, vas deferens, and seminal vesicles.
- pubertal growth spurt.
- cessation of pubertal growth spurt (epiphyseal closure).
- libido.
- spermatogenesis in Sertoli cells (paracrine effect).
- deepening of voice.
- increased muscle mass.
- growth of penis and seminal vesicles.
- negative feedback on anterior pituitary.

2. **Actions of dihydrotestosterone**

- differentiation of penis, scrotum, and prostate.
- male hair pattern.
- male pattern baldness.
- sebaceous gland activity.
- growth of prostate.

3. **Androgen insensitivity disorder** (testicular feminizing syndrome)

- is caused by **deficiency of androgen receptors** in target tissues of males.
- Testosterone and dihydrotestosterone actions in target tissues are absent.
- There are **female external genitalia** ("default"), and there is no internal genital tract.
- Testosterone levels are elevated due to lack of testosterone receptors in the anterior pituitary (lack of feedback inhibition).

D. Puberty (male and female)

- is initiated by the onset of **pulsatile GnRH** release from the hypothalamus.
- FSH and LH are, in turn, secreted in pulsatile fashion.
- GnRH **up-regulates** its own receptor in the anterior pituitary.

E. Variation in FSH and LH levels over the life span (male and female)

1. In childhood, hormone levels are lowest and FSH > LH.
2. At puberty and during the reproductive years, hormone levels increase and LH > FSH.
3. In senescence, hormone levels are highest and FSH > LH.

X. FEMALE REPRODUCTION

A. Synthesis of estrogen and progesterone (Figure 7.18)

- **Theca cells** produce androstenedione (stimulated at the first step by LH). Androstenedione diffuses to the nearby **granulosa cells**, which contain 17β-hydroxysteroid dehydrogenase, which converts androstenedione to testosterone, and aromatase, which converts testosterone to 17β-estradiol (stimulated by FSH).

B. Regulation of the ovary

1. **Hypothalamic control—GnRH**

- As in the male, pulsatile GnRH stimulates the anterior pituitary to secrete FSH and LH.

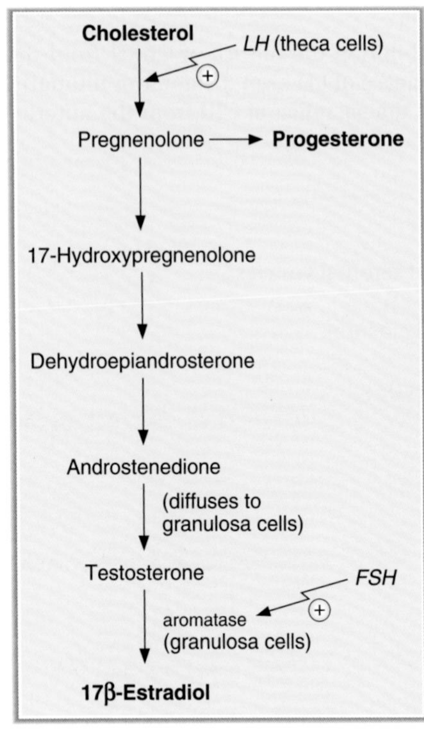

FIGURE 7.18. Synthesis of estrogen and progesterone. FSH = follicle-stimulating hormone; LH = luteinizing hormone.

2. **Anterior lobe of the pituitary—FSH and LH**
   - FSH and LH stimulate the following in the ovaries:

   a. steroidogenesis in the ovarian follicle and corpus luteum.
   b. follicular development beyond the antral stage.
   c. ovulation.
   d. luteinization.

3. **Negative and positive feedback control—estrogen and progesterone** (Table 7.13)
   - Granulosa cells secrete inhibin, which inhibits FSH secretion.
   - Granulosa cells secrete activan, which stimulates FSH secretion.

C. Actions of estrogen

1. has both negative and positive feedback effects on FSH and LH secretion.
2. causes maturation and maintenance of the fallopian tubes, uterus, cervix, and vagina.
3. causes the development of female secondary sex characteristics at puberty.
4. causes the development of the breasts.
5. up-regulates estrogen, LH, and progesterone receptors.
6. causes proliferation and development of ovarian granulosa cells.
7. during pregnancy, stimulates growth of myometrium and growth of ductal system in the breast.
8. during pregnancy, stimulates prolactin secretion (but then blocks its action on the breast).
9. increases uterine contractility.

table 7.13 Negative and Positive Feedback Control of the Menstrual Cycle

Phase of Menstrual Cycle	Hormone	Type of Feedback and Site
Follicular	Estrogen	Negative; anterior pituitary
Midcycle	Estrogen	Positive; anterior pituitary
Luteal	Estrogen Progesterone	Negative; anterior pituitary Negative; anterior pituitary

D. Actions of progesterone

1. has negative feedback effects on FSH and LH secretion during luteal phase.
2. maintains secretory activity of the uterus during the luteal phase.
3. during pregnancy, maintains endometrial lining.
4. decreases uterine contractility.
5. participates in development of the breasts.

E. Menstrual cycle (Figure 7.19)

1. Follicular phase (days 0 to 14)

- A **primordial follicle develops** to the graafian stage, with atresia of neighboring follicles.
- LH and FSH receptors are up-regulated in theca and granulosa cells.
- **Estradiol levels increase** and cause **proliferation of the uterus.**
- **FSH and LH levels are suppressed** by the negative feedback effect of estradiol on the anterior pituitary.
- Progesterone levels are low.

2. Ovulation (day 14)

- occurs 14 days before menses, regardless of cycle length. Thus, in a 28-day cycle, ovulation occurs on day 14; in a 35-day cycle, ovulation occurs on day 22.

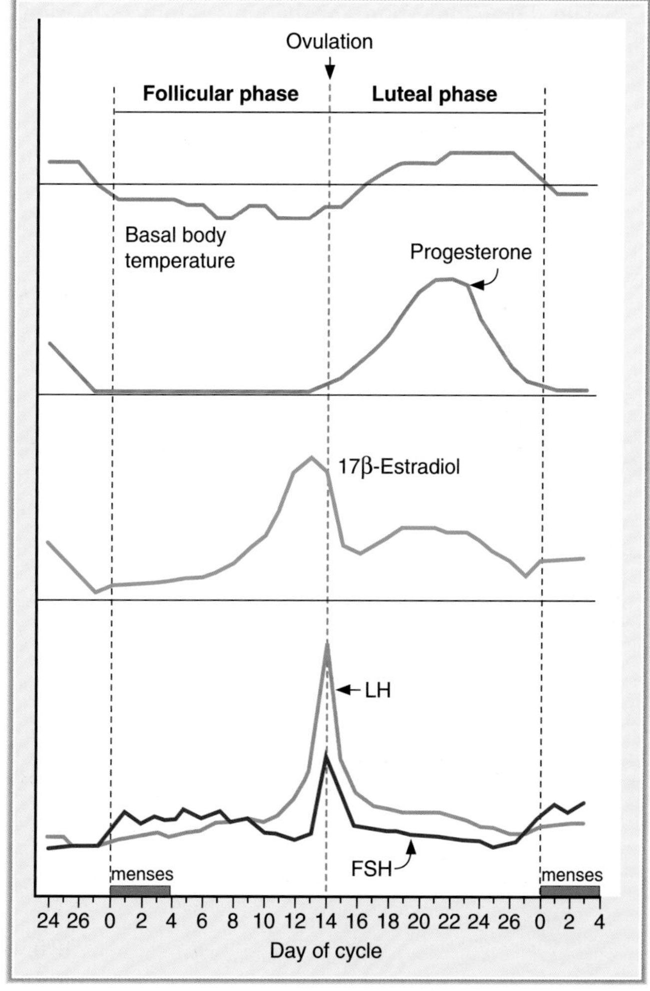

FIGURE 7.19. The menstrual cycle. FSH = follicle-stimulating hormone; LH = luteinizing hormone.

- A burst of estradiol synthesis at the end of the follicular phase has a **positive feedback** effect on the secretion of FSH and LH **(LH surge)**.
- **Ovulation** occurs as a result of the **estrogen-induced LH surge**.
- Estrogen levels decrease just after ovulation (but rise again during the luteal phase).
- **Cervical mucus increases** in quantity; it becomes less viscous and more penetrable by sperm.

3. Luteal phase (days 14 to 28)

- The **corpus luteum** begins to develop, and it **synthesizes estrogen and progesterone**.
- **Vascularity and secretory activity of the endometrium increase** to prepare for receipt of a fertilized egg.
- **Basal body temperature increases** because of the effect of progesterone on the hypothalamic thermoregulatory center.
- If fertilization does not occur, the **corpus luteum regresses** at the end of the luteal phase. As a result, estradiol and progesterone levels decrease abruptly.

4. Menses (days 0 to 4)

- **The endometrium is sloughed** because of the abrupt withdrawal of estradiol and progesterone.

F. Pregnancy (Figure 7.20)

- is characterized by steadily increasing levels of estrogen and progesterone, which maintain the endometrium for the fetus, suppress ovarian follicular function (by inhibiting FSH and LH secretion), and stimulate development of the breasts.

1. Fertilization

- If **fertilization occurs**, the corpus luteum is rescued from regression by **human chorionic gonadotropin (HCG)**, which is produced by the placenta.

2. First trimester

- The corpus luteum (stimulated by **HCG**) is responsible for the production of estradiol and progesterone.
- Peak levels of HCG occur at gestational week 9 and then decline.

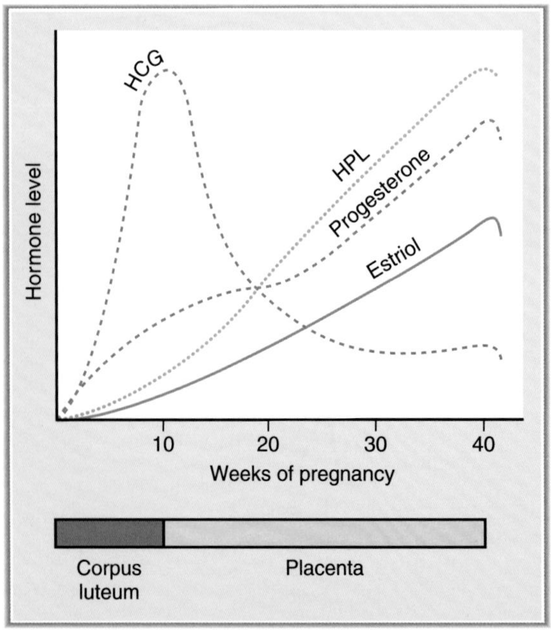

FIGURE 7.20. Hormone levels during pregnancy. HCG = human chorionic gonadotropin; HPL = human placental lactogen.

3. Second and third trimesters

- **Progesterone** is produced by the placenta.
- **Estrogens** are produced by the interplay of the **fetal adrenal gland** and the placenta. The fetal adrenal gland synthesizes dehydroepiandrosterone sulfate (DHEA-S), which is then hydroxylated in the fetal liver. These intermediates are transferred to the placenta, where enzymes remove sulfate and aromatize to estrogens. **The major placental estrogen is estriol.**
- **Human placental lactogen** is produced throughout pregnancy. Its actions are similar to those of growth hormone and prolactin.

4. Parturition

- Throughout pregnancy, estrogen increases uterine contractility and progesterone decreases uterine contractility.
- Near term, the estrogen/progesterone ratio increases, which makes the uterus more sensitive to contractile stimuli.
- Near term, increased estrogen levels increase production of local prostaglandins. **Prostaglandins** increase uterine contractility, increase gap junctions between uterine smooth muscle cells, and cause softening, thinning, and dilation of the cervix.
- The **initiating event in parturition is unknown**. (Although oxytocin is a powerful stimulant of uterine contractions, blood levels of oxytocin do not change before labor.)

5. Lactation

- Estrogens and progesterone stimulate the growth and development of the breasts throughout pregnancy.
- **Prolactin levels increase steadily during pregnancy** because estrogen stimulates prolactin secretion from the anterior pituitary.
- **Lactation does not occur during pregnancy because estrogen and progesterone block the action of prolactin on the breast.**
- After parturition, estrogen and progesterone levels decrease abruptly and lactation occurs.
- Lactation is maintained by suckling, which stimulates both oxytocin and prolactin secretion.
- **Ovulation is suppressed** as long as lactation continues because prolactin has the following effects:

a. inhibits hypothalamic GnRH secretion.

b. inhibits the action of GnRH on the anterior pituitary and consequently inhibits LH and FSH secretion.

c. antagonizes the actions of LH and FSH on the ovaries.

Review Test

QUESTIONS 1–5

Use the graph below, which shows changes during the menstrual cycle, to answer Questions 1–5.

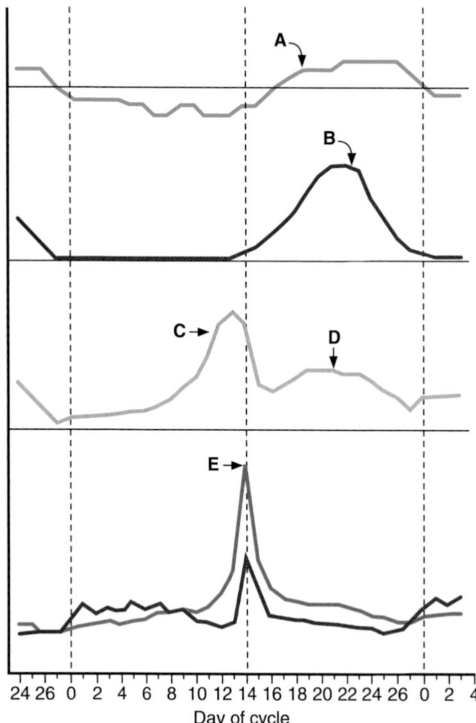

1. The increase shown at point A is caused by the effect of

(A) estrogen on the anterior pituitary
(B) progesterone on the hypothalamus
(C) follicle-stimulating hormone (FSH) on the ovary
(D) luteinizing hormone (LH) on the anterior pituitary
(E) prolactin on the ovary

2. Blood levels of which substance are described by curve B?

(A) Estradiol
(B) Estriol
(C) Progesterone
(D) Follicle-stimulating hormone (FSH)
(E) Luteinizing hormone (LH)

3. The source of the increase in concentration indicated at point C is the

(A) hypothalamus
(B) anterior pituitary
(C) corpus luteum
(D) ovary
(E) adrenal cortex

4. The source of the increase in concentration at point D is the

(A) ovary
(B) adrenal cortex
(C) corpus luteum
(D) hypothalamus
(E) anterior pituitary

5. The cause of the sudden increase shown at point E is

(A) negative feedback of progesterone on the hypothalamus
(B) negative feedback of estrogen on the anterior pituitary
(C) negative feedback of follicle-stimulating hormone (FSH) on the ovary
(D) positive feedback of FSH on the ovary
(E) positive feedback of estrogen on the anterior pituitary

6. A 41-year-old woman has hypocalcemia, hyperphosphatemia, and decreased urinary phosphate excretion. Injection of parathyroid hormone (PTH) causes an increase in urinary cyclic adenosine monophosphate (cAMP). The most likely diagnosis is

(A) primary hyperparathyroidism
(B) vitamin D intoxication
(C) vitamin D deficiency
(D) hypoparathyroidism after thyroid surgery
(E) pseudohypoparathyroidism

7. Which of the following hormones acts on its target tissues by a steroid hormone mechanism of action?

(A) Thyroid hormone
(B) Parathyroid hormone (PTH)

(C) Antidiuretic hormone (ADH) on the collecting duct
(D) β_1-adrenergic agonists
(E) Glucagon

8. A 38-year-old man who has galactorrhea is found to have a prolactinoma. His physician treats him with bromocriptine, which eliminates the galactorrhea. The basis for the therapeutic action of bromocriptine is that it

(A) antagonizes the action of prolactin on the breast
(B) enhances the action of prolactin on the breast
(C) inhibits prolactin release from the anterior pituitary
(D) inhibits prolactin release from the hypothalamus
(E) enhances the action of dopamine on the anterior pituitary

9. Which of the following hormones originates in the anterior pituitary?

(A) Dopamine
(B) Growth hormone–releasing hormone (GHRH)
(C) Somatostatin
(D) Gonadotropin-releasing hormone (GnRH)
(E) Thyroid-stimulating hormone (TSH)
(F) Oxytocin
(G) Testosterone

10. Which of the following functions of the Sertoli cells mediates negative feedback control of follicle-stimulating hormone (FSH) secretion?

(A) Synthesis of inhibin
(B) Synthesis of testosterone
(C) Aromatization of testosterone
(D) Maintenance of the blood–testes barrier

11. Which of the following substances is derived from proopiomelanocortin (POMC)?

(A) Adrenocorticotropic hormone (ACTH)
(B) Follicle-stimulating hormone (FSH)
(C) Melatonin
(D) Cortisol
(E) Dehydroepiandrosterone

12. Which of the following inhibits the secretion of growth hormone by the anterior pituitary?

(A) Sleep
(B) Stress
(C) Puberty
(D) Somatomedins
(E) Starvation
(F) Hypoglycemia

13. Selective destruction of the zona glomerulosa of the adrenal cortex would produce a deficiency of which hormone?

(A) Aldosterone
(B) Androstenedione
(C) Cortisol
(D) Dehydroepiandrosterone
(E) Testosterone

14. Which of the following explains the suppression of lactation during pregnancy?

(A) Blood prolactin levels are too low for milk production to occur
(B) Human placental lactogen levels are too low for milk production to occur
(C) The fetal adrenal gland does not produce sufficient estriol
(D) Blood levels of estrogen and progesterone are high
(E) The maternal anterior pituitary is suppressed

15. Which step in steroid hormone biosynthesis, if inhibited, blocks the production of all androgenic compounds but does not block the production of glucocorticoids?

(A) Cholesterol → pregnenolone
(B) Progesterone → 11-deoxycorticosterone
(C) 17-Hydroxypregnenolone → dehydroepiandrosterone
(D) Testosterone → estradiol
(E) Testosterone → dihydrotestosterone

16. A 46-year-old woman has hirsutism, hyperglycemia, obesity, muscle wasting, and increased circulating levels of adrenocorticotropic hormone (ACTH). The most likely cause of her symptoms is

(A) primary adrenocortical insufficiency (Addison disease)
(B) pheochromocytoma
(C) primary overproduction of ACTH (Cushing disease)
(D) treatment with exogenous glucocorticoids
(E) hypophysectomy

17. Which of the following decreases the conversion of 25-hydroxycholecalciferol to 1,25-dihydroxycholecalciferol?

(A) A diet low in Ca^{2+}
(B) Hypocalcemia
(C) Hyperparathyroidism
(D) Hypophosphatemia
(E) Chronic renal failure

18. Increased adrenocorticotropic hormone (ACTH) secretion would be expected in patients

(A) with chronic adrenocortical insufficiency (Addison disease)
(B) with primary adrenocortical hyperplasia
(C) who are receiving glucocorticoid for immunosuppression after a renal transplant
(D) with elevated levels of angiotensin II

19. Which of the following would be expected in a patient with Graves disease?

(A) Cold sensitivity
(B) Weight gain
(C) Decreased O_2 consumption
(D) Decreased cardiac output
(E) Drooping eyelids
(F) Atrophy of the thyroid gland
(G) Increased thyroid-stimulating hormone (TSH) levels
(H) Increased triiodothyronine (T_3) levels

20. Blood levels of which of the following substances is decreased in Graves disease?

(A) Triiodothyronine (T_3)
(B) Thyroxine (T_4)
(C) Diiodotyrosine (DIT)
(D) Thyroid-stimulating hormone (TSH)
(E) Iodide (I^-)

21. Which of the following hormones acts by an inositol 1,4,5-triphosphate (IP_3)-Ca^{2+} mechanism of action?

(A) 1,25-Dihydroxycholecalciferol
(B) Progesterone
(C) Insulin
(D) Parathyroid hormone (PTH)
(E) Gonadotropin-releasing hormone (GnRH)

22. Which step in steroid hormone biosynthesis is stimulated by adrenocorticotropic hormone (ACTH)?

(A) Cholesterol → pregnenolone
(B) Progesterone → 11-deoxycorticosterone
(C) 17-Hydroxypregnenolone → dehydroepiandrosterone
(D) Testosterone → estradiol
(E) Testosterone → dihydrotestosterone

23. The source of estrogen during the second and third trimesters of pregnancy is the

(A) corpus luteum
(B) maternal ovaries
(C) fetal ovaries
(D) placenta
(E) maternal ovaries and fetal adrenal gland
(F) maternal adrenal gland and fetal liver
(G) fetal adrenal gland, fetal liver, and placenta

24. Which of the following causes increased aldosterone secretion?

(A) Decreased blood volume
(B) Administration of an inhibitor of angiotensin-converting enzyme (ACE)
(C) Hyperosmolarity
(D) Hypokalemia

25. Secretion of oxytocin is increased by

(A) milk ejection
(B) dilation of the cervix
(C) increased prolactin levels
(D) increased extracellular fluid (ECF) volume
(E) increased serum osmolarity

26. A 61-year-old woman with hyperthyroidism is treated with propylthiouracil. The drug reduces the synthesis of thyroid hormones because it inhibits oxidation of

(A) triiodothyronine (T_3)
(B) thyroxine (T_4)
(C) diiodotyrosine (DIT)
(D) thyroid-stimulating hormone (TSH)
(E) iodide (I^-)

27. A 39-year-old man with untreated diabetes mellitus type I is brought to the emergency room. An injection of insulin would be expected to cause an increase in his

(A) urine glucose concentration
(B) blood glucose concentration
(C) blood K^+ concentration
(D) blood pH
(E) breathing rate

28. Which of the following results from the action of parathyroid hormone (PTH) on the renal tubule?

(A) Inhibition of 1α-hydroxylase
(B) Stimulation of Ca^{2+} reabsorption in the distal tubule
(C) Stimulation of phosphate reabsorption in the proximal tubule
(D) Interaction with receptors on the luminal membrane of the proximal tubular cells
(E) Decreased urinary excretion of cyclic adenosine monophosphate (cAMP)

29. Which step in steroid hormone biosynthesis occurs in the accessory sex target tissues of the male and is catalyzed by 5α-reductase?

(A) Cholesterol → pregnenolone
(B) Progesterone → 11-deoxycorticosterone
(C) 17-Hydroxypregnenolone → dehydroepiandrosterone
(D) Testosterone → estradiol
(E) Testosterone → dihydrotestosterone

30. Which of the following pancreatic secretions has a receptor with four subunits, two of which have tyrosine kinase activity?

(A) Insulin
(B) Glucagon
(C) Somatostatin
(D) Pancreatic lipase

31. A 16-year-old, seemingly normal female is diagnosed with androgen insensitivity disorder. She has never had a menstrual cycle and is found to have a blind-ending vagina; no uterus, cervix, or ovaries; a 46 XY genotype; and intra-abdominal testes. Her serum testosterone is elevated. Which of the following characteristics is caused by lack of androgen receptors?

(A) 46 XY genotype
(B) Testes
(C) Elevated serum testosterone
(D) Lack of uterus and cervix
(E) Lack of menstrual cycles

QUESTIONS 32–34

A 76–year-old man with lung cancer is lethargic and excreting large volumes of urine. He is thirsty and drinks water almost constantly. Laboratory values reveal an elevated serum Ca^{2+} concentration of 18 mg/dL, elevated serum osmolarity of 310 mOsm/L, and urine osmolarity of 90 mOsm/L. Administration of an ADH analogue does not change his serum or urine osmolarity.

32. The man's serum ADH level is

(A) decreased because excess water-drinking has suppressed ADH secretion
(B) decreased because his posterior pituitary is not secreting ADH
(C) normal
(D) increased because the elevated serum osmolarity has stimulated ADH secretion
(E) increased because his extreme thirst has directly stimulated ADH secretion

33. The cause of the patient's excess urine volume is

(A) dehydration
(B) syndrome of inappropriate ADH
(C) central diabetes insipidus
(D) nephrogenic diabetes insipidus

34. The most appropriate treatment is

(A) ADH antagonist
(B) ADH analogue
(C) PTH analogue
(D) half-normal saline
(E) pamidronate plus furosemide

Answers and Explanations

1. **The answer is B** [X E 3; Figure 7.19]. Curve A shows basal body temperature. The increase in temperature occurs as a result of elevated progesterone levels during the luteal (secretory) phase of the menstrual cycle. Progesterone increases the set-point temperature in the hypothalamic thermoregulatory center.
2. **The answer is C** [X E 3; Figure 7.19]. Progesterone is secreted during the luteal phase of the menstrual cycle.
3. **The answer is D** [X A, E 1; Figure 7.19]. The curve shows blood levels of estradiol. The source of the increase in estradiol concentration shown at point C is the ovarian granulosa cells, which contain high concentrations of aromatase and convert testosterone to estradiol.
4. **The answer is C** [X E 3; Figure 7.19]. The curve shows blood levels of estradiol. During the luteal phase of the cycle, the source of the estradiol is the corpus luteum. The corpus luteum prepares the uterus to receive a fertilized egg.
5. **The answer is E** [X E 2; Figure 7.20]. Point E shows the luteinizing hormone (LH) surge that initiates ovulation at midcycle. The LH surge is caused by increasing estrogen levels from the developing ovarian follicle. Increased estrogen, by positive feedback, stimulates the anterior pituitary to secrete LH and follicle-stimulating hormone (FSH).
6. **The answer is D** [VII B 3 b]. Low blood [Ca^{2+}] and high blood [phosphate] are consistent with hypoparathyroidism. Lack of parathyroid hormone (PTH) decreases bone resorption, decreases renal reabsorption of Ca^{2+}, and increases renal reabsorption of phosphate (causing low urinary phosphate). Because the patient responded to exogenous PTH with an increase in urinary cyclic adenosine monophosphate (cAMP), the G protein coupling the PTH receptor to adenylate cyclase is apparently normal. Consequently, pseudohypoparathyroidism is excluded. Vitamin D intoxication would cause hypercalcemia, not hypocalcemia. Vitamin D deficiency would cause hypocalcemia and hypophosphatemia.
7. **The answer is A** [II E; Table 7.2]. Thyroid hormone, an amine, acts on its target tissues by a steroid hormone mechanism, inducing the synthesis of new proteins. The action of antidiuretic hormone (ADH) on the collecting duct (V_2 receptors) is mediated by cyclic adenosine monophosphate (cAMP), although the other action of ADH (vascular smooth muscle, V_1 receptors) is mediated by inositol 1,4,5-triphosphate (IP_3). Parathyroid hormone (PTH), β_1-agonists, and glucagon all act through cAMP mechanisms of action.
8. **The answer is C** [III B 4 a (1), c (2)]. Bromocriptine is a dopamine agonist. The secretion of prolactin by the anterior pituitary is tonically inhibited by the secretion of dopamine from the hypothalamus. Thus, a dopamine agonist acts just like dopamine—it inhibits prolactin secretion from the anterior pituitary.
9. **The answer is E** [III B; Table 7.1]. Thyroid-stimulating hormone (TSH) is secreted by the anterior pituitary. Dopamine, growth hormone–releasing hormone (GHRH), somatostatin, and gonadotropin-releasing hormone (GnRH) all are secreted by the hypothalamus. Oxytocin is secreted by the posterior pituitary. Testosterone is secreted by the testes.
10. **The answer is A** [IX B 2, 3]. Inhibin is produced by the Sertoli cells of the testes when they are stimulated by follicle-stimulating hormone (FSH). Inhibin then inhibits further secretion of FSH by negative feedback on the anterior pituitary. The Leydig cells synthesize testosterone. Testosterone is aromatized in the ovaries.
11. **The answer is A** [III B 1, 2; Figure 7.5]. Proopiomelanocortin (POMC) is the parent molecule in the anterior pituitary for adrenocorticotropic hormone (ACTH), β-endorphin, α-lipotropin, and β-lipotropin (and in the intermediary lobe for melanocyte-stimulating

hormone [MSH]). Follicle-stimulating hormone (FSH) is not a member of this "family"; rather, it is a member of the thyroid-stimulating hormone (TSH) and luteinizing hormone (LH) "family." MSH, a component of POMC and ACTH, may stimulate melatonin production. Cortisol and dehydroepiandrosterone are produced by the adrenal cortex.

12. **The answer is D** [III B 3 a]. Growth hormone is secreted in pulsatile fashion, with a large burst occurring during deep sleep (sleep stage 3 or 4). Growth hormone secretion is increased by sleep, stress, puberty, starvation, and hypoglycemia. Somatomedins are generated when growth hormone acts on its target tissues; they inhibit growth hormone secretion by the anterior pituitary, both directly and indirectly (by stimulating somatostatin release).

13. **The answer is A** [V A 1; Figure 7.10]. Aldosterone is produced in the zona glomerulosa of the adrenal cortex because that layer contains the enzyme for conversion of corticosterone to aldosterone (aldosterone synthase). Cortisol is produced in the zona fasciculata. Androstenedione and dehydroepiandrosterone are produced in the zona reticularis. Testosterone is produced in the testes, not in the adrenal cortex.

14. **The answer is D** [X F 5]. Although the high circulating levels of estrogen stimulate prolactin secretion during pregnancy, the action of prolactin on the breast is inhibited by progesterone and estrogen. After parturition, progesterone and estrogen levels decrease dramatically. Prolactin can then interact with its receptors in the breast, and lactation proceeds if initiated by suckling.

15. **The answer is C** [Figure 7.11]. The conversion of 17-hydroxypregnenolone to dehydroepiandrosterone (as well as the conversion of 17-hydroxyprogesterone to androstenedione) is catalyzed by 17,20-lyase. If this process is inhibited, synthesis of androgens is stopped.

16. **The answer is C** [V A 5 b]. This woman has the classic symptoms of a primary elevation of adrenocorticotropic hormone (ACTH) (Cushing disease). Elevation of ACTH stimulates overproduction of glucocorticoids and androgens. Treatment with pharmacologic doses of glucocorticoids would produce similar symptoms, except that circulating levels of ACTH would be low because of negative feedback suppression at both the hypothalamic (corticotropin-releasing hormone [CRH]) and anterior pituitary (ACTH) levels. Addison disease is caused by primary adrenocortical insufficiency. Although a patient with Addison disease would have increased levels of ACTH (because of the loss of negative feedback inhibition), the symptoms would be of glucocorticoid deficit, not excess. Hypophysectomy would remove the source of ACTH. A pheochromocytoma is a tumor of the adrenal medulla that secretes catecholamines.

17. **The answer is E** [VII C 1]. Ca^{2+} deficiency (low Ca^{2+} diet or hypocalcemia) activates 1α-hydroxylase, which catalyzes the conversion of vitamin D to its active form, 1,25-dihydroxycholecalciferol. Increased parathyroid hormone (PTH) and hypophosphatemia also stimulate the enzyme. Chronic renal failure is associated with a constellation of bone diseases, including osteomalacia caused by failure of the diseased renal tissue to produce the active form of vitamin D.

18. **The answer is A** [V A 2 a (3); Table 7.6; Figure 7.12]. Addison disease is caused by primary adrenocortical insufficiency. The resulting decrease in cortisol production causes a decrease in negative feedback inhibition on the hypothalamus and the anterior pituitary. Both of these conditions will result in increased adrenocorticotropic hormone (ACTH) secretion. Patients who have adrenocortical hyperplasia or who are receiving exogenous glucocorticoid will have an increase in the negative feedback inhibition of ACTH secretion.

19. **The answer is H** [IV B 2; Table 7.5]. Graves disease (hyperthyroidism) is caused by overstimulation of the thyroid gland by circulating antibodies to the thyroid-stimulating hormone (TSH) receptor (which then increases the production and secretion of triiodothyronine [T_3] and thyroxine [T_4], just as TSH would). Therefore, the signs and symptoms of Graves disease are the same as those of hyperthyroidism, reflecting the actions of increased circulating levels of thyroid hormones: increased heat production, weight loss, increased O_2 consumption and cardiac output, exophthalmos (bulging eyes, not drooping

eyelids), and hypertrophy of the thyroid gland (goiter). TSH levels will be decreased (not increased) as a result of the negative feedback effect of increased T_3 levels on the anterior pituitary.

20. **The answer is D** [IV B 2; Table 7.5]. In Graves disease (hyperthyroidism), the thyroid is stimulated to produce and secrete vast quantities of thyroid hormones as a result of stimulation by thyroid-stimulating immunoglobulins (antibodies to the thyroid-stimulating hormone [TSH] receptors on the thyroid gland). Because of the high circulating levels of thyroid hormones, anterior pituitary secretion of TSH will be turned off (negative feedback).

21. **The answer is E** [Table 7.2]. Gonadotropin-releasing hormone (GnRH) is a peptide hormone that acts on the cells of the anterior pituitary by an inositol 1,4,5-triphosphate (IP_3)-Ca^{2+} mechanism to cause the secretion of follicle-stimulating hormone (FSH) and luteinizing hormone (LH). 1,25-Dihydroxycholecalciferol and progesterone are steroid hormone derivatives of cholesterol that act by inducing the synthesis of new proteins. Insulin acts on its target cells by a tyrosine kinase mechanism. Parathyroid hormone (PTH) acts on its target cells by an adenylate cyclase–cyclic adenosine monophosphate (cAMP) mechanism.

22. **The answer is A** [V A 2 a (2)]. The conversion of cholesterol to pregnenolone is catalyzed by cholesterol desmolase. This step in the biosynthetic pathway for steroid hormones is stimulated by adrenocorticotropic hormone (ACTH).

23. **The answer is G** [X F 3]. During the second and third trimesters of pregnancy, the fetal adrenal gland synthesizes dehydroepiandrosterone sulfate (DHEA-S), which is hydroxylated in the fetal liver and then transferred to the placenta, where it is aromatized to estrogen. In the first trimester, the corpus luteum is the source of both estrogen and progesterone.

24. **The answer is A** [V A 2 b]. Decreased blood volume stimulates the secretion of renin (because of decreased renal perfusion pressure) and initiates the renin–angiotensin–aldosterone cascade. Angiotensin-converting enzyme (ACE) inhibitors block the cascade by decreasing the production of angiotensin II. Hyperosmolarity stimulates antidiuretic hormone (ADH) (not aldosterone) secretion. Hyperkalemia, not hypokalemia, directly stimulates aldosterone secretion by the adrenal cortex.

25. **The answer is B** [III C 2]. Suckling and dilation of the cervix are the physiologic stimuli for oxytocin secretion. Milk ejection is the *result* of oxytocin action, not the cause of its secretion. Prolactin secretion is also stimulated by suckling, but prolactin does not directly cause oxytocin secretion. Increased extracellular fluid (ECF) volume and hyperosmolarity are the stimuli for the secretion of the other posterior pituitary hormone, antidiuretic hormone (ADH).

26. **The answer is E** [IV A 2]. For iodide (I^-) to be "organified" (incorporated into thyroid hormone), it must be oxidized to I_2, which is accomplished by a peroxidase enzyme in the thyroid follicular cell membrane. Propylthiouracil inhibits peroxidase and, therefore, halts the synthesis of thyroid hormones.

27. **The answer is D** [VI C 3; Table 7.7]. Before the injection of insulin, the woman would have had hyperglycemia, glycosuria, hyperkalemia, and metabolic acidosis with compensatory hyperventilation. The injection of insulin would be expected to decrease her blood glucose (by increasing the uptake of glucose into the cells), decrease her urinary glucose (secondary to decreasing her blood glucose), decrease her blood K^+ (by shifting K^+ into the cells), and correct her metabolic acidosis (by decreasing the production of ketoacids). The correction of the metabolic acidosis will lead to an increase in her blood pH and will reduce her compensatory hyperventilation.

28. **The answer is B** [VII B 2]. Parathyroid hormone (PTH) stimulates both renal Ca^{2+} reabsorption in the renal distal tubule and the 1α-hydroxylase enzyme. PTH inhibits (not stimulates) phosphate reabsorption in the proximal tubule, which is associated with an increase in urinary cyclic adenosine monophosphate (cAMP). The receptors for PTH are located on the basolateral membranes, not the luminal membranes.

29. **The answer is E** [IX A]. Some target tissues for androgens contain 5α-reductase, which converts testosterone to dihydrotestosterone, the active form in those tissues.

30. **The answer is A** [VI C 2]. The insulin receptor in target tissues is a tetramer. The two β subunits have tyrosine kinase activity and autophosphorylate the receptor when stimulated by insulin.

31. **The answer is C** [IX C]. The elevated serum testosterone is due to lack of androgen receptors on the anterior pituitary (which normally would mediate negative feedback by testosterone). The presence of testes is due to the male genotype. The lack of uterus and cervix is due to anti-müllerian hormone (secreted by the fetal testes), which suppressed differentiation of the müllerian ducts into the internal female genital tract. The lack of menstrual cycles is due to the absence of a female reproductive tract.

32. **The answer is D** [III C 1; also Chapter 7, VII]. The man is excreting large volumes of dilute urine, which has raised his serum osmolarity and made him very thirsty. The increase in serum osmolarity would then cause an increase in serum ADH levels. The fact that exogenous ADH administration did not change his serum or urine osmolarity suggests that the collecting duct of the nephron is unresponsive to ADH. Thirst does not directly increase ADH secretion.

33. **The answer is D** [III C 1; VII, B 3; Chapter 7, VI]. The man's urine osmolarity is very dilute, while his serum osmolarity is increased. In the face of increased serum osmolarity, there should be increased ADH secretion, which should then act on the collecting duct principal cells to increase water reabsorption and concentrate the urine. The fact that the urine is dilute, not concentrated, suggests that ADH either is absent (central diabetes insipidus) or is ineffective (nephrogenic diabetes insipidus). Administration of an exogenous ADH analogue separates these two possibilities—it was ineffective in changing serum or urine osmolarity; thus, it can be concluded that ADH unable to act on the collecting ducts, that is, nephrogenic diabetes insipidus. One cause of nephrogenic diabetes insipidus is hypercalcemia, which is present in this patent secondary to his lung cancer; he likely has humoral hypercalcemia of malignancy, due to secretion of PTH-rp by the tumor. Dehydration would cause increased ADH secretion and increased urine osmolarity. Syndrome of inappropriate ADH would cause increased urine osmolarity and subsequently decreased serum osmolarity, due to excess water reabsorption.

34. **The answer is E** [VII, B 3]. The man's nephrogenic diabetes insipidus is caused by hypercalcemia secondary to increased PTH-rp secreted by his lung tumor. PTH-rp has all of the actions of PTH, including increased bone resorption, increased renal Ca^{2+} reabsorption, and decreased renal phosphate reabsorption; all of these actions lead to increased serum Ca^{2+} concentration. The treatment should be directed at lowering serum Ca^{2+} concentration, which can be achieved by giving an inhibitor of bone resorption (e.g., pamidronate) and an inhibitor of renal Ca^{2+} reabsorption (furosemide). Giving an ADH antagonist would be ineffective because the man's nephrogenic diabetes insipidus has made his collecting ducts insensitive to ADH. Giving half-normal saline could lower his serum osmolarity temporarily, but would not address the underlying problem of hypercalcemia.

Comprehensive Examination

QUESTIONS 1 AND 2

After extensive testing, a 60-year-old man is found to have a pheochromocytoma that secretes mainly epinephrine.

1. Which of the following signs would be expected in this patient?

(A) Decreased heart rate
(B) Decreased arterial blood pressure
(C) Decreased excretion rate of 3-methoxy-4-hydroxymandelic acid (VMA)
(D) Cold, clammy skin

2. Symptomatic treatment would be best achieved in this man with

(A) phentolamine
(B) isoproterenol
(C) a combination of phentolamine and isoproterenol
(D) a combination of phentolamine and propranolol
(E) a combination of isoproterenol and phenylephrine

3. The principle of positive feedback is illustrated by the effect of

(A) PO_2 on breathing rate
(B) glucose on insulin secretion
(C) estrogen on follicle-stimulating hormone (FSH) secretion at mid-menstrual cycle
(D) blood $[Ca^{2+}]$ on parathyroid hormone (PTH) secretion
(E) decreased blood pressure on sympathetic outflow to the heart and blood vessels

4. In the graph at upper right, the response shown by the dotted line illustrates the effect of

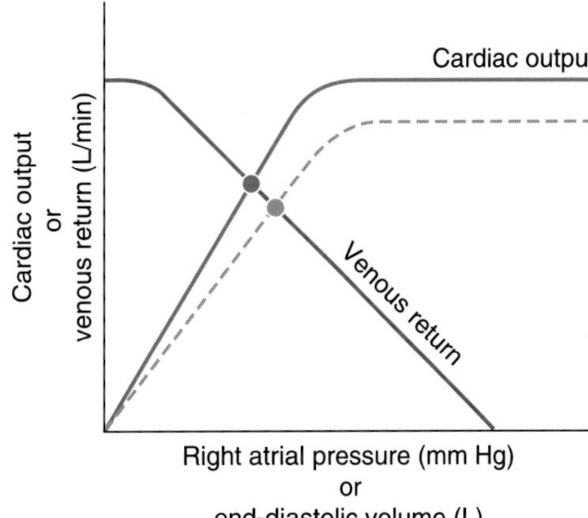

(A) administration of digitalis
(B) administration of a negative inotropic agent
(C) increased blood volume
(D) decreased blood volume
(E) decreased total peripheral resistance (TPR)

QUESTIONS 5 AND 6

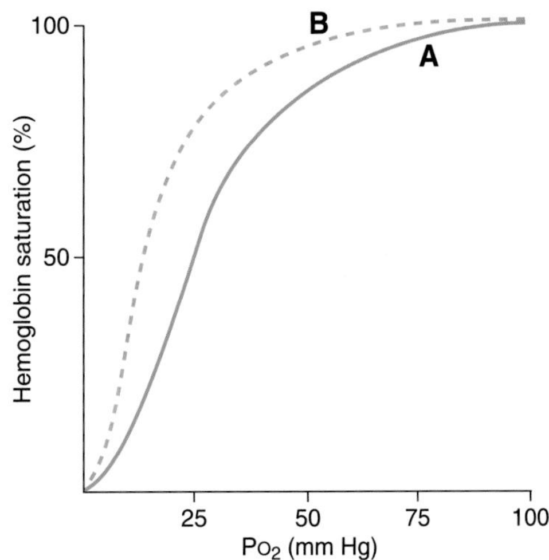

5. On the accompanying graph, the shift from curve A to curve B could be caused by

(A) fetal hemoglobin (HbF)
(B) carbon monoxide (CO) poisoning
(C) decreased pH
(D) increased temperature
(E) increased 2,3-diphosphoglycerate (DPG)

6. The shift from curve A to curve B is associated with

(A) decreased P_{50}
(B) decreased affinity of hemoglobin for O_2
(C) decreased O_2-carrying capacity of hemoglobin
(D) increased ability to unload O_2 in the tissues

7. A negative free-water clearance $\left(C_{H_2O}\right)$ would occur in a person

(A) who drinks 2 L of water in 30 minutes
(B) after overnight water restriction
(C) who is receiving lithium for the treatment of depression and has polyuria that is unresponsive to antidiuretic hormone (ADH) administration
(D) with a urine flow rate of 5 mL/min, a urine osmolarity of 295 mOsm/L, and a serum osmolarity of 295 mOsm/L
(E) with a urine osmolarity of 90 mOsm/L and a serum osmolarity of 310 mOsm/L after a severe head injury

8. CO_2 generated in the tissues is carried in venous blood primarily as

(A) CO_2 in the plasma
(B) H_2CO_3 in the plasma
(C) HCO_3^- in the plasma
(D) CO_2 in the red blood cells (RBCs)
(E) carboxyhemoglobin in the RBCs

9. In a 35-day menstrual cycle, ovulation occurs on day

(A) 12
(B) 14
(C) 17
(D) 21
(E) 28

10. Which of the following hormones stimulates the conversion of testosterone to 17β-estradiol in ovarian granulosa cells?

(A) Adrenocorticotropic hormone (ACTH)
(B) Estradiol
(C) Follicle-stimulating hormone (FSH)
(D) Gonadotropin-releasing hormone (GnRH)
(E) Human chorionic gonadotropin (HCG)
(F) Prolactin
(G) Testosterone

11. Which gastrointestinal secretion is hypotonic, has a high $[HCO_3^-]$, and has its production inhibited by vagotomy?

(A) Saliva
(B) Gastric secretion
(C) Pancreatic secretion
(D) Bile

QUESTIONS 12 AND 13

A 53-year-old man with multiple myeloma is hospitalized after 2 days of polyuria, polydipsia, and increasing confusion. Laboratory tests show an elevated serum $[Ca^{2+}]$ of 15 mg/dL, and treatment is initiated to decrease it. The patient's serum osmolarity is 310 mOsm/L.

12. The most likely reason for polyuria in this man is

(A) increased circulating levels of antidiuretic hormone (ADH)
(B) increased circulating levels of aldosterone
(C) inhibition of the action of ADH on the renal tubule
(D) stimulation of the action of ADH on the renal tubule
(E) psychogenic water drinking

13. The treatment drug is administered *in error* and produces a further increase in the patient's serum $[Ca^{2+}]$. That drug is

(A) a thiazide diuretic
(B) a loop diuretic
(C) calcitonin
(D) mithramycin
(E) etidronate disodium

14. Which of the following substances acts on its target cells via an inositol 1,4,5-triphosphate (IP_3)–Ca^{2+} mechanism?

(A) Somatomedins acting on chondrocytes
(B) Oxytocin acting on myoepithelial cells of the breast
(C) Antidiuretic hormone (ADH) acting on the renal collecting duct
(D) Adrenocorticotropic hormone (ACTH) acting on the adrenal cortex
(E) Thyroid hormone acting on skeletal muscle

15. A key difference in the mechanism of excitation–contraction coupling between the muscle of the pharynx and the muscle of the wall of the small intestine is that

(A) slow waves are present in the pharynx, but not in the small intestine
(B) adenosine triphosphate (ATP) is used for contraction in the pharynx, but not in the small intestine
(C) intracellular [Ca^{2+}] is increased after excitation in the pharynx, but not in the small intestine
(D) action potentials depolarize the muscle of the small intestine, but not of the pharynx
(E) Ca^{2+} binds to troponin C in the pharynx, but not in the small intestine, to initiate contraction

16. A 40-year-old woman has an arterial pH of 7.25, an arterial P_{CO_2} of 30 mm Hg, and serum [K^+] of 2.8 mEq/L. Her blood pressure is 100/80 mm Hg when supine and 80/50 mm Hg when standing. What is the cause of her abnormal blood values?

(A) Vomiting
(B) Diarrhea
(C) Treatment with a loop diuretic
(D) Treatment with a thiazide diuretic

17. Secretion of HCl by gastric parietal cells is needed for

(A) activation of pancreatic lipases
(B) activation of salivary lipases
(C) activation of intrinsic factor
(D) activation of pepsinogen to pepsin
(E) the formation of micelles

18. Which of the following would cause an increase in glomerular filtration rate (GFR)?

(A) Constriction of the afferent arteriole
(B) Constriction of the efferent arteriole
(C) Constriction of the ureter
(D) Increased plasma protein concentration
(E) Infusion of inulin

19. Vitamin D absorption occurs primarily in the

(A) stomach
(B) jejunum
(C) terminal ileum
(D) cecum
(E) sigmoid colon

20. Which of the following hormones causes constriction of vascular smooth muscle through an inositol 1,4,5-triphosphate (IP_3) second messenger system?

(A) Antidiuretic hormone (ADH)
(B) Aldosterone
(C) Dopamine
(D) Oxytocin
(E) Parathyroid hormone (PTH)

21. A 30-year-old woman has the anterior lobe of her pituitary gland surgically removed because of a tumor. Without hormone replacement therapy, which of the following would occur after the operation?

(A) Absence of menses
(B) Inability to concentrate the urine in response to water deprivation
(C) Failure to secrete catecholamines in response to stress
(D) Failure to secrete insulin in a glucose tolerance test
(E) Failure to secrete parathyroid hormone (PTH) in response to hypocalcemia

22. The following graph shows three relationships as a function of plasma [glucose]. At plasma [glucose] less than 200 mg/dL, curves X and Z are superimposed on each other because

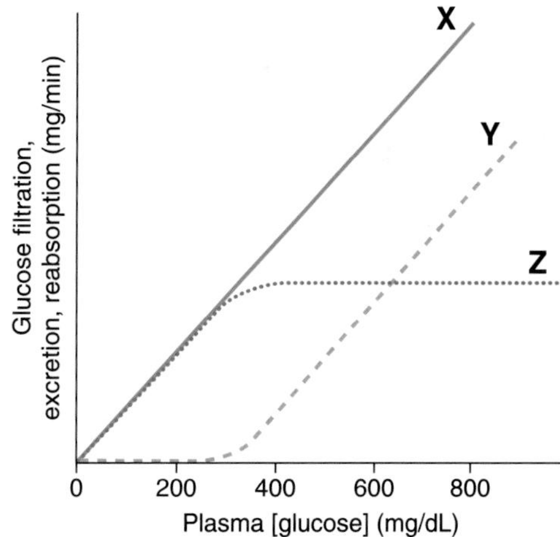

(A) the reabsorption and excretion of glucose are equal
(B) all of the filtered glucose is reabsorbed
(C) glucose reabsorption is saturated
(D) the renal threshold for glucose has been exceeded
(E) Na^+–glucose cotransport has been inhibited
(F) all of the filtered glucose is excreted

23. Which of the following responses occurs as a result of tapping on the patellar tendon?

(A) Stimulation of Ib afferent fibers in the muscle spindle
(B) Inhibition of Ia afferent fibers in the muscle spindle
(C) Relaxation of the quadriceps muscle
(D) Contraction of the quadriceps muscle
(E) Inhibition of α-motoneurons

QUESTIONS 24 AND 25

A 5-year-old boy has a severe sore throat, high fever, and cervical adenopathy.

24. It is suspected that the causative agent is *Streptococcus pyogenes*. Which of the following is involved in producing fever in this patient?

(A) Increased production of interleukin-1 (IL-1)
(B) Decreased production of prostaglandins
(C) Decreased set-point temperature in the hypothalamus
(D) Decreased metabolic rate
(E) Vasodilation of blood vessels in the skin

25. Before antibiotic therapy is initiated, the patient is given aspirin to reduce his fever. The mechanism of fever reduction by aspirin is

(A) shivering
(B) stimulation of cyclooxygenase
(C) inhibition of prostaglandin synthesis
(D) shunting of blood from the surface of the skin
(E) increasing the hypothalamic set-point temperature

26. Arterial pH of 7.52, arterial P_{CO_2} of 26 mm Hg, and tingling and numbness in the feet and hands would be observed in a

(A) patient with chronic diabetic ketoacidosis
(B) patient with chronic renal failure
(C) patient with chronic emphysema and bronchitis
(D) patient who hyperventilates on a commuter flight
(E) patient who is taking a carbonic anhydrase inhibitor for glaucoma
(F) patient with a pyloric obstruction who vomits for 5 days
(G) healthy person

27. Albuterol is useful in the treatment of asthma because it acts as an agonist at which of the following receptors?

(A) α_1 Receptor
(B) β_1 Receptor
(C) β_2 Receptor
(D) Muscarinic receptor
(E) Nicotinic receptor

28. Which of the following hormones is converted to its active form in target tissues by the action of 5α-reductase?

(A) Adrenocorticotropic hormone (ACTH)
(B) Aldosterone
(C) Estradiol
(D) Prolactin
(E) Testosterone

29. If an artery is partially occluded by an embolism such that its radius becomes one-half the preocclusion value, which of the following parameters will *increase* by a factor of 16?

(A) Blood flow
(B) Resistance
(C) Pressure gradient
(D) Capacitance

30. If heart rate increases, which phase of the cardiac cycle is decreased?

(A) Atrial systole
(B) Isovolumetric ventricular contraction
(C) Rapid ventricular ejection
(D) Reduced ventricular ejection
(E) Isovolumetric ventricular relaxation
(F) Rapid ventricular filling
(G) Reduced ventricular filling

QUESTIONS 31 AND 32

A 17-year-old boy is brought to the emergency department after being injured in an automobile accident and sustaining significant blood loss. He is given a transfusion of 3 units of blood to stabilize his blood pressure.

31. Before the transfusion, which of the following was true about his condition?

(A) His total peripheral resistance (TPR) was decreased
(B) His heart rate was decreased
(C) The firing rate of his carotid sinus nerves was increased
(D) Sympathetic outflow to his heart and blood vessels was increased

32. Which of the following is a consequence of the decrease in blood volume in this patient?

(A) Increased renal perfusion pressure
(B) Increased circulating levels of angiotensin II
(C) Decreased renal Na^+ reabsorption
(D) Decreased renal K^+ secretion

33. A 37-year-old woman suffers a severe head injury in a skiing accident. Shortly thereafter, she becomes polydipsic and polyuric. Her urine osmolarity is 75 mOsm/L, and her serum osmolarity is 305 mOsm/L. Treatment with 1-deamino-8-D-arginine vasopressin (dDAVP) causes an increase in her urine osmolarity to 450 mOsm/L. Which diagnosis is correct?

(A) Primary polydipsia
(B) Central diabetes insipidus
(C) Nephrogenic diabetes insipidus
(D) Water deprivation
(E) Syndrome of inappropriate antidiuretic hormone (SIADH)

34. Which diuretic inhibits Na^+ reabsorption and K^+ secretion in the distal tubule by acting as an aldosterone antagonist?

(A) Acetazolamide
(B) Chlorothiazide
(C) Furosemide
(D) Spironolactone

35. Which gastrointestinal secretion has a component that is required for the intestinal absorption of vitamin B_{12}?

(A) Saliva
(B) Gastric secretion
(C) Pancreatic secretion
(D) Bile

36. Secretion of which of the following hormones is stimulated by extracellular fluid volume expansion?

(A) Antidiuretic hormone (ADH)
(B) Aldosterone
(C) Atrial natriuretic peptide (ANP)
(D) 1,25-Dihydroxycholecalciferol
(E) Parathyroid hormone (PTH)

37. Which enzyme in the steroid hormone synthetic pathway is stimulated by angiotensin II?

(A) Aldosterone synthase
(B) Aromatase
(C) Cholesterol desmolase
(D) 17,20-Lyase
(E) 5α-Reductase

QUESTIONS 38–41

Use the diagram of an action potential to answer the following questions.

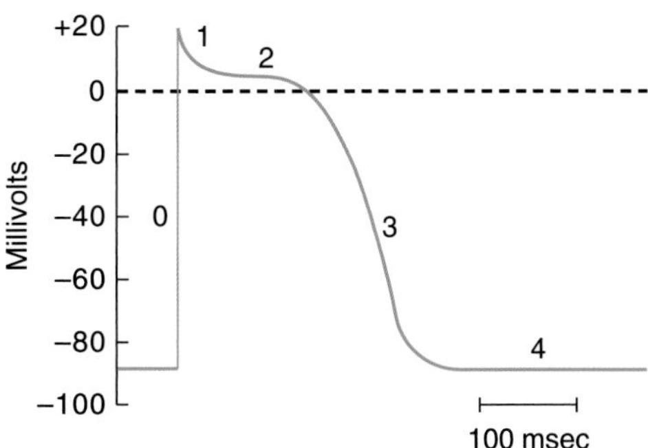

38. The action potential shown is from

(A) a skeletal muscle cell
(B) a smooth muscle cell
(C) a sinoatrial (SA) cell
(D) an atrial muscle cell
(E) a ventricular muscle cell

39. Phase 0 of the action potential shown is produced by an

(A) inward K^+ current
(B) inward Na^+ current
(C) inward Ca^{2+} current
(D) outward Na^+ current
(E) outward Ca^{2+} current

40. Phase 2, the plateau phase, of the action potential shown

(A) is the result of Ca^{2+} flux out of the cell
(B) increases in duration as heart rate increases
(C) corresponds to the effective refractory period
(D) is the result of approximately equal inward and outward currents
(E) is the portion of the action potential when another action potential can most easily be elicited

41. The action potential shown corresponds to which portion of an electrocardiogram (ECG)?

(A) P wave
(B) PR interval
(C) QRS complex
(D) ST segment
(E) QT interval

42. Which of the following is the *first* step in the biosynthetic pathway for thyroid hormones that is inhibited by propylthiouracil?

(A) Iodide (I^-) pump
(B) $I^- \rightarrow I_2$
(C) I_2 + tyrosine
(D) Diiodotyrosine (DIT) + DIT
(E) Thyroxine (T_4)→triiodothyronine (T_3)

43. Arterial pH of 7.29, arterial [HCO_3^-] of 14 mEq/L, increased urinary excretion of NH_4^+, and hyperventilation would be observed in a

(A) patient with chronic diabetic ketoacidosis
(B) patient with chronic renal failure
(C) patient with chronic emphysema and bronchitis
(D) patient who hyperventilates on a commuter flight
(E) patient who is taking a carbonic anhydrase inhibitor for glaucoma
(F) patient with a pyloric obstruction who vomits for 5 days
(G) healthy person

44. Activation of which of the following receptors increases total peripheral resistance (TPR)?

(A) α_1 Receptor
(B) β_1 Receptor
(C) β_2 Receptor
(D) Muscarinic receptor
(E) Nicotinic receptor

45. The receptor for this hormone has intrinsic tyrosine kinase activity.

(A) Adrenocorticotropic hormone (ACTH)
(B) Antidiuretic hormone (ADH)
(C) Aldosterone
(D) Insulin
(E) Parathyroid hormone (PTH)
(F) Somatostatin
(G) Growth hormone

46. If an artery is partially occluded by an embolism such that its radius becomes one-half the preocclusion value, which of the following parameters will *decrease* by a factor of 16?

(A) Blood flow
(B) Resistance
(C) Pressure gradient
(D) Capacitance

47. Which phase of the cardiac cycle is absent if there is no P wave on the electrocardiogram (ECG)?

(A) Atrial systole
(B) Isovolumetric ventricular contraction
(C) Rapid ventricular ejection
(D) Reduced ventricular ejection
(E) Isovolumetric ventricular relaxation
(F) Rapid ventricular filling
(G) Reduced ventricular filling

48. A receptor potential in the pacinian corpuscle

(A) is all-or-none
(B) has a stereotypical size and shape
(C) is the action potential of this sensory receptor
(D) if hyperpolarizing, increases the likelihood of action potential occurrence
(E) if depolarizing, brings the membrane potential closer to threshold

49. Compared with the base of the lung, in a person who is standing, the apex of the lung has

(A) a higher ventilation rate
(B) a higher perfusion rate
(C) a higher ventilation/perfusion (V/Q) ratio
(D) the same V/Q ratio
(E) a lower pulmonary capillary PO_2

50. A 54-year-old man with a lung tumor has high circulating levels of antidiuretic hormone (ADH), a serum osmolarity of 260 mOsm/L, and a negative free water clearance $\left(C_{H_2O}\right)$. Which diagnosis is correct?

(A) Primary polydipsia
(B) Central diabetes insipidus
(C) Nephrogenic diabetes insipidus
(D) Water deprivation
(E) Syndrome of inappropriate antidiuretic hormone (SIADH)

51. End-organ resistance to which of the following hormones results in polyuria and elevated serum osmolarity?

(A) Antidiuretic hormone (ADH)
(B) Aldosterone
(C) 1,25-Dihydroxycholecalciferol
(D) Parathyroid hormone (PTH)
(E) Somatostatin

52. Which diuretic causes increased urinary excretion of Na^+ and K^+ and decreased urinary excretion of Ca^{2+}?

(A) Acetazolamide
(B) Chlorothiazide
(C) Furosemide
(D) Spironolactone

53. Arterial P_{CO_2} of 72 mm Hg, arterial $[HCO_3^-]$ of 38 mEq/L, and increased H^+ excretion would be observed in a

(A) patient with chronic diabetic ketoacidosis
(B) patient with chronic renal failure
(C) patient with chronic emphysema and bronchitis
(D) patient who hyperventilates on a commuter flight
(E) patient who is taking a carbonic anhydrase inhibitor for glaucoma
(F) patient with a pyloric obstruction who vomits for 5 days
(G) healthy person

54. In a skeletal muscle capillary, the capillary hydrostatic pressure (P_c) is 32 mm Hg, the capillary oncotic pressure (π_c) is 27 mm Hg, and the interstitial hydrostatic pressure (P_i) is 2 mm Hg. Interstitial oncotic pressure (π_i) is negligible. What is the driving force across the capillary wall, and will it favor filtration or absorption?

(A) 3 mm Hg, favoring absorption
(B) 3 mm Hg, favoring filtration
(C) 7 mm Hg, favoring absorption
(D) 7 mm Hg, favoring filtration
(E) 9 mm Hg, favoring filtration

55. Which of the following substances has the lowest renal clearance?

(A) Creatinine
(B) Glucose
(C) K^+
(D) Na^+
(E) Para-aminohippuric acid (PAH)

56. Atropine causes dry mouth by inhibiting which of the following receptors?

(A) α_1 Receptor
(B) β_1 Receptor
(C) β_2 Receptor
(D) Muscarinic receptor
(E) Nicotinic receptor

57. Which of the following transport mechanisms is inhibited by furosemide in the thick ascending limb?

(A) Na^+ diffusion via Na^+ channels
(B) Na^+–glucose cotransport (symport)
(C) Na^+–K^+–$2Cl^-$ cotransport (symport)
(D) Na^+–H^+ exchange (antiport)
(E) Na^+,K^+-adenosine triphosphatase (ATPase)

58. Which of the following conditions decreases the likelihood of edema formation?

(A) Arteriolar constriction
(B) Venous constriction
(C) Standing
(D) Nephrotic syndrome
(E) Inflammation

59. Which of the following conditions causes hypoventilation?

(A) Strenuous exercise
(B) Ascent to high altitude
(C) Anemia
(D) Diabetic ketoacidosis
(E) Chronic obstructive pulmonary disease (COPD)

60. A 28-year-old man who is receiving lithium treatment for bipolar disorder becomes polyuric. His urine osmolarity is 90 mOsm/L; it remains at that level when he is given a nasal spray of dDAVP. Which diagnosis is correct?

(A) Primary polydipsia
(B) Central diabetes insipidus
(C) Nephrogenic diabetes insipidus
(D) Water deprivation
(E) Syndrome of inappropriate antidiuretic hormone (SIADH)

61. Inhibition of which enzyme in the steroid hormone synthetic pathway blocks the production of all androgenic compounds in the adrenal cortex, but not the production of glucocorticoids or mineralocorticoids?

(A) Aldosterone synthase
(B) Aromatase
(C) Cholesterol desmolase
(D) 17,20-Lyase
(E) 5α-Reductase

62. Arterial pH of 7.54, arterial $[HCO_3]$ of 48 mEq/L, hypokalemia, and hypoventilation would be observed in a

(A) patient with chronic diabetic ketoacidosis
(B) patient with chronic renal failure

(C) patient with chronic emphysema and bronchitis
(D) patient who hyperventilates on a commuter flight
(E) patient who is taking a carbonic anhydrase inhibitor for glaucoma
(F) patient with a pyloric obstruction who vomits for 5 days
(G) healthy person

63. Somatostatin inhibits the secretion of which of the following hormones?

(A) Antidiuretic hormone (ADH)
(B) Insulin
(C) Oxytocin
(D) Prolactin
(E) Thyroid hormone

64. Which of the following substances is converted to a more active form after its secretion?

(A) Testosterone
(B) Triiodothyronine (T_3)
(C) Reverse triiodothyronine (rT_3)
(D) Angiotensin II
(E) Aldosterone

65. Levels of which of the following hormones are high during the first trimester of pregnancy and decline during the second and third trimesters?

(A) Adrenocorticotropic hormone (ACTH)
(B) Estradiol
(C) Follicle-stimulating hormone (FSH)
(D) Gonadotropin-releasing hormone (GnRH)
(E) Human chorionic gonadotropin (HCG)
(F) Oxytocin
(G) Prolactin
(H) Testosterone

THE FOLLOWING DIAGRAM APPLIES TO QUESTIONS 66 AND 67

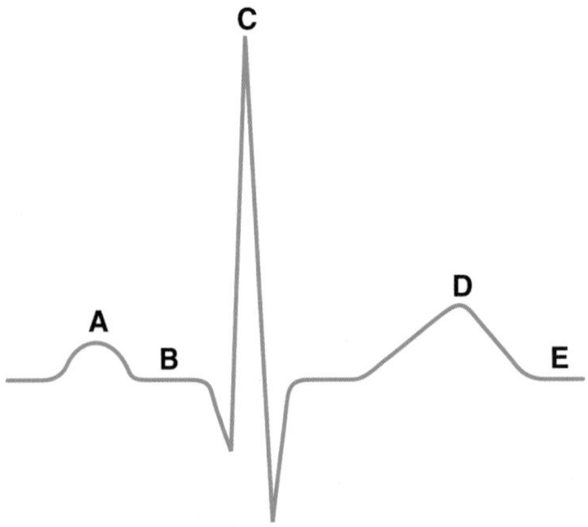

66. During which labeled wave or segment of the electrocardiogram (ECG) are both the atria and the ventricles completely repolarized?

(A) A
(B) B
(C) C
(D) D
(E) E

67. During which labeled wave or segment of the electrocardiogram (ECG) is aortic pressure at its lowest value?

(A) A
(B) B
(C) C
(D) D
(E) E

THE FOLLOWING DIAGRAM APPLIES TO QUESTIONS 68–74

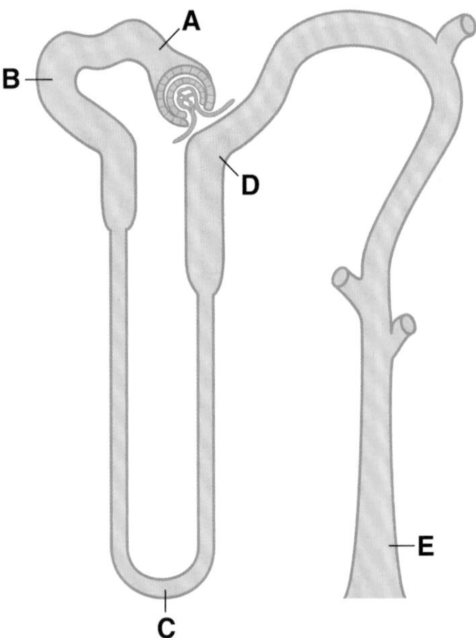

68. At which site is the amount of para-aminohippuric acid (PAH) in tubular fluid lowest?

(A) Site A
(B) Site B
(C) Site C
(D) Site D
(E) Site E

69. At which site is the creatinine concentration highest in a person who is deprived of water?

(A) Site A
(B) Site B
(C) Site C
(D) Site D
(E) Site E

70. At which site is the tubular fluid [HCO_3^-] highest?

(A) Site A
(B) Site B
(C) Site C
(D) Site D
(E) Site E

71. At which site is the amount of K^+ in tubular fluid lowest in a person who is on a very low-K^+ diet?

(A) Site A
(B) Site B
(C) Site C
(D) Site D
(E) Site E

72. At which site is the composition of tubular fluid closest to that of plasma?

(A) Site A
(B) Site B
(C) Site C
(D) Site D
(E) Site E

73. At which site is about one-third of the filtered water remaining in the tubular fluid?

(A) Site A
(B) Site B
(C) Site C
(D) Site D
(E) Site E

74. At which site is the tubular fluid osmolarity lower than the plasma osmolarity in a person who is deprived of water?

(A) Site A
(B) Site B
(C) Site C
(D) Site D
(E) Site E

75. A patient's electrocardiogram (ECG) shows periodic QRS complexes that are not preceded by P waves and that have a bizarre shape. These QRS complexes originated in the

(A) sinoatrial (SA) node
(B) atrioventricular (AV) node
(C) His–Purkinje system
(D) ventricular muscle

76. Which of the following substances would be expected to cause an increase in arterial blood pressure?

(A) Saralasin
(B) V_1 agonist
(C) Acetylcholine (ACh)
(D) Spironolactone
(E) Phenoxybenzamine

77. A decrease in which of the following parameters in an artery will produce an increase in pulse pressure?

(A) Blood flow
(B) Resistance
(C) Pressure gradient
(D) Capacitance

78. Which of the following changes occurs during moderate exercise?

(A) Increased total peripheral resistance (TPR)
(B) Increased stroke volume
(C) Decreased pulse pressure
(D) Decreased venous return
(E) Decreased arterial PO_2

79. Plasma renin activity is lower than normal in patients with

(A) hemorrhagic shock
(B) essential hypertension
(C) congestive heart failure
(D) hypertension caused by aortic constriction above the renal arteries

80. Inhibition of which enzyme in the steroid hormone synthetic pathway reduces the size of the prostate?

(A) Aldosterone synthase
(B) Aromatase
(C) Cholesterol desmolase
(D) 17,20-Lyase
(E) 5α-Reductase

81. During which phase of the cardiac cycle does ventricular pressure rise but ventricular volume remain constant?

(A) Atrial systole
(B) Isovolumetric ventricular contraction
(C) Rapid ventricular ejection
(D) Reduced ventricular ejection
(E) Isovolumetric ventricular relaxation
(F) Rapid ventricular filling
(G) Reduced ventricular filling

82. Which of the following lung volumes or capacities includes the residual volume?

(A) Tidal volume (TV)
(B) Vital capacity (VC)
(C) Inspiratory capacity (IC)
(D) Functional residual capacity (FRC)
(E) Inspiratory reserve volume (IRV)

83. Arterial [HCO_3^-] of 18 mEq/L, P_{CO_2} of 34 mm Hg, and increased urinary HCO_3^- excretion would be observed in a

(A) patient with chronic diabetic ketoacidosis
(B) patient with chronic renal failure
(C) patient with chronic emphysema and bronchitis
(D) patient who hyperventilates on a commuter flight
(E) patient who is taking a carbonic anhydrase inhibitor for glaucoma
(F) patient with a pyloric obstruction who vomits for 5 days
(G) healthy person

84. A 36-year-old woman with galactorrhea is treated with bromocriptine. The basis for bromocriptine's action is by acting as an agonist for

(A) dopamine
(B) estradiol
(C) follicle-stimulating hormone (FSH)
(D) gonadotropin-releasing hormone (GnRH)
(E) human chorionic gonadotropin (HCG)
(F) oxytocin
(G) prolactin

85. A 32-year-old woman who is thirsty has a urine osmolarity of 950 mOsm/L and a serum osmolarity of 297 mOsm/L. Which diagnosis is correct?

(A) Primary polydipsia
(B) Central diabetes insipidus
(C) Nephrogenic diabetes insipidus
(D) Water deprivation
(E) Syndrome of inappropriate antidiuretic hormone (SIADH)

86. Hypoxia causes vasoconstriction in which of the following vascular beds?

(A) Cerebral
(B) Coronary
(C) Muscle
(D) Pulmonary
(E) Skin

87. A new diuretic is developed for the treatment of acute mountain sickness and causes an increase in the pH of urine. This diuretic is in the same class as

(A) acetazolamide
(B) chlorothiazide
(C) furosemide
(D) spironolactone

88. Arterial pH of 7.25, arterial P_{CO_2} of 30 mm Hg, and decreased urinary excretion of NH_4^+ would be observed in a

(A) patient with chronic diabetic ketoacidosis
(B) patient with chronic renal failure
(C) patient with chronic emphysema and bronchitis
(D) patient who hyperventilates on a commuter flight
(E) patient who is taking a carbonic anhydrase inhibitor for glaucoma
(F) patient with a pyloric obstruction who vomits for 5 days
(G) healthy person

89. In which of the following situations will arterial PO_2 be closest to 100 mm Hg?

(A) A person who is having a severe asthmatic attack
(B) A person who lives at high altitude
(C) A person who has a right-to-left cardiac shunt
(D) A person who has a left-to-right cardiac shunt
(E) A person who has pulmonary fibrosis

90. Which of the following is an example of a primary active transport process?

(A) Na^+–glucose transport in small intestinal epithelial cells
(B) Na^+–alanine transport in renal proximal tubular cells
(C) Insulin-dependent glucose transport in muscle cells
(D) H^+–K^+ transport in gastric parietal cells
(E) Na^+–Ca^{2+} exchange in nerve cells

91. Which gastrointestinal secretion is inhibited when the pH of the stomach contents is 1.0?

(A) Saliva
(B) Gastric secretion
(C) Pancreatic secretion
(D) Bile

92. Which of the following would be expected to increase after surgical removal of the duodenum?

(A) Gastric emptying
(B) Secretion of cholecystokinin (CCK)
(C) Secretion of secretin
(D) Contraction of the gallbladder
(E) Absorption of lipids

93. Which of the following hormones causes contraction of vascular smooth muscle?

(A) Antidiuretic hormone (ADH)
(B) Aldosterone
(C) Atrial natriuretic peptide (ANP)
(D) 1,25-Dihydroxycholecalciferol
(E) Parathyroid hormone (PTH)

94. Which of the following is absorbed by facilitated diffusion?

(A) Glucose in duodenal cells
(B) Fructose in duodenal cells
(C) Dipeptides in duodenal cells
(D) Vitamin B_1 in duodenal cells
(E) Cholesterol in duodenal cells
(F) Bile acids in ileal cells

95. Which of the following hormones acts on the anterior lobe of the pituitary to inhibit secretion of growth hormone?

(A) Dopamine
(B) Gonadotropin-releasing hormone (GnRH)
(C) Insulin
(D) Prolactin
(E) Somatostatin

96. Which enzyme in the steroid hormone synthetic pathway is required for the development of female secondary sex characteristics, but not male secondary sex characteristics?

(A) Aldosterone synthase
(B) Aromatase
(C) Cholesterol desmolase
(D) 17,20-Lyase
(E) 5α-Reductase

97. At the beginning of which phase of the cardiac cycle does the second heart sound occur?

(A) Atrial systole
(B) Isovolumetric ventricular contraction
(C) Rapid ventricular ejection
(D) Reduced ventricular ejection
(E) Isovolumetric ventricular relaxation
(F) Rapid ventricular filling
(G) Reduced ventricular filling

98. Which of the following actions occurs when light strikes a photoreceptor cell of the retina?

(A) Transducin is inhibited
(B) The photoreceptor depolarizes
(C) Cyclic guanosine monophosphate (cGMP) levels in the cell decrease
(D) All-*trans* retinal is converted to 11-*cis* retinal
(E) Increased release of glutamate

99. Which step in the biosynthetic pathway for thyroid hormones produces thyroxine (T_4)?

(A) Iodide (I^-) pump
(B) $I^- \rightarrow I_2$
(C) I_2 + tyrosine
(D) Diiodotyrosine (DIT) + DIT
(E) DIT + monoiodotyrosine (MIT)

100. A 44-year-old woman is diagnosed with central diabetes insipidus following a head injury. Which of the following sets of values is consistent with her disorder?

	Urine osmolarity	Serum osmolarity	Serum ADH
(A)	↑	↑	↑
(B)	↓	↓	↓
(C)	↓	↑	Normal
(D)	↑	↓	↓
(E)	↓	↑	↓

101. A 58-year-old man with lung cancer is found to have hypercalcemia and is diagnosed with humoral hypercalcemia of malignancy. Which of the following sets of values is consistent with his disorder?

	Bone resorption	Serum phosphate	Serum PTH
(A)	↑	↑	↑
(B)	↑	↓	↑
(C)	↑	↓	↓
(D)	↓	↑	↓
(E)	↓	↓	↑

Answers and Explanations

1. **The answer is D** [Chapter 2, I C; Table 2.2]. Increased circulating levels of epinephrine from the adrenal medullary tumor stimulate both α-adrenergic and β-adrenergic receptors. Thus, heart rate and contractility are increased and, as a result, cardiac output is increased. Total peripheral resistance (TPR) is increased because of arteriolar vasoconstriction, which leads to decreased blood flow to the cutaneous circulation and causes cold, clammy skin. Together, the increases in cardiac output and TPR increase arterial blood pressure. 3-Methoxy-4-hydroxymandelic acid (VMA) is a metabolite of both norepinephrine and epinephrine; increased VMA excretion occurs in pheochromocytomas.
2. **The answer is D** [Chapter 2, I; Table 2.3]. Treatment is directed at blocking both the α-stimulatory and β-stimulatory effects of catecholamines. Phentolamine is an α-blocking agent; propranolol is a β-blocking agent. Isoproterenol is a β_1 and β_2 agonist. Phenylephrine is an α_1 agonist.
3. **The answer is C** [Chapter 7, I D; X E 2]. The effect of estrogen on the secretion of follicle-stimulating hormone (FSH) and luteinizing hormone (LH) by the anterior lobe of the pituitary gland at midcycle is one of the few examples of positive feedback in physiologic systems—increasing estrogen levels at midcycle cause *increased* secretion of FSH and LH. The other options illustrate negative feedback. Decreased arterial PO_2 causes an increase in breathing rate (via peripheral chemoreceptors). Increased blood glucose stimulates insulin secretion. Decreased blood [Ca^{2+}] causes an increase in parathyroid hormone (PTH) secretion. Decreased blood pressure decreases the firing rate of carotid sinus nerves (via the baroreceptors) and ultimately increases sympathetic outflow to the heart and blood vessels to return blood pressure to normal.
4. **The answer is B** [Chapter 3, IV F 3 a; Figures 3.8 and 3.12]. A downward shift of the cardiac output curve is consistent with decreased myocardial contractility (negative inotropism); for any right atrial pressure or end-diastolic volume, the force of contraction is decreased. Digitalis, a positive inotropic agent, would produce an upward shift of the cardiac output curve. Changes in blood volume alter the venous return curve rather than the cardiac output curve. Changes in total peripheral resistance (TPR) alter both the cardiac output and venous return curves.
5. **The answer is A** [Chapter 4, IV A 2, C; Figure 4.7]. Because fetal hemoglobin (HbF) has a greater affinity for O_2 than does adult hemoglobin, the O_2–hemoglobin dissociation curve would *shift to the left*. Carbon monoxide poisoning would cause a shift to the left but would also cause a decrease in total O_2-carrying capacity (decreased percent saturation) because CO occupies O_2-binding sites. Decreased pH, increased temperature, and increased 2,3-diphosphoglycerate (DPG) all would shift the curve to the right.
6. **The answer is A** [Chapter 4, IV C 2]. A shift to the left of the O_2–hemoglobin dissociation curve represents an increased affinity of hemoglobin for O_2. Accordingly, at any given level of PO_2, the percent saturation is increased, the P_{50} is decreased (read the PO_2 at 50% saturation), and the ability to unload O_2 to the tissues is impaired (because of the higher affinity of hemoglobin for O_2). The O_2-carrying capacity is determined by hemoglobin concentration and is unaffected by the shift from curve A to curve B.
7. **The answer is B** [Chapter 5, VII D; Table 5.6]. A person with a negative free-water clearance $\left(C_{H_2O}\right)$ would, by definition, be producing urine that is hyperosmotic to blood $\left(C_{H_2O} = V - C_{osm}\right)$. After overnight water restriction, serum osmolarity increases. This increase, via hypothalamic osmoreceptors, stimulates the release of antidiuretic hormone (ADH) from the posterior lobe of the pituitary. This ADH circulates to the collecting ducts of the kidney and causes reabsorption of water, which results in the production of

hyperosmotic urine. Drinking large amounts of water inhibits the secretion of ADH and causes excretion of dilute urine and a positive C_{H_2O}. Lithium causes nephrogenic diabetes insipidus by blocking the response of ADH on the collecting duct cells, resulting in dilute urine and a positive C_{H_2O}. In Option D, the calculated value of C_{H_2O} is zero. In Option E, the calculated value of C_{H_2O} is positive.

8. **The answer is C** [Chapter 4, V B; Figure 4.9]. CO_2 generated in the tissues enters venous blood and, in the red blood cells (RBCs), combines with H_2O in the presence of carbonic anhydrase to form H_2CO_3. H_2CO_3 dissociates into H^+ and HCO_3^-. The H^+ remains in the RBCs to be buffered by deoxyhemoglobin, and the HCO_3^- moves into plasma in exchange for Cl^-. Thus, CO_2 is carried in venous blood to the lungs as HCO_3^-. In the lungs, the reactions occur in reverse: CO_2 is regenerated and expired.

9. **The answer is D** [Chapter 7, X E 2]. Menses occurs 14 days after ovulation, regardless of cycle length. Therefore, in a 35-day menstrual cycle, ovulation occurs on day 21. Ovulation occurs at the midpoint of the menstrual cycle only if the cycle length is 28 days.

10. **The answer is C** [Chapter 7, X A]. Testosterone is synthesized from cholesterol in ovarian theca cells and diffuses to ovarian granulosa cells, where it is converted to estradiol by the action of aromatase. Follicle-stimulating hormone (FSH) stimulates the aromatase enzyme and increases the production of estradiol.

11. **The answer is A** [Chapter 6, IV A 2–4 a]. Saliva has a high [HCO_3^-] because the cells lining the salivary ducts secrete HCO_3^-. Because the ductal cells are relatively impermeable to water and because they reabsorb more solute (Na^+ and Cl^-) than they secrete (K^+ and HCO_3^-), the saliva is rendered hypotonic. Vagal stimulation increases saliva production, so vagotomy (or atropine) inhibits it and produces dry mouth.

12. **The answer is C** [Chapter 5, VII D 3; Table 5.6]. The most likely explanation for this patient's polyuria is hypercalcemia. With severe hypercalcemia, Ca^{2+} accumulates in the inner medulla and papilla of the kidney and inhibits adenylate cyclase, blocking the effect of ADH on water permeability. Because ADH is ineffective, the urine cannot be concentrated and the patient excretes large volumes of dilute urine. His polydipsia is secondary to his polyuria and is caused by the increased serum osmolarity. Psychogenic water drinking would also cause polyuria, but the serum osmolarity would be lower than normal, not higher than normal.

13. **The answer is A** [Chapter 5, VI C]. Thiazide diuretics would be contraindicated in a patient with severe hypercalcemia because these drugs cause increased Ca^{2+} reabsorption in the renal distal tubule. On the other hand, loop diuretics inhibit Ca^{2+} and Na^+ reabsorption and produce calciuresis. When given with fluid replacement, loop diuretics can effectively and rapidly lower the serum [Ca^{2+}]. Calcitonin, mithramycin, and etidronate disodium inhibit bone resorption and, as a result, decrease serum [Ca^{2+}].

14. **The answer is B** [Chapter 7; Table 7.2]. Oxytocin causes contraction of the myoepithelial cells of the breast by an inositol 1,4,5-triphosphate (IP_3)–Ca^{2+} mechanism. Somatomedins (insulin-like growth factor [IGF]), like insulin, act on target cells by activating tyrosine kinase. Antidiuretic hormone (ADH) acts on the V_2 receptors of the renal collecting duct by a cyclic adenosine monophosphate (cAMP) mechanism (although in vascular smooth muscle it acts on V_1 receptors by an IP_3 mechanism).

    Adrenocorticotropic hormone (ACTH) also acts via a cAMP mechanism. Thyroid hormone induces the synthesis of new protein (e.g., Na^+,K^+-adenosine triphosphatase [ATPase]) by a steroid hormone mechanism.

15. **The answer is E** [Chapter 1, VI B; VII B; Table 1.3]. The pharynx is skeletal muscle, and the small intestine is unitary smooth muscle. The difference between smooth and skeletal muscle is the mechanism by which Ca^{2+} initiates contraction. In smooth muscle, Ca^{2+} binds to calmodulin, and in skeletal muscle, Ca^{2+} binds to troponin C. Both types of muscle are excited to contract by action potentials. Slow waves are present in smooth muscle but not skeletal muscle. Both smooth and skeletal muscle require an increase in intracellular [Ca^{2+}] as the important linkage between excitation (the action potential) and contraction, and both consume adenosine triphosphate (ATP) during contraction.

16. **The answer is B** [Chapter 5, IX D; Table 5.9]. The arterial blood values and physical findings are consistent with metabolic acidosis, hypokalemia, and orthostatic hypotension. Diarrhea is associated with the loss of HCO_3^- and K^+ from the gastrointestinal (GI) tract, consistent with the laboratory values. Hypotension is consistent with extracellular fluid (ECF) volume contraction. Vomiting would cause metabolic alkalosis and hypokalemia. Treatment with loop or thiazide diuretics *could* cause volume contraction and hypokalemia but *would* cause metabolic alkalosis rather than metabolic acidosis.

17. **The answer is D** [Chapter 6, V B 1 c]. Pepsinogen is secreted by the gastric chief cells and is activated to pepsin by the low pH of the stomach (created by secretion of HCl by the gastric parietal cells). Lipases are *inactivated* by low pH.

18. **The answer is B** [Chapter 5, II C 6; Table 5.3]. Glomerular filtration rate (GFR) is determined by the balance of Starling forces across the glomerular capillary wall. Constriction of the efferent arteriole increases the glomerular capillary hydrostatic pressure (because blood is restricted in leaving the glomerular capillary), thus favoring filtration. Constriction of the afferent arteriole would have the opposite effect and would reduce the glomerular capillary hydrostatic pressure. Constriction of the ureter would increase the hydrostatic pressure in the tubule and, therefore, oppose filtration. Increased plasma protein concentration would increase the glomerular capillary oncotic pressure and oppose filtration. Infusion of inulin is used to measure the GFR and does not alter the Starling forces.

19. **The answer is B** [Chapter 6, V C 1, 2]. Vitamin D is fat-soluble and absorbed along with dietary fat. First, fat absorption requires the breakdown of dietary lipids to fatty acids, monoglycerides, and cholesterol in the duodenum by pancreatic lipases. Second, fat absorption requires the presence of bile acids, which are secreted into the small intestine by the gallbladder. These bile acids form micelles around the products of lipid digestion and deliver them to the absorbing surface of the small intestinal cells. Because the bile acids are recirculated to the liver from the ileum, fat absorption must be complete before the chyme reaches the terminal ileum.

20. **The answer is A** [Chapter 7, III C 1 b]. Antidiuretic hormone (ADH) causes constriction of vascular smooth muscle by activating a V_1 receptor that uses the inositol 1,4,5-triphosphate (IP_3) and Ca^{2+} second messenger system. When hemorrhage or extracellular fluid (ECF) volume contraction occurs, ADH secretion by the posterior pituitary is stimulated via volume receptors. The resulting increase in ADH levels causes increased water reabsorption by the collecting ducts (V_2 receptors) and vasoconstriction (V_1 receptors) to help restore blood pressure.

21. **The answer is A** [Chapter 7, III B]. Normal menstrual cycles depend on the secretion of follicle-stimulating hormone (FSH) and luteinizing hormone (LH) from the anterior pituitary. Concentration of urine in response to water deprivation depends on the secretion of antidiuretic hormone (ADH) by the posterior pituitary. Catecholamines are secreted by the adrenal medulla in response to stress, but anterior pituitary hormones are not involved. Anterior pituitary hormones are not involved in the direct effect of glucose on the beta cells of the pancreas or in the direct effect of Ca^{2+} on the chief cells of the parathyroid gland.

22. **The answer is B** [Chapter 5, III B]. Curves X, Y, and Z show glucose filtration, glucose excretion, and glucose reabsorption, respectively. Below a plasma [glucose] of 200 mg/dL, the carriers for glucose reabsorption are unsaturated, so all of the filtered glucose can be reabsorbed, and none will be excreted in the urine.

23. **The answer is D** [Chapter 2, III C 1; Figure 2.9]. When the patellar tendon is stretched, the quadriceps muscle also stretches. This movement activates Ia afferent fibers of the muscle spindles, which are arranged in parallel formation in the muscle. These Ia afferent fibers form synapses on α-motoneurons in the spinal cord. In turn, the pool of α-motoneurons is activated and causes reflex contraction of the quadriceps muscle to return it to its resting length.

24. **The answer is A** [Chapter 2, VI C]. *Streptococcus pyogenes* causes increased production of interleukin-1 (IL-1) in macrophages. IL-1 acts on the anterior hypothalamus to increase

the production of prostaglandins, which increase the hypothalamic set-point temperature. The hypothalamus then "reads" the core temperature as being lower than the new set-point temperature and activates various heat-generating mechanisms that increase body temperature (fever). These mechanisms include shivering and vasoconstriction of blood vessels in the skin.

25. The answer is C [Chapter 2, VI C 2]. By inhibiting cyclooxygenase, aspirin inhibits the production of prostaglandins and lowers the hypothalamic set-point temperature to its original value. After aspirin treatment, the hypothalamus "reads" the body temperature as being higher than the set-point temperature and activates heat-loss mechanisms, including sweating and vasodilation of skin blood vessels. This vasodilation shunts blood toward the surface skin. When heat is lost from the body by these mechanisms, body temperature is reduced.

26. The answer is D [Chapter 5, IX D 4; Table 5.9]. The blood values are consistent with acute respiratory alkalosis from hysterical hyperventilation. The tingling and numbness are symptoms of a reduction in serum ionized [Ca^{2+}] that occurs secondary to alkalosis. Because of the reduction in [H^+], fewer H^+ ions will bind to negatively charged sites on plasma proteins, and more Ca^{2+} binds (decreasing the free ionized [Ca^{2+}]).

27. The answer is C [Chapter 2, I C 1 d]. Albuterol is an adrenergic β_2 agonist. When activated, the β_2 receptors in the bronchioles produce bronchodilation.

28. The answer is E [Chapter 7, IX A; Figure 7.16]. Testosterone is converted to its active form, dihydrotestosterone, in some target tissues by the action of 5α-reductase.

29. The answer is B [Chapter 3, II C, D]. A decrease in radius causes an increase in resistance, as described by the Poiseuille relationship (resistance is inversely proportional to r^4). Thus, if radius decreases twofold, the resistance will increase by $(2)^4$ or 16-fold.

30. The answer is G [Chapter 3, V; Figure 3.15]. When heart rate increases, the time between ventricular contractions (for refilling of the ventricles with blood) decreases. Because most ventricular filling occurs during the "reduced" phase, this phase is the most compromised by an increase in heart rate.

31. The answer is D [Chapter 3, IX C; Table 3.6; Figure 3.21]. The blood loss that occurred in the accident caused a decrease in arterial blood pressure. The decrease in arterial pressure was detected by the baroreceptors in the carotid sinus and caused a decrease in the firing rate of the carotid sinus nerves. As a result of the baroreceptor response, sympathetic outflow to the heart and blood vessels increased, and parasympathetic outflow to the heart decreased. Together, these changes caused an increased heart rate, increased contractility, and increased total peripheral resistance (TPR) (in an attempt to restore the arterial blood pressure).

32. The answer is B [Chapter 3, IX C; Table 3.6; Figure 3.21; Chapter 5 IV C 3 b (1)]. The decreased blood volume causes decreased renal perfusion pressure, which initiates a cascade of events, including increased renin secretion, increased circulating angiotensin II, increased aldosterone secretion, increased Na^+ reabsorption, and increased K^+ secretion by the renal tubule.

33. The answer is B [Chapter 5, VII C; Table 5.6]. A history of head injury with production of dilute urine accompanied by elevated serum osmolarity suggests central diabetes insipidus. The response of the kidney to exogenous antidiuretic hormone (ADH) (1-deamino-8-D-arginine vasopressin [dDAVP]) eliminates nephrogenic diabetes insipidus as the cause of the concentrating defect.

34. The answer is D [Chapter 5, IV C 3 b (1); Table 5.11]. Spironolactone inhibits distal tubule Na^+ reabsorption and K^+ secretion by acting as an aldosterone antagonist.

35. The answer is B [Chapter 6, V E 1 c; Table 6.3]. Gastric parietal cells secrete intrinsic factor, which is required for the intestinal absorption of vitamin B_{12}.

36. The answer is C [Chapter 3, VI C 4]. Atrial natriuretic peptide (ANP) is secreted by the atria in response to extracellular fluid volume expansion and subsequently acts on the kidney to cause increased excretion of Na^+ and H_2O.

37. **The answer is A** [Chapter 7, V A 2 b; Figure 7.11]. Angiotensin II increases production of aldosterone by stimulating aldosterone synthase, the enzyme that catalyzes the conversion of corticosterone to aldosterone.

38. **The answer is E** [Chapter 3, III B; Figures 3.4 and 3.5]. The action potential shown is characteristic of ventricular muscle, with a stableresting membrane potential and a long plateau phase of almost 300 msec. Action potentials in skeletal cells are much shorter (only a few milliseconds). Smooth muscle action potentials would be superimposed on fluctuating baseline potentials (slow waves). Sinoatrial (SA) cells of the heart have spontaneous depolarization (pacemaker activity) rather than a stableresting potential. Atrial muscle cells of the heart have a much shorter plateau phase and a much shorter overall duration.

39. **The answer is B** [Chapter 3, III C 1 a]. Depolarization, as in phase 0, is caused by an inward current (defined as the movement of positive charge into the cell). The inward current during phase 0 of the ventricular muscle action potential is caused by opening of Na^+ channels in the ventricular muscle cell membrane, movement of Na^+ into the cell, and depolarization of the membrane potential toward the Na^+ equilibrium potential (approximately +65 mV). In sinoatrial (SA) cells, phase 0 is caused by an inward Ca^{2+} current.

40. **The answer is D** [Chapter 3, III B 1 c]. Because the plateau phase is a period of stable-membrane potential, by definition, the inward and outward currents are equal and balance each other. Phase 2 is the result of opening of Ca^{2+} channels and inward, not outward, Ca^{2+} current. In this phase, the cells are refractory to the initiation of another action potential. Phase 2 corresponds to the absolute refractory period, rather than the effective refractory period (which is longer than the plateau). As heart rate increases, the duration of the ventricular action potential decreases, primarily by decreasing the duration of phase 2.

41. **The answer is E** [Chapter 3, III A 4; Figure 3.3]. The action potential shown represents both depolarization and repolarization of a ventricular muscle cell. Therefore, on an electrocardiogram (ECG), it corresponds to the period of depolarization (beginning with the Q wave) through repolarization (completion of the T wave). That period is defined as the QT interval.

42. **The answer is B** [Chapter 7, IV A 2]. The oxidation of I^- to I_2 is catalyzed by peroxidase and inhibited by propylthiouracil, which can be used in the treatment of hyperthyroidism. Later steps in the pathway that are catalyzed by peroxidase and inhibited by propylthiouracil are iodination of tyrosine, coupling of diiodotyrosine (DIT) and DIT, and coupling of DIT and monoiodotyrosine (MIT).

43. **The answer is A** [Chapter 5, IX D 1; Table 5.9]. The blood values are consistent with metabolic acidosis, as would occur in diabetic ketoacidosis. Hyperventilation is the respiratory compensation for metabolic acidosis. Increased urinary excretion of NH_4^+ reflects the adaptive increase in NH_3 synthesis that occurs in chronic acidosis. Patients with metabolic acidosis secondary to chronic renal failure would have reduced NH_4^+ excretion (because of diseased renal tissue).

44. **The answer is A** [Chapter 2, I C 1 a]. When adrenergic α_1 receptors on the vascular smooth muscle are activated, they cause vasoconstriction and increased total peripheral resistance (TPR).

45. **The answer is D** [Chapter 7; Table 7.2]. Hormone receptors with intrinsic tyrosine kinase activity include those for insulin and for insulin-like growth factors (IGFs). The β subunits of the insulin receptor have tyrosine kinase activity and, when activated by insulin, the receptors autophosphorylate. The phosphorylated receptors then phosphorylase intracellular proteins; this process ultimately results in the physiologic actions of insulin. Growth hormone utilizes tyrosine kinase-associated receptors.

46. **The answer is A** [Chapter 3, II C, D]. Blood flow through the artery is proportional to the pressure difference and inversely proportional to the resistance ($Q = \Delta P/R$). Because resistance increased 16-fold when the radius decreased twofold, blood flow must decrease 16-fold.

47. The answer is A [Chapter 3, V; Figure 3.15]. The P wave represents electrical activation (depolarization) of the atria. Atrial contraction is always preceded by electrical activation.

48. The answer is E [Chapter 2, II A 4; Figure 2.2]. Receptor potentials in sensory receptors (such as the pacinian corpuscle) are not action potentials and therefore do not have the stereotypical size and shape or the all-or-none feature of the action potential. Instead, they are graded potentials that vary in size depending on the stimulus intensity. A hyperpolarizing receptor potential would take the membrane potential away from threshold and decrease the likelihood of action potential occurrence. A depolarizing receptor potential would bring the membrane potential toward threshold and increase the likelihood of action potential occurrence.

49. The answer is C [Chapter 4, VII C; Table 4.5]. In a person who is standing, both ventilation and perfusion are greater at the base of the lung than at the apex. However, because the regional differences for perfusion are greater than those for ventilation, the ventilation/perfusion (V/Q) ratio is higher at the apex than at the base. The pulmonary capillary Po_2 therefore is higher at the apex than at the base because the higher V/Q ratio makes gas exchange more efficient.

50. The answer is E [Chapter 5, VII D 4]. A negative value for free-water clearance $\left(C_{H_2O}\right)$ means that "free water" (generated in the diluting segments of the thick ascending limb and early distal tubule) is reabsorbed by the collecting ducts. A negative C_{H_2O} is consistent with high circulating levels of antidiuretic hormone (ADH). Because ADH levels are high at a time when the serum is very dilute, ADH has been secreted "inappropriately" by the lung tumor.

51. The answer is A [Chapter 5, VII C; Table 5.6]. End-organ resistance to antidiuretic hormone (ADH) is called nephrogenic diabetes insipidus. It may be caused by lithium intoxication (which inhibits the G_s protein in collecting duct cells) or by hypercalcemia (which inhibits adenylate cyclase). The result is inability to concentrate the urine, polyuria, and increased serum osmolarity (resulting from the loss of free water in the urine).

52. The answer is B [Chapter 5, IV C 3 a; VI C 2; Table 5.11]. Thiazide diuretics act on the early distal tubule (cortical diluting segment) to inhibit Na^+ reabsorption. At the same site, they enhance Ca^{2+} reabsorption so that urinary excretion of Na^+ is increased while urinary excretion of Ca^{2+} is decreased. K^+ excretion is increased because the flow rate is increased at the site of distal tubular K^+ secretion.

53. The answer is C [Chapter 5, IX D 3; Table 5.9]. The blood values are consistent with respiratory acidosis with renal compensation. The renal compensation involves increased reabsorption of HCO_3^- (associated with increased H^+ secretion), which raises the serum [HCO_3^-].

54. The answer is B [Chapter 3, VII C]. The driving force is calculated from the Starling forces across the capillary wall. The net pressure = $(P_c - P_i) - (\pi_c - \pi_i)$. Therefore, net pressure = (32 mm Hg − 2 mm Hg) − (27 mm Hg) = +3 mm Hg. Because the sign of the net pressure is positive, filtration is favored.

55. The answer is B [Chapter 5, III D]. Glucose has the lowest renal clearance of the substances listed, because at normal blood concentrations, it is filtered and completely reabsorbed. Na^+ is also extensively reabsorbed, and only a fraction of the filtered Na^+ is excreted. K^+ is reabsorbed but also secreted. Creatinine, once filtered, is not reabsorbed at all. Para-aminohippuric acid (PAH) is filtered and secreted; therefore, it has the highest renal clearance of the substances listed.

56. The answer is D [Chapter 2, I C 2 b]. Atropine blocks cholinergic muscarinic receptors. Because saliva production is increased by stimulation of the parasympathetic nervous system, atropine treatment reduces saliva production and causes dry mouth.

57. The answer is C [Chapter 5, IV C 2]. Na^+–K^+–$2Cl^-$ cotransport is the mechanism in the luminal membrane of the thick ascending limb cells that is inhibited by loop diuretics such as furosemide. Other loop diuretics that inhibit this transporter are bumetanide and ethacrynic acid.

58. The answer is A [Chapter 3, VII C; Table 3.2]. Constriction of arterioles causes decreased capillary hydrostatic pressure and, as a result, decreased net pressure (Starling forces) across the capillary wall; filtration is reduced, as is the tendency for edema. Venous constriction and standing cause increased capillary hydrostatic pressure and tend to cause increased filtration and edema. Nephrotic syndrome results in the excretion of plasma proteins in the urine and a decrease in the oncotic pressure of capillary blood, which also leads to increased filtration and edema. Inflammation causes local edema by dilating arterioles.

59. The answer is E [Chapter 4, IX A, B; Chapter 5 IX D]. Chronic obstructive pulmonary disease (COPD) causes hypoventilation. Strenuous exercise increases the ventilation rate to provide additional oxygen to the exercising muscle. Ascent to high altitude and anemia cause hypoxemia, which subsequently causes hyperventilation by stimulating peripheral chemoreceptors. The respiratory compensation for diabetic ketoacidosis is hyperventilation.

60. The answer is C [Chapter 5, VII C]. Lithium inhibits the G protein that couples the antidiuretic hormone (ADH) receptor to adenylate cyclase. The result is inability to concentrate the urine. Because the defect is in the target tissue for ADH (nephrogenic diabetes insipidus), exogenous ADH administered by nasal spray will not correct it.

61. The answer is D [Chapter 7, V A 1; Figure 7.11]. 17,20-Lyase catalyzes the conversion of glucocorticoids to the androgenic compounds dehydroepiandrosterone and androstenedione. These androgenic compounds are the precursors of testosterone in both the adrenal cortex and the testicular Leydig cells.

62. The answer is F [Chapter 5, IX D 2; Table 5.9]. The blood values and history of vomiting are consistent with metabolic alkalosis. Hypoventilation is the respiratory compensation for metabolic alkalosis. Hypokalemia results from the loss of gastric K^+ and from hyperaldosteronism (resulting in increased renal K^+ secretion) secondary to volume contraction.

63. The answer is B [Chapter 6, II B 1; Chapter 7 III B 3 a (1), VI D]. The actions of somatostatin are diverse. It is secreted by the hypothalamus to inhibit the secretion of growth hormone by the anterior lobe of the pituitary. It is secreted by cells of the gastrointestinal (GI) tract to inhibit the secretion of the GI hormones. It is also secreted by the delta cells of the endocrine pancreas and, via paracrine mechanisms, inhibits the secretion of insulin and glucagon by the beta cells and alpha cells, respectively. Prolactin secretion is inhibited by a different hypothalamic hormone, dopamine.

64. The answer is A [Chapter 7, IX A; Figure 7.16]. Testosterone is converted to a more active form (dihydrotestosterone) in some target tissues. Triiodothyronine (T_3) is the active form of thyroid hormone; reverse triiodothyronine (rT_3) is an inactive alternative form of T_3. Angiotensin I is converted to its active form, angiotensin II, by the action of angiotensin-converting enzyme (ACE). Aldosterone is unchanged after it is secreted by the zona glomerulosa of the adrenal cortex.

65. The answer is E [Chapter 7, X F 2; Figure 7.20]. During the first trimester of pregnancy, the placenta produces human chorionic gonadotropin (HCG), which stimulates estrogen and progesterone production by the corpus luteum. Peak levels of HCG occur at about the 9th gestational week and then decline. At the time of the decline in HCG, the placenta assumes the responsibility for steroidogenesis for the remainder of the pregnancy.

66. The answer is E [Chapter 3, V; Figure 3.15]. The atria depolarize during the P wave and then repolarize. The ventricles depolarize during the QRS complex and then repolarize during the T wave. Thus, both the atria and the ventricles are fully repolarized at the completion of the T wave.

67. The answer is C [Chapter 3, V; Figure 3.15]. Aortic pressure is lowest just before the ventricles contract.

68. The answer is A [Chapter 5, III C]. Para-aminohippuric acid (PAH) is filtered across the glomerular capillaries and then secreted by the cells of the late proximal tubule. The sum of filtration plus secretion of PAH equals its excretion rate. Therefore, the smallest amount of PAH present in tubular fluid is found in the glomerular filtrate before the site of secretion.

69. The answer is E [Chapter 5, III C; IV A 2]. Creatinine is a glomerular marker with characteristics similar to inulin. The creatinine concentration in tubular fluid is an indicator of water reabsorption along the nephron. The creatinine concentration increases as water is reabsorbed. In a person who is deprived of water (antidiuresis), water is reabsorbed throughout the nephron, including the collecting ducts, and the creatinine concentration is greatest in the final urine.

70. The answer is A [Chapter 5, IX C 1 a]. HCO_3^- is filtered and then extensively reabsorbed in the early proximal tubule. Because this reabsorption exceeds that for H_2O, the $[HCO_3^-]$ of proximal tubular fluid decreases. Therefore, the highest concentration of $[HCO_3^-]$ is found in the glomerular filtrate.

71. The answer is E [Chapter 5, V B]. K^+ is filtered and then reabsorbed in the proximal tubule and loop of Henle. In a person on a diet that is very low in K^+, the distal tubule continues to reabsorb K^+ so that the amount of K^+ present in tubular fluid is lowest in the final urine. If the person were on a high-K^+ diet, then K^+ would be secreted, not reabsorbed, in the distal tubule.

72. The answer is A [Chapter 5, II C 4 b]. In the glomerular filtrate, tubular fluid closely resembles plasma; there, its composition is virtually identical to that of plasma, except that it does not contain plasma proteins. These proteins cannot pass across the glomerular capillary because of their molecular size. Once the tubular fluid leaves Bowman space, it is extensively modified by the cells lining the tubule.

73. The answer is B [Chapter 5, IV C 1]. The proximal tubule reabsorbs about two-thirds of the glomerular filtrate isosmotically. Therefore, one-third of the glomerular filtrate remains at the end of the proximal tubule.

74. The answer is D [Chapter 5, VII B, C]. Under conditions of either water deprivation (antidiuresis) or water loading, the thick ascending limb of the loop of Henle performs its basic function of reabsorbing salt without water (owing to the water impermeability of this segment). Thus, fluid leaving the loop of Henle is dilute with respect to plasma, even when the final urine is more concentrated than plasma.

75. The answer is C [Chapter 3, III A]. Because there are no P waves associated with the bizarre QRS complex, activation could not have begun in the sinoatrial (SA) node. If the beat had originated in the atrioventricular (AV) node, the QRS complex would have had a "normal" shape because the ventricles would activate in their normal sequence. Therefore, the beat must have originated in the His–Purkinje system, and the bizarre shape of the QRS complex reflects an improper activation sequence of the ventricles. Ventricular muscle does not have pacemaker properties.

76. The answer is B [Chapter 3, III E; VI B]. V_1 agonists simulate the vasoconstrictor effects of antidiuretic hormone (ADH). Because saralasin is an angiotensin-converting enzyme (ACE) inhibitor, it blocks the production of the vasoconstrictor substance angiotensin II. Spironolactone, an aldosterone antagonist, blocks the effects of aldosterone to increase distal tubule Na^+ reabsorption and consequently reduces extracellular fluid (ECF) volume and blood pressure. Phenoxybenzamine, an α-blocking agent, inhibits the vasoconstrictor effect of α-adrenergic stimulation. Acetylcholine (ACh), via production of endothelium-derived relaxing factor (EDRF), causes vasodilation of vascular smooth muscle and reduces blood pressure.

77. The answer is D [Chapter 3, II E]. A decrease in the capacitance of the artery means that for a given volume of blood in the artery, the pressure will be increased. Thus, for a given stroke volume ejected into the artery, both the systolic pressure and pulse pressure will be greater.

78. The answer is B [Chapter 3, IX B; Table 3.5]. During moderate exercise, sympathetic outflow to the heart and blood vessels is increased. The sympathetic effects on the heart cause increased heart rate and contractility, and the increased contractility results in increased stroke volume. Pulse pressure increases as a result of the increased stroke volume. Venous return also increases because of muscular activity; this increased venous return further contributes to increased stroke volume by the Frank-Starling mechanism. Total peripheral resistance (TPR) might be expected to increase because of sympathetic stimulation of the blood vessels. *However*, the buildup of local metabolites in the exercising muscle causes local vasodilation, which overrides the sympathetic vasoconstrictor effect, thus decreasing TPR. Arterial Po_2 does not decrease during moderate exercise, although O_2 consumption increases.

79. The answer is B [Chapter 3, VI B]. Patients with essential hypertension have decreased renin secretion as a result of increased renal perfusion pressure. Patients with congestive heart failure and hemorrhagic shock have increased renin secretion because of reduced intravascular volume, which results in decreased renal perfusion pressure. Patients with aortic constriction above the renal arteries are hypertensive because decreased renal perfusion pressure causes increased renin secretion, followed by increased secretion of angiotensin II and aldosterone.

80. The answer is E [Chapter 7, IX A]. 5α-Reductase catalyzes the conversion of testosterone to dihydrotestosterone. Dihydrotestosterone is the active androgen in several male accessory sex tissues (e.g., prostate).

81. The answer is B [Chapter 3, V; Figure 3.15]. Because the ventricles are contracting during isovolumetric contraction, ventricular pressure increases. Because all of the valves are closed, the contraction is isovolumetric. No blood is ejected into the aorta until ventricular pressure increases enough to open the aortic valve.

82. The answer is D [Chapter 4, I A, B]. Residual volume is the volume present in the lungs after maximal expiration, or expiration of the vital capacity (VC). Therefore, residual volume is not included in the tidal volume (TV), VC, inspiratory reserve volume (IRV), or inspiratory capacity (IC). The functional residual capacity (FRC) is the volume remaining in the lungs after expiration of a normal TV and, therefore, includes the residual volume.

83. The answer is E [Chapter 5, IX D 1; Table 5.9]. The blood values are consistent with metabolic acidosis (calculate pH = 7.34). Treatment with a carbonic anhydrase inhibitor causes metabolic acidosis because it increases HCO_3^- excretion.

84. The answer is A [Chapter 7, III B 4 a, c (2)]. Prolactin secretion by the anterior pituitary is tonically inhibited by dopamine secreted by the hypothalamus. If this inhibition is disrupted (e.g., by interruption of the hypothalamic–pituitary tract), then prolactin secretion will be increased, causing galactorrhea. The dopamine agonist bromocriptine simulates the tonic inhibition by dopamine and inhibits prolactin secretion.

85. The answer is D [Chapter 5, VII A 1; Table 5.6; Figure 5.14]. The description is of a normal person who is deprived of water. Serum osmolarity is slightly higher than normal because insensible water loss is not being replaced by drinking water. The increase in serum osmolarity stimulates (via osmoreceptors in the anterior hypothalamus) the release of antidiuretic hormone (ADH) from the posterior pituitary. ADH then circulates to the kidney and stimulates water reabsorption from the collecting ducts to concentrate the urine.

86. The answer is D [Chapter 3, VIII C-F; Table 3.3]. Both the pulmonary and coronary circulations are regulated by Po_2. However, the critical difference is that hypoxia causes vasodilation in the coronary circulation and vasoconstriction in the pulmonary circulation. The cerebral and muscle circulations are regulated primarily by local metabolites, and the skin circulation is regulated primarily by sympathetic innervation (for temperature regulation).

87. The answer is A [Chapter 5, IX C 1; Tables 5.9 and 5.11]. Acetazolamide, a carbonic anhydrase inhibitor, is used to treat respiratory alkalosis caused by ascent to high altitude. It acts on the renal proximal tubule to inhibit the reabsorption of filtered HCO_3^- so that the person excretes alkaline urine and develops mild metabolic acidosis.

88. The answer is B [Chapter 5, IX D 1; Table 5.9]. The blood values are consistent with metabolic acidosis with respiratory compensation. Because the urinary excretion of NH_4^+ is decreased, chronic renal failure is a likely cause.

89. **The answer is D** [Chapter 3, VI D]. In a person with a left-to-right cardiac shunt, arterial blood from the left ventricle is mixed with venous blood in the right ventricle. Therefore, Po_2 in pulmonary arterial blood is higher than normal, but systemic arterial blood would be expected to have a normal Po_2 value or 100 mm Hg. During an asthmatic attack, Po_2 is reduced because of increased resistance to airflow. At high altitude, arterial Po_2 is reduced because the inspired air has reduced Po_2. Persons with a right-to-left cardiac shunt have decreased arterial Po_2 because blood is shunted from the right ventricle to the left ventricle without being oxygenated or "arterialized." In pulmonary fibrosis, the diffusion of O_2 across the alveolar membrane is decreased.

90. **The answer is D** [Chapter 1, II]. H^+–K^+ transport occurs via H^+, K^+-adenosine triphosphatase (ATPase) in the luminal membrane of gastric parietal cells, a primary active transport process that is energized directly by ATP. Na^+–glucose and Na^+–alanine transport are examples of cotransport (symport) that are secondary active transport processes and do not use ATP directly. Glucose uptake into muscle cells occurs via facilitated diffusion. Na^+–Ca^{2+} exchange is an example of countertransport (antiport) and is a secondary active transport process.

91. **The answer is B** [Chapter 6, II A 1 c; IV B 4 a]. When the pH of the stomach contents is very low, secretion of gastrin by the G cells of the gastric antrum is inhibited. When gastrin secretion is inhibited, further gastric HCl secretion by the parietal cells is also inhibited. Pancreatic secretion is *stimulated* by low pH of the duodenal contents.

92. **The answer is A** [Chapter 6, II A 2 a]. Removal of the duodenum would remove the source of the gastrointestinal (GI) hormones, cholecystokinin (CCK), and secretin. Because CCK stimulates contraction of the gallbladder (and, therefore, ejection of bile acids into the intestine), lipid absorption would be impaired. CCK also inhibits gastric emptying, so removing the duodenum should accelerate gastric emptying (or decrease gastric emptying time).

93. **The answer is A** [Chapter 7, III C 1 b]. Antidiuretic hormone (ADH) not only produces increased water reabsorption in the renal collecting ducts (V_2 receptors) but also causes constriction of vascular smooth muscle (V_1 receptors).

94. **The answer is B** [Chapter 6, V A 2 b]. Monosaccharides (glucose, galactose, and fructose) are the absorbable forms of carbohydrates. Glucose and galactose are absorbed by Na^+-dependent cotransport; fructose is absorbed by facilitated diffusion. Dipeptides and water-soluble vitamins are absorbed by cotransport in the duodenum, and bile acids are absorbed by Na^+-dependent cotransport in the ileum (which recycles them to the liver). Cholesterol is absorbed from micelles by simple diffusion across the intestinal cell membrane.

95. **The answer is E** [Chapter 7, III B 3 a (1)]. Somatostatin is secreted by the hypothalamus and inhibits the secretion of growth hormone by the anterior pituitary. Notably, much of the feedback inhibition of growth hormone secretion occurs by stimulating the secretion of somatostatin (an inhibitory hormone). Both growth hormone and somatomedins stimulate the secretion of somatostatin by the hypothalamus.

96. **The answer is B** [Chapter 7, X A]. Aromatase catalyzes the conversion of testosterone to estradiol in the ovarian granulosa cells. Estradiol is required for the development of female secondary sex characteristics.

97. **The answer is E** [Chapter 3, V; Figure 3.15]. Closure of the aortic and pulmonic valves creates the second heart sound. The closure of these valves corresponds to the end of ventricular ejection and the beginning of ventricular relaxation.

98. **The answer is C** [Chapter 2, II C 4; Figure 2.5]. Light striking a photoreceptor cell causes the conversion of 11-*cis* retinal to all-*trans* retinal; activation of a G protein called transducin; activation of phosphodiesterase, which catalyzes the conversion of cyclic guanosine monophosphate (cGMP) to 5′-GMP so that cGMP levels decrease; closure of Na^+ channels by the decreased cGMP levels; hyperpolarization of the photoreceptor; and decreased release of glutamate, an excitatory neurotransmitter.

99. The answer is D [Chapter 7, IV A 4]. The coupling of two molecules of diiodotyrosine (DIT) results in the formation of thyroxine (T_4). The coupling of DIT to monoiodotyrosine (MIT) produces triiodothyronine (T_3).

100. The answer is E [Chapter 7, III C 1]. The initiating event in central diabetes insipidus is decreased antidiuretic hormone (ADH) secretion by the posterior pituitary (head injury severs the axons that carry ADH from the hypothalamus to the posterior pituitary). Decreased ADH leads to decreased water reabsorption by the late distal tubule and collecting ducts, increased water excretion and, consequently, decreased urine osmolarity. The increased water excretion leads to increased serum osmolarity.

101. The answer is C [Chapter 7, VII B 3]. The initiating event in humoral hypercalcemia of malignancy is increased levels of parathyroid hormone–related peptide (PTH-rp), secreted by the man's lung cancer. PTH-rp has the same physiologic actions as PTH. Thus, the high levels of PTH-rp cause increased bone resorption and increased renal Ca^{2+} reabsorption, which together result in hypercalcemia. High levels of PTH-rp also cause decreased renal phosphate reabsorption, leading to decreased serum phosphate concentration. Finally, the hypercalcemia inhibits PTH secretion by the chief cells of the parathyroid gland.

Appendix A Key Physiology Topics for USMLE Step I

CELL PHYSIOLOGY

Transport mechanisms
Ionic basis for action potential
Excitation–contraction coupling in skeletal, cardiac, and smooth muscle
Neuromuscular transmission

AUTONOMIC PHYSIOLOGY

Cholinergic receptors
Adrenergic receptors
Effects of autonomic nervous system on organ system function

CARDIOVASCULAR PHYSIOLOGY

Events of cardiac cycle
Pressure, flow, resistance relationships
Frank-Starling law of the heart
Ventricular pressure–volume loops
Ionic basis for cardiac action potentials
Starling forces in capillaries
Regulation of arterial pressure (baroreceptors and renin–angiotensin II–aldosterone system)
Cardiovascular and pulmonary responses to exercise
Cardiovascular responses to hemorrhage
Cardiovascular responses to changes in posture

RESPIRATORY PHYSIOLOGY

Lung and chest-wall compliance curves
Breathing cycle
Hemoglobin–O_2 dissociation curve
Causes of hypoxemia and hypoxia
V/Q, PO_2, and PCO_2 in upright lung
V/Q defects
Peripheral and central chemoreceptors in control of breathing
Responses to high altitude

RENAL AND ACID–BASE PHYSIOLOGY

Fluid shifts between body fluid compartments
Starling forces across glomerular capillaries
Transporters in various segments of nephron (Na^+, Cl^-, HCO_3^-, H^+, K^+, and glucose)
Effects of hormones on renal function
Simple acid–base disorders
Common mixed acid–base disorders

GASTROINTESTINAL PHYSIOLOGY

Gastrointestinal hormones
Gastrointestinal motility
Salivary, gastric, pancreatic, and biliary secretions
Digestion and absorption of carbohydrates, proteins, and lipids

ENDOCRINE AND REPRODUCTIVE PHYSIOLOGY

Mechanisms of hormone action
ADH actions and pathophysiology

Thyroid: steps in synthesis, pathophysiology of hypothyroidism and hyperthyroidism

Adrenal cortex: hormone synthesis, pathophysiology of Addison disease, Cushing syndrome, and adrenogenital syndromes

Insulin: secretion, insulin receptors and actions, type I and II diabetes mellitus

PTH: actions, hyperparathyroidism, hypoparathyroidism, PTH-rp, pseudohypoparathryoidism

Actions of testosterone and dihydrotestosterone

Menstrual cycle

Hormones of pregnancy

Appendix B Key Physiology Equations for USMLE Step I

Cardiac Output (Ohm law)	$CO = \frac{P_a - RAP}{TPR}$
Resistance	$R = \frac{8\eta l}{\pi r^4}$
Compliance	$C = \frac{V}{P}$
Cardiac Output	$CO = SV \times HR$
Ejection fraction	$EF = \frac{SV}{EDV}$
Cardiac output (measurement)	$CO = \frac{O_2 \text{ consumption}}{[O_2]_{\text{pulmonary vein}} - [O_2]_{\text{pulmonary artery}}}$
Starling equation	$J_v = K_f[(P_c - P_i) - (\pi_c - \pi_i)]$
Physiologic dead space	$V_D = V_T \times \frac{P_{A_{CO_2}} - P_{E_{CO_2}}}{P_{A_{CO_2}}}$
Alveolar ventilation	$V_A = (V_T - V_D) \times \text{Breaths/min}$
Renal clearance	$C = \frac{UV}{P}$
Glomerular filtration rate	$GFR = \frac{[U]_{\text{inulin}} V}{[P]_{\text{inulin}}}$
Free water clearance	$C_{H_2O} = V - C_{osm}$
Henderson-Hasselbalch equation	$pH = pK + \log \frac{A^-}{HA}$
Serum anion gap	$\text{Anion Gap} = Na^+ - (Cl^- + HCO_3^-)$

Appendix C

Normal Blood Values

PLASMA, SERUM, OR BLOOD CONCENTRATIONS

Substance	Average Normal Value	Range	Comments
Bicarbonate (HCO_3^-)	24 mEq/L	22–26 mEq/L	Venous blood; measured as total CO_2
Calcium (Ca^{2+}), ionized	5 mg/dL		
Calcium (Ca^{2+}), total	10 mg/dL		
Chloride (Cl^-)	100 mEq/L	98–106 mEq/L	
Creatinine	1.2 mg/dL	0.5–1.5 mg/dL	
Glucose	80 mg/dL	70–100 mg/dL	Fasting
Hematocrit	0.45	0.4–0.5	Men, 0.47; women, 0.41
Hemoglobin	15 g/dL		
Hydrogen ion (H^+)	40 nEq/L		Arterial blood
Magnesium (Mg^{2+})	0.9 mmol/L		
Osmolarity	287 mOsm/L	280–298 mOsm/L	Osmolality is mOsm/kg H_2O
O_2 saturation	98%	96%–100%	Arterial blood
P_{CO_2}, arterial	40 mm Hg		
P_{CO_2}, venous	46 mm Hg		
P_{O_2}, arterial	100 mm Hg		
P_{O_2}, venous	40 mm Hg		
pH, arterial	7.4	7.37–7.42	
pH, venous	7.37		
Phosphate	1.2 mmol/L		
Potassium (K^+)	4.5 mEq/L		
Protein, albumin	4.5 g/dL		
Protein, total	7 g/dL	6–8 g/dL	
Sodium (Na^+)	140 mEq/L		
Urea nitrogen (BUN)	12 mg/dL	9–18 mg/dL	Varies with dietary protein
Uric acid	5 mg/dL		

Index

(*Note*: Page numbers followed by "*f*" denote figures; those followed by "*t*" denote tables; those followed by "Q" denote questions; and those followed by "E" denote explanations.)

B

P

U